Fractional Order Analysis

Fractional Order Analysis

Theory, Methods and Applications

Edited by
Hemen Dutta, Ahmet Ocak Akdemir, and
Abdon Atangana

This edition first published 2020

Registered Office
John Wiley & Sons, Inc., 111 River Street, Hoboken, NJ 07030, USA

Editorial Office
111 River Street, Hoboken, NJ 07030, USA

For details of our global editorial offices, customer services, and more information about Wiley products visit us at www.wiley.com.

Wiley also publishes its books in a variety of electronic formats and by print-on-demand. Some content that appears in standard print versions of this book may not be available in other formats.

Library of Congress Cataloging-in-Publication Data

Names: Dutta, Hemen, 1981- editor. | Akdemir, Ahmet Ocak, 1985- editor. | Atangana, Abdon, editor.
Title: Fractional order analysis : theory, methods and applications / edited by Hemen Dutta, Ahmet Ocak Akdemir, Abdon Atangana.
Description: Hoboken, NJ : Wiley, [2020] | Includes bibliographical references and index.
Identifiers: LCCN 2020015384 (print) | LCCN 2020015385 (ebook) | ISBN 9781119654162 (cloth) | ISBN 9781119654209 (adobe pdf) | ISBN 9781119654230 (epub)
Subjects: LCSH: Fractional calculus.
Classification: LCC QA314 .F735 2020 (print) | LCC QA314 (ebook) | DDC 515/.83–dc23
LC record available at https://lccn.loc.gov/2020015384
LC ebook record available at https://lccn.loc.gov/2020015385

Cover Design: Wiley
Cover Image: © oxygen/Getty Images

Set in 9.5/12.5pt STIXTwoText by SPi Global, Chennai, India

10 9 8 7 6 5 4 3 2 1

Contents

Preface

The book covers several new research findings in the area of fractional-order analysis and its applications. Different tools and techniques of fractional-order analysis are presented in the chapters, and several practical applications have also been demonstrated by means of different mathematical methods and models. Readers should find several useful, relevant, and connected topics in the area of fractional-order analysis those are necessary for crucial understanding of various research problems in science and technology. This book should be useful for graduate and PhD students, researchers, and educators interested in fractional-order analysis and its diverse applications. There are 11 chapters in the book, and they are organized as follows:

Chapter **"On the Fractional Derivative and Integral Operators"** first discussed interesting developments of fractional calculus. Then, it presented the properties of Grünwald–Letnikov, Riemann–Liouville and Caputo fractional derivative and integral operators. Also, comparisons of these operators have been made in detail. The Caputo–Fabrizio derivative operator is obtained by using the exponential function and its features have also been given. Atangana–Baleanu fractional derivative with nonlocal and nonsingular kernel is obtained by using the generalized form of the Mittag-Leffler function. Finally, the Keller–Segel and cancer treatment models were compared by expanding to Caputo, Caputo–Fabrizio, and Atangana–Baleanu fractional derivative operators.

Chapter **"Generalized Conformable Fractional Operators and Their Applications"** aims to present new generalized fractional integral and derivative operators along with their applications. These operators are known as left-sided and right-sided generalized conformable fractional operators, and they contain several operators of fractional calculus. Then some basic properties such as linearity, continuity, boundedness, etc. of such operators are presented. A nonlinear generalized conformable fractional differential equation is also formed. It is shown that this equation is equivalent to a Volterra integral equation and, then the existence and uniqueness of the solution is demonstrated. Finally, some Hermite–Hadamard-type inequalities for conformable integrals have been presented and discussed their applications for Trapezoidal formula and means.

Chapter **"Analysis of New Trends of Fractional Differential Equations"** discussed several results related to fractal–fractional derivatives in Caputo sense when the kernels are power law, exponential decay law the generalized Mittag-Leffler functions and presented numerical approximation for each case. It considered partial differential equations with these new differential operators. It also presented numerical analysis in detail and the stability for each case. Numerical simulations have been incorporated to explain the efficiency of the numerical scheme adopted, and also to see the effect of the fractal dimension and fractional order.

Chapter **"New Estimations for Exponentially Convexity via Conformable Fractional Operators"** discussed different fractional integral operators and their basic properties. It also established some new Hadamard-type integral inequalities for exponentially convex functions via conformable fractional integrals.

Chapter **"Lyapunov-type Inequalities for Local Fractional Proportional Derivatives"** reviewed some Lyapunov-type inequalities for certain local and nonlocal fractional derivatives and presented a Lyapunov-type inequality for the sequential local fractional proportional derivatives with constant references as the special case $\alpha = 2$ of the nonlocal fractional proportional derivative ${}_aD^{\alpha,\rho}$. An open problem is also presented for a more general sequential local fractional proportional boundary value problem. Then, it presented a higher order extension in order to investigate the ability of proving a Lyapunov inequality, which cannot obtained from the nonlocal one, for the local fractional proportional derivative of order $1 < \rho \leq 2$.

Chapter **"Minkowski-type Inequalities for Mixed Conformable Fractional Integrals"** first presented some fractional integrals and Minkowski-type inequalities obtained for these integrals. Then, the reverse Minkowski inequality and related inequalities for mixed conformable fractional integrals have been presented.

Chapter **"New Estimations for Different Kinds of Convex Functions via Conformable Integrals and Riemann–Liouville Fractional Integral Operators"** discussed several new fractional bounds involving the functions having geometrical convexity and co-ordinated convexity properties via conformable integrals and Riemann–Liouville fractional integrals. In order to obtain main results, it also derived new fractional integral identities. Then, it generalized some new integral inequalities for *GG*-convex functions whose second derivative at certain powers are established via conformable integrals. Several new results were further derived by choosing different values of n and α.

Chapter "**Legendre-Spectral Algorithms for Solving Some Fractional Differential Equations**" discussed some algorithms for treating some kinds of fractional differential equations by utilizing suitable spectral methods. A Galerkin method is employed for solving time-fractional telegraph equation, and a double shifted Legendre expansion is proposed as an approximating polynomial. Further, the spectral methods Petrov–Galerkin and collocation, respectively, are applied for obtaining spectral solutions of space fractional linear diffusion problem. The two suggested algorithms have been built by using a certain double shifted Legendre basis. Investigation for the convergence and error analysis of the two suggested approximate double expansions have also been performed. Numerical results were provided to justify the efficiency of the proposed algorithms.

Chapter "**Mathematical Modeling of an Autonomous Nonlinear Dynamical System for Malaria Transmission Using Caputo Derivative**" proposed a seven-dimensional Caputo-type fractional-order model describing the dynamics for the spread of malaria virus transmitted to humans (host) by the bite of mosquito (vector), affecting the latter itself. Considering that both underlying populations may not behave exactly the same, the first group of population (human) is assigned the fractional-order ε whereas the fractional-order κ is for the second group (mosquito). The model is shown to have globally asymptotically stable steady-state solution with $R_0 < 1$ (disease cannot spread) and an unstable endemic equilibrium point for $R_0 > 1$ (disease can spread), where R_0 is the basic reproductive number. Fixed point theory is used for discussing the existence and uniqueness of the solution of the model. Further, it is proved that the non-negative hyperoctant R_+^7 is a positively invariant region for the model. Numerical simulation is also presented to show that the model under Caputo differentiation is more accurate than its classical version to describe the complexity of the dynamics of the disease's transmission.

Chapter "**MHD-free Convection Flow Over a Vertical Plate with Ramped Wall Temperature and Chemical Reaction in view of Nonsingular Kernel**" aims to study the unsteady MHD-free convection flow of an electrically conducting incompressible generalized Maxwell fluid over an infinite vertical plate with ramped temperature and constant concentration. Fractional-order Caputo, Caputo–Fabrizio, and Atangana–Baleanu time derivatives are used to study the effect of fractional parameters on the dynamics of fluid. The motion of plate is rectilinear translation with an arbitrary time-dependent velocity. It observed that fractional-order model is best to explain the memory effect and flow behavior of the fluid. The influence of transverse magnetic fields is also studied. Moreover, the effects of system parameters on the filed velocity are analyzed through numerical simulation and graphs.

Chapter **"Comparison of the Different Fractional Derivatives for the Dynamics of Zika Virus"** presented a comparative study of the Zika model with two different operators, viz., the Caputo–Fabrizio (CF) and the Atangana–Baleanu (AB) derivative. It considered a latest mathematical model that considers Zika dynamics with mutation. Then, it presented some basic mathematical results for the model and then applied the CF and AB derivative, and also presented their analysis. Some key results related to fractional order are further incorporated for the appropriateness of CF and AB for modeling purposes.

The editors are grateful to the contributors for their contribution and co-operation throughout the whole process of editing the book. The editors have benefited from the remarks and comments of several experts on the topics of this book. The editors would also like to thank the editors at Wiley and production staff for their support and help. Finally, the editors offer sincere thanks to all those who contributed in some way to complete this book project.

February, 2020

Hemen Dutta
Ahmet Ocak Akdemir
Abdon Atangana

List of Contributors

Waleed Mohammed Abd-Elhameed
Department of Mathematics
Faculty of Science
Cairo University
Giza 12613
Egypt

Thabet Abdeljawad
Department of Mathematics and Physical Sciences
Prince Sultan University
Riyadh 11586
Saudi Arabia

and

Department of Medical Research
China Medical University
Taichung
Taiwan

and

Department of Computer Science and Information Engineering
Asia University
Taichung
Taiwan

Ahmet Ocak Akdemir
Department of Mathematics
Faculty of Science and Letters
Ağrı İbrahim Çeçen University
Ağrı
Turkey

Ali Akgül
Department of Mathematics
Art and Science Faculty
Siirt University
Siirt
Turkey

Abdon Atangana
Institute for Groundwater Studies
Faculty of Natural and Agricultural Science
University of Free State
Bloemfontein 9300
South Africa

and

Department of Medical Research
China Medical University Hospital
China Medical University
Taichung
Taiwan

Mustafa Ali Dokuyucu
Department of Mathematics
Faculty of Science and Letters
Ağrı İbrahim Çeçen University
Ağrı-Turkey

Sever Silvestru Dragomir
Theory of Inequality
School of Engineering and Science
Victoria University
PO Box 14428
Melbourne VIC 8001
Australia

Hemen Dutta
Department of Mathematics
Gauhati University
Guwahati 781014
Assam
India

Alper Ekinci
Department of International Trade
Bandirma Vocational High School
Bandirma Onyedi Eylul University
Balikesir
Turkey

Muhammad Adil Khan
Department of Mathematics
University of Peshawar
Peshawar, Khyber Pakhtunkhwa
Pakistan

Muhammad Altaf Khan
Institute of Ground water studies
Faculty of Natural and Agricultural Sciences
University of the Free State
Bloemfontein
South Africa

Tahir Ullah Khan
Department of Mathematics
University of Peshawar
Peshawar, Khyber Pakhtunkhwa
Pakistan

Muhamet Emin Özdemir
Department of Mathematics Education
Uludağ University
Görükle Campus
Bursa-Turkey

Sania Qureshi
Department of Basic Sciences and Related Studies
Mehran University of Engineering and Technology
Jamshoro 76062
Pakistan

Muhammad Bilal Riaz
Department of Mathematics
University of Management and Technology
Lahore
Pakistan

and

Institute for Groundwater Studies (IGS)
University of the Free State
Bloemfontein
South Africa

Syed Tauseef Saeed
Department of Mathematics
National University of Computer & Emerging Sciences
Lahore Campus
Pakistan

Erhan Set
Faculty of Science and Letters
Department of Mathematics
Ordu University
Ordu-Turkey

Youssri Hassan Youssri
Department of Mathematics
Faculty of Science
Cairo University
Giza 12613
Egypt

About the Editors

Hemen Dutta has been serving the Department of Mathematics at Gauhati University as a faculty member. He did his Master of Science in Mathematics, Post Graduate Diploma in Computer Application, Master of Philosophy in Mathematics and Doctor of Philosophy in Mathematics. His primary research areas include diverse topics of mathematical analysis and their applications. He has more than 120 published items as research papers in reputed journals and chapters in books published by top class publishers. He has published 12 books as text book, monograph and edited book and four conference proceedings with leading publishers. He has delivered several invited talks in national and international events, and visited some foreign institutions on invitations. He has organized five academic events so far. He is the recipient of some grants for conference organization, research project, and travel. He has reviewed papers for journals and databases, and also associated with editing of issues in journals. He has also authored several popular articles in newspaper, magazine, and science portal.

Ahmet Ocak Akdemir is an Associate Professor in the Department of Mathematics at Ağrı Ibrahim Çeçen University, Ağrı, Turkey. He has teaching and research experiences at the same university. His present position is Head of Mathematics Department and also carrying out many administrative tasks. He received his Ph.D. degree in Analysis in 2012 from Atatürk University, Erzurum, Turkey. His research interest focuses mainly in inequality theory, convex analysis, and real functions of two variables, especially fractional calculus and integral operators. He has several research papers published with indexed and pioneer journals and delivered several talks at international conferences and meetings. He is the Editor-In-Chief of Turkish Journal of Science and guest editor of some journals. He has organized several international conferences as chairman and member of organizing committee. He is the recipient of numerous publication encouragement awards given by his own university and some private institutions. Also, he has carried out four projects at national level and published its results. His papers have been cited by several researchers and his current *h*-index is 7 according to Web of Science record. He has collaborations with several mathematicians around the world.

Abdon Atangana is Full Professor at the Faculty of Natural and Agricultural Science, Institute for Groundwater Studies, University of the Free State (UFS), Bloemfontein, South Africa. He obtained his honors and master's degrees in Applied Mathematics from the Department of Applied Mathematics at the UFS with distinction. He serves as a reviewer for more than 200 accredited international journals, and has been awarded the world champion of peer review in 2016 and in 2017. He also serves on the editorial boards of more than 20 journals of international repute. His research interests are in the methods and applications of partial and ordinary differential equations, fractional differential equations, perturbation methods, asymptotic methods, iterative methods, and groundwater modeling. He is the founder of fractional calculus with non-local and non-singular kernels, popular in applied mathematics today. Since 2013, he has published over 225 research articles in several accredited journals of applied mathematics, applied physics, geohydrology, and biomathematics. He is the author of two books, *Fractional Operators with Constant and Variable Order with Application to Geo-hydrology*; and *Derivative with a New Parameter: Theory, Methods and Applications*.

1

On the Fractional Derivative and Integral Operators

Mustafa A. Dokuyucu

Department of Mathematics, Ağrı İbrahim Çeçen University, Ağrı, Turkey

1.1 Introduction

It is formulated using mathematical expressions to solve problems in engineering, science, and many other fields. These formulas are obtained and functions are formed to solve the problems under certain initial and boundary conditions. The resulting equation generally contains derivatives of fractional, first or higher order. Expressions containing such equations are solved by some known methods. While making these solutions, the known methods of classical analysis are not always sufficient. In this case, fractional calculus tools are activated. As many researchers know, the story of the fractional calculus that began with that letter in 1695 was answered by Leibniz [1, 2].

Although fractional calculus tools have been known and used in different fields for a long time, the theory of fractional differential equations has recently begun to be studied. Many important books have been written on this subject in the literature. The subject of fractional calculus can still be improved and is of great importance in helping other fields. In this section, chronologically fractional derivative and integral operators will be introduced and important properties of these operators will be given [3]. Firstly, the derivative of Grünwald–Letnikov developed with the classical derivative half will be given. Later, Riemann and Liouville developed the definition of fractional derivative of Grünwald–Letnikov and introduced a new operator to the literature [4]. This operator has had an important place for a long time as it is today. Later, in 1967, Caputo made a significant development in this regard and introduced an operator to be used until the early 2000s [5]. In 2015, Caputo and Fabrizio changed the kernel of the Caputo derivative definition to a definition. The kernel they use is very important in terms of singularity. It also gives good results in solving real world problems [6]. Finally, Atangana and Baleanu put forward a definition of both nonsingular and nonlocal. The

Fractional Order Analysis: Theory, Methods and Applications, First Edition.
Edited by Hemen Dutta, Ahmet Ocak Akdemir, and Abdon Atangana.

core obtained by using the generalized form of the Mittag-Leffler function has an important place, although it has recently been discovered [7].

It is tried to explain the important features of all the operators given above by supporting them with definitions, theorems and lemmas. In the last chapter, the application of two important models such as Keller–Segel and Cancer Treatment to these derivatives is shown [8, 9].

1.2 Fractional Derivative and Integral Operators

In this section, fractional derivative and integral operators will be introduced chronologically. Important theorems and lemmas will be given about these operators.

1.2.1 Properties of the Grünwald–Letnikov Fractional Derivative and Integral

Let $y = f(t)$ is a continuous function. According to the definition, the first-order derivative of the function $f(t)$ is defined by

$$f'(t) = \frac{df}{dt} = \lim_{h\to 0} \frac{f(t) - f(t-h)}{h}. \tag{1.1}$$

The second-order derivative using Eq. (1.1), then

$$\begin{aligned} f''(t) = \frac{d^2f}{dt^2} &= \lim_{h\to 0} \frac{f'(t) - f'(t-h)}{h} \\ &= \lim_{h\to 0} \frac{1}{h}\left[\frac{f(t)-f(t-h)}{h} - \frac{f(t-h)-f(t-2h)}{h}\right] \\ &= \lim_{h\to 0} \frac{f(t) - 2f(t-h) + f(t-2h)}{h^2}. \end{aligned} \tag{1.2}$$

Similarly, third-order derivative as:

$$f'''(t) = \frac{d^3f}{dt^3} = \lim_{h\to 0} \frac{f(t) - 3f(t-h) + 3f(t-2h) - f(t-3h)}{h^3}. \tag{1.3}$$

When this situation is generalized, Eq. (1.14) is obtained

$$f^n(t) = \frac{d^nf}{dt^n} = \lim_{h\to 0} \frac{1}{h^n} \sum_{r=0}^{n} (-1)^n \binom{n}{r} f(t - rh), \tag{1.4}$$

where

$$\binom{n}{r} = \frac{n(n-1)(n-2)\cdots(n-r+1)}{r!}. \tag{1.5}$$

Let now examine the following expression generalizing the fractions (1.2)–(1.15):

$$f_h^{(p)}(t) = \frac{1}{h^p} \sum_{r=0}^{n} (-1)^k \binom{p}{r} f(t - rh), \tag{1.6}$$

where p is an arbitrary integer number, n is also integer, as above.

Obviously, for $p \leq n$ we have,

$$\lim_{h\to 0} f_h^{(p)}(t) = f^{(p)}(t) = \frac{d^p f}{dt^p} \tag{1.7}$$

because in such a case, as follows from (1.15), all the coefficients in the numerator after $\binom{p}{p}$ are equal to 0.

Let us consider negative values of p. For convenience, let us denote

$$\begin{bmatrix} p \\ r \end{bmatrix} = \frac{p(p+1)\cdots(p+r-1)}{r!}. \tag{1.8}$$

Then we have

$$\binom{-p}{r} = \frac{-p(-p-1)\cdots(-p-r+1)}{r!} = (1-)^r \begin{bmatrix} p \\ r \end{bmatrix} \tag{1.9}$$

and replacing p in (1.6) with $-p$ we can write

$$f_h^{(-p)}(t) = \frac{1}{h^p} \sum_{r=0}^{n} \begin{bmatrix} p \\ r \end{bmatrix} f(t - rh), \tag{1.10}$$

where p is a positive integer number.

If n is fixed, then $f_h^{(-p)}(t)$ tends to the uninteresting limit 0 as $h \to 0$. To arrive at a nonzero limit, we have to suppose that $n \to \infty$ as $h \to 0$. We can take $h = \frac{t-a}{n}$, where a is a real constant, and consider the limit value, either finite or infinite, of $f_h^{(-p)}(t)$, which we will denote as

$$\lim_{\substack{h\to 0 \\ nh=t-a}} f_h^{(-p)}(t) = {}_a D_h^{(-p)}(t).$$

Here $D_h^{(-p)}(t)$ denotes, in fact, a certain operation performed on the function $f(t)$; a and t are the terminals – the limits relating to this operation.

Let us consider several particular cases.

For $p = 1$, we have

$$f_h^{(-1)}(t) = h \sum_{r=0}^{n} f(t - rh). \tag{1.11}$$

Taking into account that $t - nh = a$ and that the function $f(t)$ is assumed to be continuous, we conclude that

$$\lim_{\substack{h\to 0 \\ nh=t-a}} f_h^{(-1)}(t) = {}_a D_t^{(-1)} f(t) = \int_0^{t-a} f(t - z)dt = \int_a^t f(\tau)d\tau. \tag{1.12}$$

Let us take $p = 2$. In this case

$$\begin{bmatrix} 2 \\ r \end{bmatrix} = \frac{2, 3, \cdots, (2 + r - 1)}{r!} = r + 1$$

and we have

$$f_h^{(-2)}(t) = h \sum_{r=0}^{n} (rh) f(t - rh). \tag{1.13}$$

Denoting $t + h = y$, we can write

$$f_h^{(-2)}(t) = h \sum_{r=1}^{n+1} (rh) f(t - rh) \tag{1.14}$$

and taking $h \to 0$, we obtain

$$\lim_{\substack{h\to 0 \\ nh=t-a}} f_h^{(-2)}(t) = {}_a D_t^{(-2)} f(t) = \int_0^{t-a} z f(t - z) dt = \int_a^t (t - \tau) f(\tau) d\tau, \tag{1.15}$$

because $y \to t$ as $h \to 0$. Relationships (11)–(15) suggest the following general expression:

$${}_a D_t^{(-p)} f(t) = \lim_{\substack{h\to 0 \\ nh=t-a}} h^p \begin{bmatrix} p \\ r \end{bmatrix} f(t - rh) = \frac{1}{(p-1)!} \int_a^t (t - \tau)^{p-1} f(\tau) d\tau. \tag{1.16}$$

To prove the formula (1.16) by induction, we have to show that f holds for some p, then it holds also for $p + 1$.

Let us introduce the function

$$f_1(t) = \int_a^t f(\tau) d\tau, \tag{1.17}$$

which has the obvious property $f_1(a) = 0$, and consider

$$\begin{aligned} {}_a D_t^{(-p-1)} f(t) &= \lim_{\substack{h\to 0 \\ nh=t-a}} h^{p+1} \sum_{r=0}^{n} \begin{bmatrix} p+1 \\ r \end{bmatrix} f(t - rh) \\ &= \lim_{\substack{h\to 0 \\ nh=t-a}} h^{p} \sum_{r=0}^{n} \begin{bmatrix} p+1 \\ r \end{bmatrix} f_1(t - rh) \\ &\quad - \lim_{\substack{h\to 0 \\ nh=t-a}} h^{p} \sum_{r=0}^{n} \begin{bmatrix} p+1 \\ r \end{bmatrix} f_1(t - (r+1)h). \end{aligned} \tag{1.18}$$

Using (1.8), it is easy to verify that

$$\begin{bmatrix} p+1 \\ r \end{bmatrix} = \begin{bmatrix} p \\ r \end{bmatrix} + \begin{bmatrix} p+1 \\ r-1 \end{bmatrix}, \tag{1.19}$$

where we must put

$$\begin{bmatrix} p+1 \\ -1 \end{bmatrix} = 0.$$

Relationship (1.19) applied to the first sum in (1.18) and the replacement of r by $r-1$ in the second sum gives:

$$\begin{aligned}
{}_aD_t^{(-p-1)}f(t) &= \lim_{\substack{h\to 0\\ nh=t-a}} h^p \sum_{r=0}^{n} \begin{bmatrix} p+1 \\ r \end{bmatrix} f(t-rh) \\
&\quad + \lim_{\substack{h\to 0\\ nh=t-a}} h^p \sum_{r=0}^{n} \begin{bmatrix} p+1 \\ r-1 \end{bmatrix} f_1(t-rh) \\
&\quad - \lim_{\substack{h\to 0\\ nh=t-a}} h^p \sum_{r=1}^{n+1} \begin{bmatrix} p+1 \\ r-1 \end{bmatrix} f_1(t-rh) \\
&= {}_a D_t^{(-p)} f_1(t) - \lim_{\substack{h\to 0\\ nh=t-a}} h^p \begin{bmatrix} p+1 \\ n \end{bmatrix} f(t-(n+1)h) \\
&= {}_a D_t^{(-p)} f_1(t) - (t-a)^p \lim_{n\to\infty} \begin{bmatrix} p+1 \\ n \end{bmatrix} \tfrac{1}{n^p} f_1\left(a - \tfrac{t-a}{n}\right).
\end{aligned} \tag{1.20}$$

It follows from the definition (1.16) of the function $f_1(t)$ that

$$\lim_{n\to\infty} f_1\left(a - \frac{t-a}{n}\right) = 0.$$

Taking into account the known limit

$$\lim_{n\to\infty} \begin{bmatrix} p+1 \\ n \end{bmatrix} \frac{1}{n^p} = \lim_{n\to\infty} \frac{(p+1)(p+2)\cdots(p+n)}{n^p n!} = \frac{1}{\Gamma(p+1)},$$

we obtain

$$\begin{aligned}
{}_aD_t^{(-p-1)}f(t) &= {}_a D_t^{(-p)} f_1(t) = \frac{1}{(p-1)!}\int_a^t (t-\tau)^{p-1} f_1(\tau)d\tau \\
&= -\frac{(t-\tau)^p f_1(\tau)}{p!}\Big|_{\tau=a}^{\tau=t} + \frac{1}{p!}\int_a^t (t-\tau)^p f(\tau)d\tau \\
&= \frac{1}{p!}\int_a^t (t-\tau)^p f(\tau)d\tau,
\end{aligned} \tag{1.21}$$

which ends the proof of formula (1.16) by induction.

Now let us show that formula (1.16) is a representation of a p-fold integral. Integrating the relationship

$$\frac{d}{dt}({}_aD_t^{-p} f(t)) = \frac{1}{(p-2)!}\int_a^t (t-\tau)^{p-2} f(\tau)d\tau = {}_a D_t^{-p+1} f(t)$$

from a to t, we obtain:

$${}_aD_t^{-p} f(t) = \int_a^t ({}_aD_t^{-p+1} f(t))dt,$$

$${}_aD_t^{-p+1} f(t) = \int_a^t ({}_aD_t^{-p+2} f(t))dt, \text{etc.},$$

and therefore

$$
\begin{aligned}
{}_aD_t^{-p}f(t) &= \int_a^t dt \int_a^t ({}_aD_t^{-p+2}f(t)) \\
&= \int_a^t dt \int_a^t dt \int_a^t ({}_aD_t^{-p+3}f(t))dt \\
&= \underbrace{\int_a^t dt \int_a^t dt \cdots \int_a^t}_{p\ \text{times}} f(t)dt.
\end{aligned}
\tag{1.22}
$$

We see that the derivative of an integer order n (1.14) and the p-fold integral (1.16) of the continuous function $f(t)$ are particular cases of the general expression

$$
{}_aD_t^p f(t) = \lim_{\substack{h\to 0\\ nh=t-a}} h^{-p} \sum_{r=0}^{n} (-1)^r \binom{p}{r} f(t-rh),
\tag{1.23}
$$

which represent the derivative of order m if $p = m$ and the m-fold integral if $p = -m$ [4].

1.2.1.1 Integral of Arbitrary Order

Let us consider the case of $p < 0$. For convenience let us replace p by $-p$ in the expression (1.23). Then (1.23) takes the form

$$
{}_aD_t^{-p} f(t) = \lim_{\substack{h\to 0\\ nh=t-a}} h^{p} \sum_{r=0}^{n} (-1)^r \begin{bmatrix} p \\ r \end{bmatrix} f(t-rh),
\tag{1.24}
$$

where, as above, the values of h and n relate as $nh = t - a$.

To prove the existence of the limit (1.24) and evaluate that limit, we need the following theorem [3].

Theorem 1.1 *Let us take a sequence β_k, $(k = 1, 2, \ldots)$ and suppose that*

$$
\lim_{k\to\infty} \beta_k = 1;
\tag{1.25}
$$

$$
\lim_{n\to\infty} v_{n,k} = 0 \quad \textit{for all } k,
\tag{1.26}
$$

$$
\lim_{n\to\infty} \sum_{k=1}^{n} v_{n,k} = A \quad \textit{for all } k,
\tag{1.27}
$$

$$
\sum_{k=1}^{n} |v_{n,k}| < K \quad \textit{for all } n.
\tag{1.28}
$$

Then

$$
\lim_{n\to\infty} \sum_{k=1}^{n} v_{n,k}\beta_k = A.
\tag{1.29}
$$

1.2.1.2 Derivatives of Arbitrary Order

Let us consider the case of $p > 0$. Our aim is, as above, to evaluate the limit [4]

$$ {}_aD_t^{-p}f(t) = \lim_{\substack{h\to 0\\ nh=t-a}} h^{-p}\sum_{r=0}^{n}(-1)^r\binom{p}{r}f(t-rh) = \lim_{\substack{h\to 0\\ nh=t-a}} f_h^{(p)}(t), \tag{1.30} $$

where

$$ f_h^{(p)}(t) = h^{-p}\sum_{r=0}^{n}(-1)^r\binom{p}{r}f(t-rh). \tag{1.31} $$

To evaluate the limit (1.30), let us first transform the expression for $f_h^{(p)}(t)$ in the following way. Using the known property of the binomial coefficient

$$ \binom{p}{r} = \binom{p-1}{r} + \binom{p-1}{r-1}. \tag{1.32} $$

We can write

$$ \begin{aligned} f_h^{(p)}(t) &= h^{-p}\sum_{r=0}^{n}(-1)^r\binom{p-1}{r}f(t-rh)\\ &\quad +h^{-p}\sum_{r=1}^{n}(-1)^r\binom{p-1}{r-1}f(t-rh)\\ &= h^{-p}\sum_{r=0}^{n}(-1)^r\binom{p-1}{r}f(t-rh)\\ &\quad +h^{-p}\sum_{r=0}^{n-1}(-1)^{r+1}\binom{p-1}{r}f(t-(r+1)h)\\ &= (-1)^n\binom{p-1}{n}h^{-p}f(a)\\ &\quad +h^{-p}\sum_{r=0}^{n-1}(-1)^r\binom{p-1}{r}\Delta f(t-rh), \end{aligned} \tag{1.33} $$

where we denote

$$ \Delta f(t-rh) = f(t-rh) - f(t-(r+1)h). $$

Obviously, $\Delta f(t-rh)$ is a first-order backward difference of the function $f(\tau)$ at the point $\tau = t - rh$.

Applying the property (1.32) of the binomial coefficients repeatedly m times, we obtain starting from (1.33):

$$ \begin{aligned} f_h^{(p)}(t) &= (-1)^n\binom{p-1}{n}h^{-p}f(a) + (-1)^{n-1}\binom{p-2}{n-1}h^{-p}\Delta f(a+h)\\ &\quad + h^{-p}\sum_{r=0}^{n-2}(-1)^r\binom{p-2}{r}\Delta^2 f(t-rh)\\ &= (-1)^n\binom{p-1}{n}h^{-p}f(a) + (-1)^{n-1}\binom{p-2}{n-1}h^{-p}\Delta f(a+h) \end{aligned} $$

$$
\begin{aligned}
&+(-1)^{n-2}\binom{p-3}{n-3}h^{-p}\Delta^2 f(a+2h) \\
&+h^{-p}\sum_{r=0}^{n-3}(-1)^r\binom{p-3}{r}\Delta^3 f(t-rh) \\
&=\cdots \\
&=\sum_{r=0}^{m}(-1)^{n-k}\binom{p-k-1}{n-k}h^{-p}\Delta^k f(a+kh) \\
&+h^{-p}\sum_{r=0}^{n-m-1}(-1)^r\binom{p-m-1}{r}\Delta^{m+1}f(t-rh). \qquad (1.34)
\end{aligned}
$$

Let us evaluate the limit of the kth term in the first sum in (1.34):

$$
\begin{aligned}
&\lim_{\substack{h\to 0\\ nh=t-a}}(-1)^{n-k}\binom{p-k-1}{n-k}h^{-p}\Delta^k f(a+kh) \\
&=\lim_{\substack{h\to 0\\ nh=t-a}}(-1)^{n-k}\binom{p-k-1}{n-k}(n-k)^{p-k} \\
&\quad\times\left(\frac{n}{n-k}\right)^{p-k}(nh)^{-p+k}\frac{\Delta^k f(a+kh)}{h^k} \\
&=(t-a)^{-p+k}\lim_{n\to\infty}(-1)^{n-k}\binom{p-k-1}{n-k}(n-k)^{p-k} \\
&\quad\times\lim_{n\to\infty}\left(\frac{n}{n-k}\right)^{p-k}\times\lim_{h\to\infty}\frac{\Delta^k f(a+kh)}{h^k} \\
&=\frac{f^k(a)(t-a)^{-p+k}}{\Gamma(-p+k+1)}.
\end{aligned} \qquad (1.35)
$$

Using the property of the gamma function limits,

$$
\lim_{n\to\infty}(-1)^{n-k}\binom{p-k-1}{n-k}(n-k)^{p-k}
$$

$$
=\lim_{n\to\infty}\frac{(-p+k+1)(-p+k+2)\cdots(-p+n)}{(n-k)^{-p+k}(n-k)!}=\frac{1}{\Gamma(-p+k+1)}
$$

and

$$
\lim_{n\to\infty}\left(\frac{n}{n-k}\right)^{p-k}=1,
$$

$$
\lim_{h\to 0}\frac{\Delta^k f(a+kh)}{h^k}=f^k(a).
$$

Knowing the limit (1.35), we can easily write the limit of the first sum in (1.34). To evaluate the limit of the second sum in (1.34), let us write it in the form

$$\frac{1}{\Gamma(-p+m+1)}\sum_{r=0}^{n-m-1}(-1)^r\Gamma(-p+m+1)\binom{p-m-1}{r}r^{-m+p} \times h(rh)^{m-p}\frac{\Delta^{m+1}f(t-rh)}{h^{m+1}}. \tag{1.36}$$

Using the property of the gamma function limits, we verify that

$$\lim_{r\to\infty} = \lim_{r\to\infty}(-1)^r\Gamma(-p+m+1)\binom{p-m-1}{r}r^{-m+p} = 1. \tag{1.37}$$

In addition, if $m-p>-1$, then

$$\lim_{n\to\infty}\sum_{r=0}^{n-m-1}v_{n,r} = \lim_{\substack{h\to 0\\ nh=t-a}}\sum_{r=0}^{n-m-1}h(rh)^{m-p}\frac{\Delta^{m+1}f(t-rh)}{h^{m+1}} = \int_a^t (t-\tau)^{m-p}f^{(m+1)}(\tau)d\tau. \tag{1.38}$$

Taking into account (1.37) and (1.38) and applying Theorem 1.1, we conclude that

$$\lim_{\substack{h\to 0\\ nh=t-a}} h^{-p}\sum_{r=0}^{n-m-1}(-1)^r\binom{p-m-1}{r}\Delta^{m+1}f(t-rh) = \frac{1}{\Gamma(-p+m+1)}\int_a^t (t-\tau)^{m-p}f^{(m+1)}(\tau)d\tau. \tag{1.39}$$

Using (1.35) and (1.39), we finally obtain the limit (1.30):

$${}_aD_t^p = \lim_{\substack{h\to 0\\ nh=t-a}} f_h^{(p)}(t) = \sum_{k=0}^{m}\frac{f^{(k)}(a)(t-a)^{-p+k}}{\Gamma(-p+k+1)} + \frac{1}{\Gamma(-p+m+1)}\int_a^t (t-\tau)^{m-p}f^{(m+1)}(\tau)d\tau. \tag{1.40}$$

The formula (1.40) has been obtained under the assumption that the derivatives $f^{(k)}(t)$, $(k=1,2,\dots,m+1)$ are continuous in the closed interval $[a,t]$ and that m is an integer number satisfying the condition $m>p-1$. The smallest possible value for m is determined by the inequality

$$m<p<m+1.$$

1.2.2 Properties of Riemann–Liouville Fractional Derivative and Integral

Manipulation with the Grünwald–Letnikov fractional derivatives defined as a limit of a fractional-order backward difference is not convenient. The obtained expression (1.40) looks better because of the presence of the integral in it; but what about the nonintegral terms? The answer is simple and elegant: to consider the expression (1.40) as a particular case of the integro-differential expression

$$ {}_aD_t^p f(t) = \left(\frac{d}{dt}\right)^{m+1} \int_a^t (t-\tau)^{m-p} f(\tau)d\tau, \quad (m \le p \le m+1). \tag{1.41} $$

The expression (1.41) is the most widely known definition of the fractional derivative; it is usually called the Riemann–Liouville definition.

Obviously, the expression (1.40), which has been obtained for the Grünwald–Letnikov fractional derivative under the assumption that the function $f(t)$ must be $m+1$ times continuously differentiable, can be obtained from (1.41) under the same assumption by performing repeatedly integration by parts and differentiation. This gives

$$ \begin{aligned} {}_aD_t^p &= \left(\frac{d}{dt}\right)^{m+1} \int_a^t (t-\tau)^{m-p} f(\tau)d\tau \\ &= \sum_{k=0}^{m} \frac{f^{(k)}(a)(t-a)^{-p+k}}{\Gamma(-p+k+1)} \\ &\quad + \frac{1}{\Gamma(-p+m+1)} \int_a^t (t-\tau)^{m-p} f^{(m+1)}(\tau)d\tau \\ &= {}_aD_t^p f(t), \quad (m \le p \le m+1). \end{aligned} \tag{1.42} $$

Therefore, if we consider a class of functions $f(t)$ having $m+1$ continuous derivatives for $t \ge 0$, then the Grünwald–Letnikov definition (1.30) (or, what is in this case the same, its integral form (1.40)) is equivalent to the Riemann–Liouville definition (1.41).

From the pure mathematical point of view, such a class of functions is narrow, however, this class of functions is very important for applications, because the character of the majority of dynamical processes is smooth enough and does not allow discontinuities. Understanding this fact is important for the proper use of the methods of the fractional calculus in applications, especially because of the fact that the Riemann–Liouville definition (1.41) provides an excellent opportunity to weaken the conditions on the function $f(t)$. Namely, it is enough to require the integrability of $f(t)$; then the integral (1.41) exists for $t > a$ and can be differentiated $m+1$ times. The weak conditions on the function $f(t)$ in (1.41) are necessary, for example, for obtaining the solution of the Abel integral equation.

Let us look at how the Riemann–Liouville definition (1.41) appears as the result of the unification of the notions of integer-order integration and differentiation [4].

1.2.2.1 Unification of Integer-Order Derivatives and Integrals

Let us suppose that the function $f(\tau)$ is continuous and integrable in every finite interval (a, t); the function $f(t)$ may have an integrable singularity of order $r < 1$ at the point $\tau = a$:

$$\lim_{\tau \to a} (\tau - a)^r f(t) = \text{const}(\neq 0). \tag{1.43}$$

Then the integral

$$f^{-1}(t) = \int_a^t f(\tau)d\tau \tag{1.44}$$

exists and has a finite value, namely equal to 0, as $t \to a$. Indeed, performing the substitution $\tau = a + y(t - a)$ and the denoting $\epsilon = t - a$, we obtain

$$\begin{aligned} \lim_{t \to a} f^{(-1)}(t) &= \lim_{t \to a} \int_a^t f(\tau)d\tau \\ &= \lim_{t \to a}(t - a) \int_0^1 f(a + y(t - a))dy \\ &= \lim_{\epsilon \to 0} \epsilon^{1-r} \int_0^1 (\epsilon y)^r f(a + y\epsilon)y^{-r}dy = 0, \end{aligned} \tag{1.45}$$

because $r < 1$. Therefore, we can consider the twofold integral

$$\begin{aligned} f^{-2}(t) &= \int_a^t d\tau_1 \int_a^{\tau_1} f(\tau)d\tau = \int_a^t f(\tau)d\tau \int_\tau^t d\tau_1 \\ &\int_a^t (t - \tau)f(\tau)d\tau. \end{aligned} \tag{1.46}$$

Integration of (1.46) gives the threefold integral of $f(\tau)$:

$$\begin{aligned} f^{-3}(t) &= \int_a^t d\tau_1 \int_a^{\tau_1} d\tau_2 \int_a^{\tau_2} f(\tau_3)d\tau_3 \\ &= \int_a^t d\tau_1 \int_a^{\tau_1} (\tau_1 - \tau)f(\tau)d\tau \\ &= \frac{1}{2} \int_a^t (t - \tau)^2 f(\tau)d\tau \end{aligned} \tag{1.47}$$

and by induction in the general case, we have the Cauchy formula

$$f^{(-n)}(t) = \frac{1}{\Gamma(n)} \int_a^t (t - \tau)^{n-1} f(\tau)d\tau. \tag{1.48}$$

Let us suppose that $n \geq 1$ is fixed and take integer $k \geq 0$. Obviously, we will obtain

$$f^{(-k-n)}(t) = \frac{1}{\Gamma(n)} D^{-k} \int_a^t (t - \tau)^{n-1} f(\tau)d\tau, \tag{1.49}$$

where the symbol $D^{-k}(k \geq 0)$ denotes k iterated integrations.

On the other hand, for a fixed n $geq1$ and integer $k \geq n$, the $(k - n)$-th derivative of the function $f(t)$ can be written as

$$f^{(k-n)}(t) = \frac{1}{\Gamma(n)} D^k \int_a^t (t - \tau)^{n-1} f(\tau)d\tau, \tag{1.50}$$

where the symbol $D^k(k \geq 0)$ denotes k iterated differentiations.

We see that the formulas (1.49) and (1.50) can be considered as particular cases of one them, namely (1.50), in which $n(n \leq 1)$ is fixed and the symbol D^k means k integrations if $k \leq 0$ and k differentiations if $k > 0$. If $k = n-1, n-2, \dots$, then the formula (1.50) gives iterated integrals of $f(t)$; for $k = n$, it gives the function $f(t)$; for $k = n+1, n+2, n+3, \dots$, it gives derivatives of order $k - n = 1, 2, 3, \dots$ of the function $f(t)$ [4].

1.2.2.2 Integrals of Arbitrary Order

To extend the notion of n-fold integration to noninteger values of n, we can start with the Cauchy formula (1.48) and replace the integer n in it by a real $p > 0$:

$$ {}_aD_t^{-p} = \frac{1}{\Gamma(p)} \int_a^t (t-\tau)^{p-1} f(\tau)d\tau. \tag{1.51} $$

In (1.48), the integer n must satisfy the condition $n \geq 1$; the corresponding for p is weaker: for the existence of the integral (1.51), we must have $p > 0$.

Moreover, under certain reasonable assumptions

$$ \lim_{p\to 0} D_t^{-p} f(t) = f(t), \tag{1.52} $$

so we can put

$$ {}_aD_t^0 f(t) = f(t). \tag{1.53} $$

The proof of the relationship (1.52) is very simple if $f(t)$ has continuous derivatives for $t \leq 0$. In such a case, integration by parts and the use of gamma property, it gives

$$ {}_aD_t^{-p} f(t) = \frac{(t-a)^p}{f}(a)\Gamma(p+1) + \frac{1}{\Gamma(p+1)} \int_a^t (t-\tau)^p f'(\tau)d\tau, $$

and we obtain

$$ \lim_{p\to 0} D_t^{-p} = f(a) + \int_a^t f'(\tau)d\tau = f(a) + (f(t - f(a))) = f(t). $$

If $f(t)$ is only continuous for $t \geq a$, then the proof (1.52) is somewhat longer. In such case, let us write ${}_aD_t^{(-p)} f(t)$ in the form,

$$ \begin{aligned} {}_aD_t^{(-p)} f(t) &= \frac{1}{\Gamma(p)} \int_a^t (t-\tau)^{p-1}(f(\tau) - f(t))d\tau + \frac{f(t)}{\Gamma(p)} \int_a^t (t-\tau)^{p-1} d\tau \\ &= \frac{1}{\Gamma(p)} \int_a^{t-\delta} (t-\tau)^{p-1}(f(\tau) - f(t)d\tau \\ &\quad + \frac{1}{\Gamma(p)} \int_{t-\delta}^t (t-\tau)^{p-1}(f(\tau) - f(t)d\tau + \frac{f(t)(t-a)^p}{\Gamma(p+1)}. \end{aligned} \tag{1.54} $$

Let us second part of the integral (1.54). Since $f(t)$ is continuous, for every $\delta > 0$ there exists $\epsilon > 0$ such that

$$|f(\tau) - f(t)| < \epsilon.$$

Then we have following estimate of the second part of the integral (1.54):

$$|I_2| < \frac{\epsilon}{\Gamma(p)} \int_{t-\delta}^{t} (t-\tau)^{p-1} d\tau < \frac{\epsilon \delta^p}{\Gamma(p+1)}, \tag{1.55}$$

and taking into account that $\epsilon \to 0$ and $\delta \to 0$, we obtain that for all $p \geq 0$

$$\lim_{\delta \to 0} |I_2| = 0. \tag{1.56}$$

Let us now take an arbitrary $\epsilon > 0$ and choose δ such that

$$|I_2| < \epsilon \tag{1.57}$$

for all $p \geq 0$. For this fixed δ, we obtain the following estimate first part of the integral (1.54):

$$|I_1| \leq \frac{M}{\Gamma(p)} \int_{a}^{t-\delta} (t-\tau)^{p-1} d\tau \leq \frac{M}{\Gamma(p+1)} (\delta^p - (t-a)^p), \tag{1.58}$$

from which it follows that, for fixed $\delta > 0$

$$\lim_{p \to 0} |I_1| = 0. \tag{1.59}$$

Considering

$$|{}_aD_t^{(-p)} f(t) - f(t)| \leq |I_1| + |I_2| + |f(t)| \times \left| \frac{(t-a)^p}{\Gamma(p+1)} - 1 \right|$$

and taking into account the limits (1.56) the estimate (1.57), we obtain

$$\lim_{p \to 0} \sup |{}_aD_t^{(-p)} f(t) - f(t)| \leq \epsilon,$$

where ϵ can be chosen as small as we wish. Therefore,

$$\lim_{p \to 0} \sup |{}_aD_t^{(-p)} f(t) - f(t)| = 0,$$

and (1.52) holds if $f(t)$ is continuous for $t \geq a$.

If $f(t)$ is continuous for $t \geq a$, then integration of arbitrary real order defined by (1.51) has the following important property:

$$D_t^{-p}(D_t^{-q} f(t)) = D_t^{-p-q} f(t). \tag{1.60}$$

Indeed, we have

$$\begin{aligned} D_t^{-p}(D_t^{-q}f(t)) &= \frac{1}{\Gamma(q)}\int_a^t (t-\tau)^{q-1}D_\tau^{-p}d\tau \\ &= \frac{1}{\Gamma(p)\Gamma(q)}\int_a^t (t-\tau)^{q-1}d\tau \int_a^\tau (\tau-\xi)^{p-1}f(\xi)d\xi \\ &= \frac{1}{\Gamma(p)\Gamma(q)}\int_a^t f(\xi)d\xi \int_\xi^t (t-\tau)^{q-1}(\tau-\xi)^{p-1}d\tau \\ &= \frac{1}{\Gamma(p+q)}\int_a^t (t-\xi)^{p+q-1}f(\xi)d\xi \\ &= D_t^{-p-q}f(t). \end{aligned} \tag{1.61}$$

Obviously, we can interchange p and q, so we have

$$D_t^{-p}(D_t^{-q}f(t)) = D_t^{-q}(D_t^{-p}f(t)) = D_t^{-p-q}f(t). \tag{1.62}$$

One may note that the rule (1.62) is similar to the well-known property of integer-order derivatives:

$$\frac{d^m}{dt^m}\left(\frac{d^n f(t)}{dt^n}\right) = \frac{d^n}{dt^n}\left(\frac{d^m f(t)}{dt^m}\right) = \frac{d^{m+n}f(t)}{dt^{m+n}}. \tag{1.63}$$

1.2.2.3 Derivatives of Arbitrary Order

The representation (1.50) for the derivative of an integer order $k - n$ provides an opportunity for extending the notion of differentiation to noninteger order. Namely, we can leave integer k and replace integer n with a real ν so that $k - \nu > 0$. This gives

$${}_aD_t^{k-\nu}f(t) = \frac{1}{\Gamma(\nu)}\frac{d^k}{dt^k}\int_{(a)}^t (t-\tau)^{\nu-1}f(\tau)d\tau, \quad (0 < \nu \leq 1), \tag{1.64}$$

where the only substantial restriction for $\nu > 0$, which is necessary for the convergence of the integral in (1.164). This restriction, however, can be without loss of generality, this can be easily shown with the help of the property (1.62) of the integrals of arbitrary real order and the definition (1.164).

Denoting $p = k - \nu$, we can write (1.164) as

$${}_aD_t^p f(t) = \frac{1}{\Gamma(k-p)}\frac{d^k}{dt^k}\int_{(a)}^t (t-\tau)^{k-p-1}f(\tau)d\tau, \quad (k-1 \leq p < k) \tag{1.65}$$

or

$${}_aD_t^p f(t) = \frac{d^k}{dt^k}({}_aD_t^{-(k-p)}f(t)), \quad (k-1 \leq p < k). \tag{1.66}$$

If $p = k - 1$, then we obtain a conventional integer-order derivative of order $k - 1$:

$$\begin{aligned} {}_aD_t^p f(t) &= \frac{d^k}{dt^k}({}_aD_t^{-(k-(k-1))}f(t)) \\ &= \frac{d^k}{dt^k}({}_aD_t^{-1}f(t)) = f^{(k-1)}(t). \end{aligned} \tag{1.67}$$

Moreover, using (1.68) we see that for $p = k \geq 1$ and $t > a$

$$ {}_aD_t^p f(t) = \frac{d^k}{dt^k}({}_aD_t^0 f(t)) = \frac{d^k f(t)}{dt^k} = f^{(k)}(t), \tag{1.68}$$

which means that for $t > a$ the Riemann–Liouville fractional derivative (1.165) of order $p = k > 1$ coincides with the conventional derivative of order k.

Let us now consider some properties of the Riemann–Liouville fractional derivatives. The first and maybe the most important property of the Riemann–Liouville fractional derivative is that for $p > 0$ and $t > a$

$$ {}_aD_t^p({}_aD_t^{-p} f(t)) = f(t), \tag{1.69}$$

which means that the Riemann–Liouville fractional differentiation operator is a left inverse to the Riemann–Liouville fractional integration operator of the some order p.

To prove the property (1.69), let us consider the case of integer $p = n \geq 1$:

$$\begin{aligned} {}_aD_t^n({}_aD_t^{-n} f(t)) &= \frac{d^n}{dt^n}\int_a^t (t-\tau)^{n-1} f(\tau)d\tau \\ &= \frac{d}{dt}\int_a^t f(\tau)d\tau = f(t). \end{aligned} \tag{1.70}$$

Taking now $k - 1 \leq p < k$ and using the composition rule (1.62) for the Riemann–Liouville fractional integrals, we can write

$$ {}_aD_t^{-k} f(t) = {}_aD_t^{-k-p}({}_aD_t^{-p} f(t)). \tag{1.71}$$

Therefore,

$$\begin{aligned} {}_aD_t^p({}_aD_t^{-p} f(t)) &= \frac{d^k}{dt^k}\{{}_aD_t^{-(k-p)}({}_aD_t^{-p} f(t))\} \\ &= \frac{d^k}{dt^k}\{{}_aD_t^{-p} f(t)\} = f(t), \end{aligned} \tag{1.72}$$

which ends the proof of the property (1.71).

As with conventional integer-order differentiation and integration, fractional differentiation and integration do not compute.

If the fractional derivative ${}_aD_t^p f(t)$, $k - 1 \leq p < k$, of a function $f(t)$ is integrable, then

$$ {}_aD_t^p({}_aD_t^{-p} f(t)) = f(t) - \sum_{j=1}^{k} [{}_aD_t^{p-j}]_{t=a} \frac{(t-a)^{p-j}}{\Gamma(p-j+1)}. \tag{1.73}$$

Indeed, on the one hand we have

$$\begin{aligned} {}_aD_t^{-p}({}_aD_t^p f(t)) &= \frac{1}{\Gamma(p)}\int_a^t (t-\tau)^{p-1}({}_aD_\tau^p f(\tau))d\tau \\ &= \frac{d}{dt}\left\{\frac{1}{\Gamma(p+1)}\int_a^t (t-\tau)^p({}_aD_\tau^p f(\tau))d\tau\right\}. \end{aligned} \tag{1.74}$$

On the other hand, repeatedly integrating by parts and then using (1.72), we obtain

$$\begin{aligned}
&\frac{1}{\Gamma(p+1)}\int_a^t (t-\tau)^p({}_aD_\tau^p f(\tau))d\tau \\
&= \frac{1}{\Gamma(p+1)}\int_a^t (t-\tau)^p \frac{d^k}{d\tau^k}({}_aD_\tau^{-(k-p)} f(\tau))d\tau \\
&= \frac{1}{\Gamma(p-k+1)}\int_a^t (t-\tau)^{p-k}({}_aD_\tau^{-(k-p)} f(\tau))d\tau \\
&\quad - \sum_{j=1}^{k}\left[\frac{d^{k-j}}{dt^{k-j}}\,{}_aD_t^{-(k-p)} f(t)\right]_{t=a}\frac{(t-a)^{p-j+1}}{\Gamma(2+p-j)} \\
&= \frac{1}{\Gamma(p-k+1)}\int_a^t (t-\tau)^{p-k}({}_aD_\tau^{-(k-p)} f(\tau))d\tau \\
&\quad - \sum_{j=1}^{k}[{}_aD_t^{p-j} f(t)]_{t=a}\frac{(t-a)^{p-j+1}}{\Gamma(2+p-j)} \\
&= {}_aD_t^{-(p-k+1)}({}_aD_t^{-(k-p)} f(t)) \\
&\quad - \sum_{j=1}^{k}[{}_aD_t^{p-j} f(t)]_{t=a}\frac{(t-a)^{p-j+1}}{\Gamma(2+p-j)} \\
&= {}_aD_t^{-1} f(t) - \sum_{j=1}^{k}[{}_aD_t^{p-j} f(t)]_{t=a}\frac{(t-a)^{p-j+1}}{\Gamma(2+p-j)}.
\end{aligned} \tag{1.75}$$

The existence of all terms in (1.75) follows from the integrability of ${}_aD_t^p f(t)$, because due to this condition the fractional derivatives ${}_aD_t^{p-j} f(t)$, $(j = 1, 2, \dots, k)$ are all bounded at $t = a$.

Combining (1.74) and (1.75) ends the proof of the relationship (1.73). An important particular case must be mentioned. If $0 < p < 1$, then

$${}_aD_t^{-p}({}_aD_t^p f(t)) = f(t) - [{}_aD_t^{p-1}]_{t=a}\frac{(t-a)^{p-1}}{\Gamma(p)}. \tag{1.76}$$

The property (1.69) is a particular case of a more general property

$${}_aD_t^p({}_aD_t^{-p} f(t)) = {}_aD_t^{p-q} f(t), \tag{1.77}$$

where we assume that $f(t)$ is continuous and, if $p \geq q \geq 0$, that the derivative ${}_aD_t^{p-q} f(t)$ exists.

Two cases must be considered: $q \geq p \geq 0$ and $p > q \geq 0$.

If $q \geq p \geq 0$, then using the properties (1.69) and (1.77), we obtain

$$\begin{aligned}
{}_aD_t^p({}_aD_t^{-q} f(t)) &= {}_aD_t^p({}_aD_t^{-p}\,{}_aD_t^{-(q-p)} f(t)) \\
&= {}_aD_t^{-(q-p)} f(t) = {}_aD_t^{p-q} f(t).
\end{aligned} \tag{1.78}$$

Now let us consider the case $p > q \geq 0$. Let us denote by m and n integers such that $0 \leq m-1 \leq p < m$ and $0 \leq n \leq p-q < n$. Obviously, $n \leq m$. Then, using the definition (1.165) and the property (1.77), we obtain

$$
\begin{aligned}
{}_aD_t^p({}_aD_t^{-q}f(t)) &= \frac{d^m}{dt^m}\{{}_aD_t^{-(m-p)}({}_aD_t^{-q}f(t))\} \\
&= \frac{d^m}{dt^m}\{{}_aD_t^{p-q-m}f(t)\} \\
&= \frac{d^n}{dt^n}\{{}_aD_t^{p-q-n}f(t)\} = {}_aD_t^{p-q}f(t).
\end{aligned} \tag{1.79}
$$

The above-mentioned property (1.73) is a particular case of the more

$$
{}_aD_t^{-p}({}_aD_t^{q}f(t)) = {}_aD_t^{q-p}f(t) - \sum_{j=1}^{k}[{}_aD_t^{q-j}]_{t=a}\frac{(t-a)^{p-j}}{\Gamma(p-j+1)}. \tag{1.80}
$$

To prove the formula (1.80), we first use property (1.77) (if $q \leq p$) or property (1.79) (if $q \geq p$) and then property (1.73). This gives

$$
\begin{aligned}
{}_aD_t^{-p}({}_aD_t^{q}f(t)) &= {}_aD_t^{q-p}\{{}_aD_t^{-q}({}_aD_t^{q}f(t))\} \\
&= {}_aD_t^{q-p}\left\{f(t) - \sum_{j=1}^{k}[{}_aD_t^{q-j}f(t)]_{t=a}\frac{(t-a)^{q-j}}{\Gamma(p-j+1)}\right\} \\
&= {}_aD_t^{q-p}f(t) - \sum_{j=1}^{k}[{}_aD_t^{q-j}f(t)]_{t=a}\frac{(t-a)^{p-j}}{\Gamma(p-j+1)},
\end{aligned} \tag{1.81}
$$

where we used the known derivative of the power function [4].

1.3 Properties of Caputo Fractional Derivative and Integral

The definition (1.165) of the fractional differentiation of the Riemann–Liouville type played an important role in the development of the theory of fractional derivatives and integrals and for its applications in pure mathematics (solution of integer-order differential equations, definitions of new function classes, summation of series, etc.).

However, the demands of modern technology require a certain revision of the well-established pure mathematical approach. There have appeared a number of works, especially in the theory of viscoelasticity and in hereditary solid mechanics, where fractional derivatives are used for a better description of material properties. Mathematical modeling based on enhances rheological models naturally leads to differential equations of fractional order – and to the necessity of the formulation of initial conditions to such equations. Applied problems require definitions of fractional derivatives allowing the utilization of physically interpretable initial conditions, which contain $f(a), f'(a)$, etc. Unfortunately, the Riemann–Liouville approach leads to initial conditions containing the limit values of the Riemann–Liouville fractional derivatives at the lower terminal $t = a$, for example

$$\begin{aligned} &\lim_{t\to a}({}_aD_t^{\nu-1}f(t)) = b_1, \\ &\lim_{t\to a}({}_aD_t^{\nu-2}f(t)) = b_2, \\ &\quad \vdots, \\ &\lim_{t\to a}({}_aD_t^{\nu-n}f(t)) = b_n, \end{aligned} \tag{1.82}$$

where $b_k, k = 1, 2, \ldots, n$ are given constants.

In spite of the fact that initial value problems with such initial conditions can be successfully solved mathematically (see, for example, solutions given in [10] arid in this book), their solutions are practically useless, because there is no known physical interpretation for such types of initial conditions.

Here we observe a conflict between the well-established and polished mathematical theory and practical needs.

A certain solution to this conflict was proposed by M. Caputo first in his paper [5] and two years later in his book [11], and recently (in Banach spaces) by El-Sayed [12, 13]. Caputo's definition can be written as

$${}_a^CD_t^{\nu}f(t) = \frac{1}{\Gamma(\nu - n)}\int_a^t \frac{f^{(n)}(\tau)d\tau}{(t-\tau)^{\nu+1-n}}, \qquad (n - 1 < \nu < n). \tag{1.83}$$

Under natural conditions on the function $f(t)$, for $\nu \to n$ the Caputo derivative becomes a conventional nth derivative of the function $f(t)$. Indeed, let us assume that $0 \leq n - 1 < \nu < n$ and that the function $f(t)$ has $n + 1$ continuous bounded derivatives in $[a, T]$ for every $T > a$. Then

$$\begin{aligned} {}_a^CD_t^{\nu}f(t) &= \lim_{a\to n}\left(\frac{f^n(a)(t-a)^{n-\nu}}{\Gamma(n-\nu+1)}\right. \\ &\quad \left.+\frac{1}{\Gamma(n-\nu+1)}\int_a^t (t-\tau)^{n-\nu}f^{(n+1)}(\tau)d\tau\right) \\ &= f^n(a) + \int_a^t f^{(n+1)}(\tau)d\tau = f^{(n)}(t), \quad n = 1, 2, \ldots \end{aligned} \tag{1.84}$$

This says that, similarly to the Grünwald–Letnikov and the Riemann–Liouville approaches, the Caputo approach also provides an interpolation between integer-order derivatives.

The main advantage of Caputo's approach is that the initial conditions for fractional differential equations with Caputo derivatives take on the same form as for integer-order differential equations, i.e. contain the limit values of integer-order derivatives of unknown functions at the lower terminal $t = a$.

To underline the difference in the form of the initial conditions which must accompany fractional differential equations in terms of the Riemann–Liouville and the Caputo derivatives, let us recall the corresponding Laplace transform formulas for the case $\nu = 0$.

The formula for the Laplace transform of the Riemann-Liouville fractional derivative is

$$\int_0^\infty \{{}^0D_t^\nu f(t)\}dt = p^\nu F(p) - \sum_{k=0}^{n-1} p^k \{{}^0D_t^{\nu-k-1} f(t)\}|_{t=0}, \quad (n-1 < \nu \leq n), \tag{1.85}$$

whereas Caputo's formula, first obtained in [5], for the Laplace transform of the Caputo derivative is

$$\int_0^\infty e^{-pt}\{{}_C^0D_t^\nu f(t)\}dt = p^\nu F(p) - \sum_{k=0}^{n-1} p^{\nu-k-1} f^{(k)}(0), \quad (n-1 < \nu \leq n). \tag{1.86}$$

We see that the Laplace transform of the Riemann-Liouville fractional derivative allows utilization of initial conditions of the type (1.97), which can cause problems with their physical interpretation. On the contrary, the Laplace transform of the Caputo derivative allows utilization of initial values of classical integer-order derivatives with known physical interpretations.

The Laplace transform method is frequently used for solving applied problems. To choose the appropriate Laplace transform formula, it is very important to understand which type of definition of fractional derivative must be used.

Another difference between the Riemann–Liouville definition (1.165) and the Caputo definition (1.83) is that the Caputo derivative of a constant is 0, whereas in the cases of a finite value of the lower terminal a the Riemann–Liouville fractional derivative of a constant C is not equal to 0, but

$${}^0D_t^\nu C = \frac{Ct^{-\nu}}{\Gamma(1-\nu)}. \tag{1.87}$$

This fact led, for example, Ochmann and Makarov [14] to using the Riemann–Liouville definition with $a = -\infty$, because, on the one hand, from the physical point of view they need the fractional derivative of a constant equal to zero and on the other hand, formula (1.87) gives 0 if $a \to -\infty$. The physical meaning of this step is that the starting time of the physical process is set to $-\infty$. In such a case transient effects cannot be studied. However, taking $a = -\infty$ is the necessary abstraction for the consideration of the steady-state processes, for example for studying the response of the fractional-order dynamic system to the periodic input signal, wave propagation in viscoelastic materials, etc.

Putting $a = -\infty$ in both definitions and requiring reasonable behavior of $f(t)$ and its derivatives for $t \to -\infty$, we arrive at the same formula

$${}_{-\infty}D_t^\nu f(t) = {}_{-\infty}^{C} D_t^\nu f(t) = \frac{1}{\Gamma(n-\nu)} \int_{-\infty}^t \frac{f^{(n)}(\tau)d\tau}{(t-\tau)^{\nu+1-n}}, \quad (n-1 < \nu < n), \tag{1.88}$$

which shows that for the study of steady-state dynamical processes the Riemann–Liouville definitions and the Caputo definitions must give the same results.

There is also another difference between the Riemann–Liouville and the Caputo approaches, which we would like to mention here and which seems to be important for applications. Namely, for the Caputo derivative, we have

$$ {}_a^C D_t^{\nu} ({}_a^C D_t^{m} f(t)) = {}_a^C D_t^{\nu+m} f(t), \quad (m = 0, 1, 2, \dots ; \quad n-1 < \nu < n), \tag{1.89} $$

while for the Riemann–Liouville derivative,

$$ {}_a D_t^{m} ({}_a D_t^{\nu} f(t)) = {}_a D_t^{\nu+m} f(t), \quad (m = 0, 1, 2, \dots ; \quad n-1 < \nu < n). \tag{1.90} $$

The interchange of the differentiation operators in formulas (1.89) and (1.90) is allowed under different conditions:

$$ {}_a^C D_t^{\nu} ({}_a^C D_t^{m} f(t)) = {}^C D_t^{m} ({}_a^C D_t^{\nu} f(t)) = {}_a^C D_t^{\nu+m} f(t), \ (m = 0, 1, 2, \dots ; \ n-1 < \nu < n) \tag{1.91} $$

$$ f^{(s)}(0) = 0, \quad s = n, n+1, \dots, m $$

and

$$ {}_a D_t^{m} ({}_a D_t^{\nu} f(t)) = {}_a D_t^{\nu} ({}_a D_t^{m} f(t)) = {}_a D_t^{\nu+m} f(t), \ (m = 0, 1, 2, \dots ; \ n-1 < \nu < n) \tag{1.92} $$

$$ f^{(s)}(0) = 0, \quad s = 0, 1, 2, \dots, m. $$

We see that contrary to the Riemann–Liouville approach, in the case of the Caputo derivative there are no restrictions on the values $f^{(s)}(0)$, $(s = 0, 1, \dots, n-1)$ [4].

1.4 Properties of the Caputo–Fabrizio Fractional Derivative and Integral

Let us recall the usual Caputo fractional time derivative (UFD$_t$) of order ν, given by [6]

$$ D_t^{(\nu)} f(t) = \frac{1}{\Gamma(1-\nu)} \int_a^t \frac{f'(\tau)}{(t-\tau)^{\nu}} d\tau \tag{1.93} $$

with $\nu \in [0, 1]$ and $a \in (-\infty, t), f \in H^1(a, b), b > a$. By changing the kernel $(t-\tau)^{\nu}$ with the function $\exp(-\frac{\nu}{1-\nu} t)$ and $\frac{1}{\Gamma(1-\nu)}$ with $\frac{M(\nu)}{1-\nu}$, we obtain the following new definition of fractional time derivative NFD$_t$

$$ \mathfrak{D}_t^{(\nu)} f(t) = \frac{M(\nu)}{1-\nu} \int_a^t f'(\tau) \exp\left[-\frac{\nu(t-\tau)}{1-\nu}\right] d\tau, \tag{1.94} $$

where $M(\nu)$ is a normalization function such that $M(0) = M(1) = 1$. According to the definition (1.94), the NFD_t is zero when $f(t)$ is constant, as in the UFD_t, but, contrary to the UFD_t, the kernel does not have singularity for $t = \tau$.

The new NFD_t can also be applied to functions that do not belong to $H^1(a, b)$. Indeed, the definition (1.94) can be formulated also for $f \in L^1(-\infty, b)$ and for any $\nu \in [0, 1]$ as

$$\mathfrak{D}_t^{(\nu)} f(t) = \frac{\nu M(\nu)}{1-\nu} \int_{-\infty}^{t} (f(t) - f(\tau)) \exp\left[-\frac{\nu(t-\tau)}{1-\nu}\right] d\tau. \tag{1.95}$$

Now, it is worth to observe that if we put

$$\sigma = \frac{1-\nu}{\nu} \in [0, \infty], \quad \nu = \frac{1}{1+\sigma} \in [0, 1]$$

the definition (1.94) of NFD_t assumes the form

$$\tilde{\mathfrak{D}}_t^{(\sigma)} f(t) = \frac{N(\sigma)}{\sigma} \int_a^t f'(\tau) \exp\left[-\frac{(t-\tau)}{\sigma}\right] d\tau, \tag{1.96}$$

where $\sigma \in [0, \infty]$ and $N(\sigma)$ is the corresponding normalization term of $M(\nu)$, such that $N(0) = N(\infty) = 1$. Moreover, because

$$\lim_{\sigma \to 0} \frac{1}{\sigma} \exp\left[-\frac{(t-\tau)}{\sigma}\right] d\tau \tag{1.97}$$

and for $\nu \to 1$, we have $\sigma \to 0$. Then,

$$\begin{aligned} \lim_{\nu \to 1} \mathfrak{D}_t^{(\nu)} f(t) &= \lim_{\nu \to 1} \tfrac{M(\nu)}{1-\nu} \int_a^t f'(\tau) \exp\left[-\frac{(t-\tau)}{1-\nu}\right] d\tau \\ &= \lim_{\sigma \to 0} \tfrac{N(\sigma)}{\sigma} \int_a^t f'(\tau) \exp\left[-\frac{(t-\tau)}{\sigma}\right] d\tau = f'(t). \end{aligned} \tag{1.98}$$

Otherwise, when $\nu \to 0$, then $\sigma \to +\infty$. Hence,

$$\begin{aligned} \lim_{\nu \to 0} \mathfrak{D}_t^{(\nu)} f(t) &= \lim_{\nu \to 0} \tfrac{M(\nu)}{1-\nu} \int_a^t f'(\tau) \exp\left[-\frac{(t-\tau)}{1-\nu}\right] d\tau \\ &= \lim_{\sigma \to +\infty} \tfrac{N(\sigma)}{\sigma} \int_a^t f'(\tau) \exp\left[-\frac{(t-\tau)}{\sigma}\right] d\tau = f(t) - f(a). \end{aligned} \tag{1.99}$$

Theorem 1.2 *For NFD_t, if the function $f(t)$ is such that*

$$f^{(s)}(a) = 0, \quad s = 1, 2, \ldots, n$$

then, we have

$$\mathfrak{D}_t^{(n)} (\mathfrak{D}_t^{(\nu)} f(t)) = \mathfrak{D}_t^{(\nu)} (\mathfrak{D}_t^{(n)} f(t)). \tag{1.100}$$

Proof: We begin considering $n = 1$, then from definition (1.101) of $\mathfrak{D}_t^{(\nu+1)} f(t)$, we obtain

$$\mathfrak{D}_t^{(\nu)} (\mathfrak{D}_t^{(1)} f(t)) = \frac{M(\nu)}{1-\nu} \int_a^t f'(\tau) \exp\left[-\frac{\nu(t-\tau)}{1-\nu}\right] d\tau. \tag{1.101}$$

Hence, after an integration by parts and assuming $f'(a) = 0$, we have

$$\begin{aligned} \mathfrak{D}_t^{(\nu)}(\mathfrak{D}_t^{(1)}f(t)) &= \tfrac{M(\nu)}{1-\nu}\int_a^t \left(\frac{d}{d\tau}f'(\tau)\right)\exp\left[-\frac{\nu(t-\tau)}{1-\nu}\right]d\tau \\ &= \tfrac{M(\nu)}{1-\nu}\left[\int_a^t \frac{d}{d\tau}(f'(\tau))\exp\left(-\frac{\nu(t-\tau)}{1-\nu}\right)d\tau\right. \\ &\quad \left. -\tfrac{\nu}{1-\nu}\int_a^t f'(\tau)\exp\left(-\frac{\nu(t-\tau)}{1-\nu}\right)d\tau\right] \\ &= \tfrac{M(\nu)}{1-\nu}\left[f'(t) - \tfrac{\nu}{1-\nu}\int_a^t f'(\tau)\exp\left(-\frac{\nu(t-\tau)}{1-\nu}\right)d\tau\right], \end{aligned} \tag{1.102}$$

otherwise

$$\begin{aligned} \mathfrak{D}_t^{(1)}(\mathfrak{D}_t^{(\nu)}f(t)) &= \tfrac{d}{dt}\left(\tfrac{M(\nu)}{1-\nu}\int_a^t f'(\tau)\exp\left[-\frac{\nu(t-\tau)}{1-\nu}\right]d\tau\right) \\ &= \tfrac{M(\nu)}{1-\nu}\left[f'(t) - \tfrac{\nu}{1-\nu}\int_a^t f'(\tau)\exp\left(-\frac{\nu(t-\tau)}{1-\nu}\right)d\tau\right]. \end{aligned} \tag{1.103}$$

It is easy to generalize the proof for any $n > 1$ [6]. □

It is well known that Laplace transform plays an important role in the study of ordinary differential equations. In the case of this new fractional definition, it is also known (see [6]) that, for $0 < \nu < 1$,

$$\mathfrak{L}[{}^{CF}D_t^{\nu}f(t)](s) = \frac{(2-\nu)M(\nu)}{2(s+\nu(1-s))}(s\mathfrak{L}[f(t)](s) - f(0)), \quad s > 0, \tag{1.104}$$

where $\mathfrak{L}[g(t)]$ denotes the Laplace transform of function g. So, it is clear that if we work with Caputo–Fabrizio derivative, Laplace transform will also be a very useful tool [15].

After the notion of fractional derivative of order $0 < \nu < 1$, that of fractional integral of order $0 < \nu < 1$ becomes a natural requirement. In this section, we obtain the fractional integral associated to the Caputo–Fabrizio fractional derivative previously introduced. Let $0 < \nu < 1$. Consider now the following fractional differential equation,

$$^{CF}D_t^{\nu}f(t) = u(t), \quad t \geq 0 \tag{1.105}$$

Using Laplace transform, we obtain:

$$\mathfrak{L}[{}^{CF}D_t^{\nu}f(t)](s) = \mathfrak{L}[u(t)](s), \quad s > 0. \tag{1.106}$$

That is, using (1.106), we have that

$$\frac{(2-\nu)M(\nu)}{2(s+\nu(1-s))}(s\mathfrak{L}[f(t)](s) - f(0)) = \mathfrak{L}[u(t)](s), \quad s > 0,$$

or equivalently,

$$\mathfrak{L}[f(t)](s) = \frac{1}{s}f(0) + \frac{2\nu}{s(2-\nu)M(\nu)}\mathfrak{L}[u(t)](s) + \frac{2(1-\nu)}{(2-\nu)M(\nu)}\mathfrak{L}[u(t)](s), \ s > 0.$$

Hence, using now well-known properties of inverse Laplace transform, we deduce that

$$f(t) = \frac{2(1-\nu)}{(2-\nu)M(\nu)}u(t) + \frac{2\nu}{(2-\nu)M(\nu)}\int_0^t u(s)ds + c, \quad t \geq 0, \tag{1.107}$$

where $c \in R$ is a constant and is also a solution of (1.107).

We can also rewrite fractional differential equation (1.107) as

$$\frac{(2-\nu)M(\nu)}{2(1-\nu)}\int_0^t \exp\left(-\frac{\nu}{1-\nu}(t-s)\right)f'(s)ds = u(t), \quad t \geq 0,$$

or equivalently,

$$\int_0^t \exp\left(\frac{\nu}{1-\nu}s\right)f'(s)ds = \frac{2(1-\nu)}{(2-\nu)M(\nu)}\exp\left(\frac{\nu}{1-\nu}t\right)u(t), \quad t \geq 0.$$

Differentiating both sides of the latter equation, we obtain that,

$$f'(t) = \frac{2(1-\nu)}{(2-\nu)M(\nu)}\left(u'(t) + \frac{\nu}{1-\nu}u(t)\right), \quad t \geq 0.$$

Hence, integrating now from 0 to t, we deduce as in (1.109), that

$$f(t) = \frac{2(1-\nu)}{(2-\nu)M(\nu)}[u(t) - u(0)] + \frac{2\nu}{(2-\nu)M(\nu)}\int_0^t u(s)ds + f(0), \quad t \geq 0.$$

Thus, as consequence, we expect that the fractional integral of Caputo–Fabrizio type must be defined as follows.

Definition 1.1 Let $0 < \nu < 1$. The fractional integral of order ν of a function f is defined by,

$$^{CF}I^{\nu}f(t) = \frac{2(1-\nu)}{(2-\nu)M(\nu)}u(t) + \frac{2\nu}{(2-\nu)M(\nu)}\int_0^t u(s)ds, \quad t \geq 0. \tag{1.108}$$

Definition 1.2 Let $0 < \nu < 1$. The fractional Caputo–Fabrizio derivative of order ν of a function f is given by,

$$^{CF}D^{\nu}_{\star}f(t) = \frac{1}{1-\nu}\int_0^t \exp\left(-\frac{\nu}{1-\nu}(t-s)\right)f'(s)ds, \quad t \geq 0. \tag{1.109}$$

Lemma 1.1 *Let $0 < \nu < 1$ and f be a solution of the following fractional differential equation,*

$$^{CF}D^{\nu}f(t) = 0, \quad t \geq 0. \tag{1.110}$$

Then, f is a constant function. The converse, as indicated in the Introduction, is also true [15].

Proof: From (1.109), we obtain that the solution of (1.112) must satisfy $f(t) = f(0)$ for all $t \geq 0$. Hence, it is clear that f must be a constant function. □

Proposition 1.1 *Let $0 < \nu < 1$. Then, the unique solution of the following initial value problem* [15]

$$^{CF}D^{\nu}f(t) = \sigma(t), \quad t \geq 0, \tag{1.111}$$

$$f(0) = f_0 \in R \tag{1.112}$$

is given by

$$f(t) = f_0 + a_{\nu}(\sigma(t) - \sigma(0)) + b_{\nu}I^1\sigma(t), \quad t \geq 0, \tag{1.113}$$

where $I^1\sigma$ denotes a primitive of σ and

$$a_{\nu} = \frac{2(1-\nu)}{(2-\nu)M(\nu)}, \quad b_{\nu} = \frac{2\nu}{(2-\nu)M(\nu)}. \tag{1.114}$$

Proposition 1.2 *Let $0 < \nu < 1$. Then, initial value problem given by* [15]

$$^{CF}D^{\nu}f(t) = \lambda f(t) + u(t), \quad t \geq 0,$$

$$f(0) = f_0 \in R$$

has a unique solution for any $\lambda \in R$ [15].

1.5 Properties of the Atangana–Baleanu Fractional Derivative and Integral

We recall that the Mittag-Leffler function is the solution of the following fractional ordinary differential equation [16–18]:

$$\frac{d^{\nu}y}{dx^{\nu}} = ay, \quad 0 < \nu < 1. \tag{1.115}$$

The Mittag-Leffler function and its generalized versions are therefore considered as nonlocal functions. Let us consider the following generalized Mittag-Leffler function:

$$E_{\nu}(-t^{\nu}) = \sum_{k=0}^{\infty} \frac{(-t)^{\nu k}}{\Gamma(\nu k + 1)}. \tag{1.116}$$

The Taylor series of $\exp(-(t-y))$ at the point t is given by:

$$\exp(-a(t-y)) = \sum_{k=0}^{\infty} \frac{(-a(t-y))^k}{k!}. \tag{1.117}$$

If we chose $a = \frac{\nu}{1-\nu}$ and replace the above expression into Caputo–Fabrizio derivative, we conclude that

$$D_t^\nu(f(t)) = \frac{M(\nu)}{1-\nu}\sum_{k=0}^{\infty}\frac{(-a)^k}{k!}\int_b^t \frac{df(y)}{dy}(t-y)^k dy. \tag{1.118}$$

To solve the problem of nonlocality, we derive the following expression.

In Eq. (1.120), we replace $k!$ by $\Gamma(\nu k+1)$ also $(t-y)^k$ is replaced by $(t-y)^{\nu k}$ to obtain:

$$D_t^\nu(f(t)) = \frac{M(\nu)}{1-\nu}\sum_{k=0}^{\infty}\frac{(-a)^k}{\Gamma(\nu k+1)}\int_b^t \frac{df(y)}{dy}(t-y)^{\nu k} dy. \tag{1.119}$$

Thus, the following derivative is proposed.

Definition 1.3 Let $f \in H^1(a,b), b > a, \nu \in [0,1]$ then, the definition of the new fractional derivative is given as:

$${}_b^{ABC}D_t^\nu(f(t)) = \frac{B(\nu)}{1-\nu}\int_b^t f'(x)E_\nu\left[-\nu\frac{(t-x)^\nu}{1-\nu}\right]dx. \tag{1.120}$$

Of course $B(\nu)$ has the same properties as in Caputo and Fabrizio case. The above definition will be helpful to discuss real world problems, and it also will have a great advantage when using the Laplace transform to solve some physical problem with initial condition. However, when ν is 0 we do not recover the original function except when at the origin the function vanishes. To avoid this issue, we propose the following definition.

Definition 1.4 Let $f \in H^1(a,b), b > a, \nu \in [0,1]$ then, the definition of the new fractional derivative is given as:

$${}_b^{ABC}D_t^\nu(f(t)) = \frac{B(\nu)}{1-\nu}\frac{d}{dt}\int_b^t f(x)E_\nu\left[-\nu\frac{(t-x)^\nu}{1-\nu}\right]dx. \tag{1.121}$$

Equations (1.122) and (1.123) have a nonlocal kernel. Also in Eq. (1.122) when the function is constant, we get zero. We now show the relation between both derivatives with Laplace transform. By simple calculation, we conclude that

$$\mathfrak{L}[{}_0^{ABR}D_t^\nu f(t)](s) = \frac{B(\nu)}{1-\nu}\frac{s^\nu\mathfrak{L}\{f(t)\}(s)}{s^\nu+\frac{\nu}{1-\nu}} \tag{1.122}$$

and

$$\mathfrak{L}[{}_0^{ABC}D_t^\nu f(t)](s) = \frac{B(\nu)}{1-\nu}\frac{s^\nu\mathfrak{L}\{f(t)\}(s)-s^{\nu-1}f(0)}{s^\nu+\frac{\nu}{1-\nu}} \tag{1.123}$$

respectively.

The following theorem can therefore be established.

Theorem 1.3 *Let $f \in H^1(a,b), b > a, \nu \in [0,1]$ then, the following relation is obtained*

$$ {}_0^{ABC}D_t^{\nu}(f(t)) = {}_0^{ABR}D_t^{\nu}(f(t)) + H(t). \tag{1.124}$$

Proof: By using the definition (1.126) and the Laplace transform applied on both sides, we obtain easily the following result:

$$\mathfrak{L}[{}_0^{ABC}D_t^{\nu}f(t)](s) = \frac{B(\nu)}{1-\nu}\frac{s^{\nu}\mathfrak{L}\{f(t)\}(s)}{s^{\nu}+\frac{\nu}{1-\nu}} - \frac{s^{\nu-1}f(0)}{s^{\nu}+\frac{\nu}{1-\nu}}\frac{B(\nu)}{1-\nu}. \tag{1.125}$$

Following Eq. (1.124), we have

$$\mathfrak{L}[{}_0^{ABC}D_t^{\nu}f(t)](s) = \mathfrak{L}[{}_0^{ABR}D_t^{\nu}f(t)](s) - \frac{s^{\nu-1}f(0)}{s^{\nu}+\frac{\nu}{1-\nu}}\frac{B(\nu)}{1-\nu}. \tag{1.126}$$

Applying the inverse Laplace on both sides of Eq. (1.128), we obtain

$$\mathfrak{L}[{}_0^{ABC}D_t^{\nu}f(t)] = \mathfrak{L}[{}_0^{ABR}D_t^{\nu}f(t)] - \frac{B(\nu)}{1-\nu}f(0)E_{\nu}\left(-\frac{\nu}{1-\nu}t^{\nu}\right). \tag{1.127}$$

This completes the proof. □

Theorem 1.4 *Let f be a continuous function on a closed interval $[a,b]$. Then, the following inequality is obtained on $[a,b]$*

$$||{}_0^{ABC}D_t^{\nu}f(t)|| < \frac{B(\nu)}{1-\nu}K, \quad ||h(t)|| = \max_{a\le t\le b}|h(t)|. \tag{1.128}$$

Proof:

$$\begin{aligned} ||{}_0^{ABR}D_t^{\nu}f(t)|| &= \frac{B(\nu)}{1-\nu}\frac{d}{dt}\int_0^t f(x)E_{\nu}\left[-\nu\frac{(t-x)^{\nu}}{1-\nu}\right]dx \\ &< \frac{B(\nu)}{1-\nu}||\frac{d}{dt}\int_0^t f(x)dx|| = \frac{B(\nu)}{1-\nu}||f(x)||. \end{aligned}$$

Then taking K to be $||f(x)||$ the proof is completed. □

Theorem 1.5 *The A.B. derivative in Riemann and Caputo sense possess the Lipschitz condition, that is to say, for a given couple function f and h, the following inequalities can be established:*

$$||{}_0^{ABR}D_t^{\nu}f(t) - {}_0^{ABR}D_t^{\nu}h(t)|| \le H||f(t)-h(t)|| \tag{1.129}$$

and also

$$||{}_0^{ABC}D_t^{\nu}f(t) - {}_0^{ABC}D_t^{\nu}h(t)|| \le H||f(t)-h(t)||. \tag{1.130}$$

We present the proof of (1.131) *as the proof of* (1.132) *can be obtained similarly.*

Proof:

$$||{}_0^{ABR}D_t^{\nu}f(t) - {}_0^{ABR}D_t^{\nu}h(t)|| =$$

$$||\frac{B(\nu)}{1-\nu}\frac{d}{dt}\int_0^t f(x)E_{\nu}\left[-\nu\frac{(t-x)^{\nu}}{1-\nu}\right]dx - \frac{B(\nu)}{1-\nu}\frac{d}{dt}\int_0^t h(x)E_{\nu}\left[-\nu\frac{(t-x)^{\nu}}{1-\nu}\right]dx||$$

Using the Lipschitz condition of the first-order derivative, we can find a small positive constant such that:

$$||{}_0^{ABR}D_t^{\nu}f(t) - {}_0^{ABR}D_t^{\nu}h(t)|| < \frac{B(\nu)\theta_1}{1-\nu}E_{\nu}\left[-\nu\frac{t^{\nu}}{1-\nu}\right]||\int_0^t f(x)dx - \int_0^t h(x)dx|| \tag{1.131}$$

and then the following result is obtained:

$$||{}_0^{ABR}D_t^{\nu}f(t) - {}_0^{ABR}D_t^{\nu}h(t)|| < \frac{B(\nu)\theta_1}{1-\nu}E_{\nu}\left[-\nu\frac{t^{\nu}}{1-\nu}\right]$$
$$||f(x) - h(x)||t = H||f(x) - h(x)||, \tag{1.132}$$

which produces the requested result.

Let f be an n-times differentiable with natural number and $f^{(k)}(0) = 0$, $k = 1, 2, 3, \ldots, n$, then by inspection we obtain

$${}_0^{ABC}D_t^{\nu}\left(\frac{d^n f(t)}{dt^n}\right) = \frac{d^n}{dt^n}({}_0^{ABR}D_t^{\nu}f(t)). \tag{1.133}$$

Now, we can easily prove by taking the inverse Laplace transform and using the convolution theorem that the following time fractional ordinary differential equation:

$${}_0^{ABC}D_t^{\nu}(f(t)) = u(t) \tag{1.134}$$

has a unique solution, namely

$$f(t) = \frac{1-\nu}{B(\nu)}u(t) + \frac{\nu}{B(\nu)\Gamma(\nu)}\int_0^t u(y)(t-y)^{\nu-1}dy.$$

□

Definition 1.5 The fractional integral associate to the new fractional derivative with nonlocal kernel is defined as:

$${}_a^{AB}I_t^{\nu}(f(t)) = \frac{1-\nu}{B(\nu)}u(t) + \frac{\nu}{B(\nu)\Gamma(\nu)}\int_a^t f(y)(t-y)^{\nu-1}dy.$$

When ν is zero, we recover the initial function and if also ν is 1, we obtain the ordinary integral [7].

1.6 Applications

1.6.1 Keller–Segel Model with Caputo Derivative

The best approach in terms of the technique to be employed to research on this topic is to visit the Keller and Segel model depicted in the ground-breaking paper (1970). It predates the formal structure Keller–Segel though it is probably the first edition. They [19, 20] presented the illustration of the aggregation behavior of cellular slime mold which they said is caused by instability. Oldham and Spainer also came up with four species important to the approach in the Keller and Segel [21].

One-dimensional Keller–Segel model is given by

$$\begin{cases} \rho_t = D\rho_{xx} - \chi(\rho a_x)_x, \\ a_t = D_a a_{xx} + h\rho - ka. \end{cases} \tag{1.135}$$

The parameters D, D_a, χ are constants. D and D_a are the diffusion coefficient and a, respectively, h and k are positive constants. The first term in the first equation in (1.135) involves a Laplacian, representing the random spatial motion of the cells. The second term models the chemotactic motion of the cells. In the second equation in (1.135), the first term represent diffusion of the chemoattractant. The second term models the production of the chemoattractant by the cells, and the third term represents linear decay. The initial conditions are $\rho(x, 0) = \rho_0(x)$ and $a(x, 0) = a_0(x)$ for the system. The system (1.135) with Caputo derivative is given as below

$$\begin{cases} {}_0^C D_t^{\nu}\rho = D\rho_{xx} - \chi(\rho a_x)_x, \\ {}_0^C D_t^{\nu} a = D_a a_{xx} + h\rho - ka. \end{cases} \tag{1.136}$$

1.6.1.1 Existence and Uniqueness Solutions

We will give in this chapter the existence and uniqueness of the solutions. We will also present the uniqueness of the positive solutions. Let us present every continuous functions $G = C[a, b]$ in the Banach space defined in the closed set $[a, b]$ and consider $Z = \{\rho, a \in G, \rho(x, t) \geq 0 \quad and \quad a(x, t) \geq 0, a \leq t \leq b\}$.

Definition 1.6 Let X be a Banach space with a cone H. H initiates a restricted order $\leq$ in E in the succeeding approach.

$$y \geq x \Longrightarrow y - x \in H.$$

Now applying the fractional integral in Eq. (1.136), we obtain the following,

$$\begin{cases} \rho(x,t) - \rho(x,0) = \dfrac{1}{\Gamma(\nu)} \displaystyle\int_0^t (t-r)^{\nu-1}[D\rho(x,t)_{xx} - \chi(\rho(x,t)a(x,t)_x)_x]dr, \\ a(x,t) - a(x,0) = \dfrac{1}{\Gamma(\nu)} \displaystyle\int_0^t (t-r)^{\nu-1}[D_a a(x,t)_{xx} + h\rho(x,t) - ka(x,t)]dr. \end{cases} \tag{1.137}$$

Now we can use system (1.137) to show the existence of Eq. (1.136). Necessary lemma for the existence of the solutions are given as Lemma 1.2. We now need to define an operator which $T : K \to K$.

$$\begin{cases} T\rho(x,t) = \dfrac{1}{\Gamma(\nu)} \displaystyle\int_0^t (t-r)^{\nu-1} s(x,r,\rho(x,r))dr, \\ Ta(x,t) = \dfrac{1}{\Gamma(\nu)} \displaystyle\int_0^t (t-r)^{\nu-1} s(x,r,a(x,r))dr. \end{cases} \tag{1.138}$$

To be dealt with more easily, let us consider below

$$\begin{cases} s(x,r,\rho) = D\rho_{xx} - \chi\rho_x a_{xx}, \\ s(x,r,a) = D_a a_{xx} + h\rho - ka. \end{cases} \tag{1.139}$$

Lemma 1.2 *The mapping $T : K \to K$ is completely continuous.*

Proof: Let $M \subset K$ be bounded. There exists a constants $l, m > 0$ such that $||p|| < l$, $||a|| < m$. Let,

$$L_1 = \max_{\substack{0 \le t \le 1 \\ 0 \le \rho \le l}} s(x,t,\rho(x,t)) \quad \text{and} \quad L_2 = \max_{\substack{0 \le t \le 1 \\ 0 \le a \le m}} s(x,t,a(x,t)),$$

$\forall \rho, a \in M$, we have

$$\begin{aligned} ||T\rho(x,t)|| &\le \frac{1}{\Gamma(\nu)} \int_0^t (t-r)^{\nu-1} ||s(x,r,\rho(x,r))||dr \\ &\le \frac{L_1}{\Gamma(\nu)} \int_0^t (t-r)^{\nu-1} dr \\ &= \frac{L_1}{\Gamma(\nu+1)} t^\nu. \end{aligned} \tag{1.140}$$

So that, we can write as below,

$$||T\rho|| \le \frac{L_1}{\Gamma(\nu+1)}.$$

Similarly,

$$\begin{aligned}||Ta(x,t)|| &\leq \frac{1}{\Gamma(v)}\int_0^t (t-r)^{v-1}||s(x,r,a(x,r))||dr \\ &\leq \frac{L_2}{\Gamma(v)}\int_0^t (t-r)^{v-1}dr \\ &= \frac{L_2}{\Gamma(v+1)}t^v.\end{aligned} \tag{1.141}$$

So that, we can write as below,

$$||Ta|| \leq \frac{L_2}{\Gamma(v+1)}.$$

Hence, $T(M)$ is bounded.

Now in the following part, we will consider $t_1 < t_2$ and $\rho(x,t), a(x,t) \in M$ and then for a given $\epsilon > 0$ if $|t_2 - t_1| < \delta$, we have

$$\begin{aligned}||T\rho(x,t_2) - T\rho(x,t_1)|| &= \frac{1}{\Gamma(v)}\int_0^{t_2} (t_2-r)^{v1}||s(x,r,\rho(x,r))||dr \\ &\quad -\frac{1}{\Gamma(v)}\int_0^{t_1} (t_1-r)^{v-1}||s(x,r,\rho(x,r))dr|| \\ &= \frac{1}{\Gamma(v)}\int_0^{t_2} (t_2-r)^{v-1}||s(x,r,\rho(x,r))||dr \\ &\quad -\frac{1}{\Gamma(v)}\int_0^{t_2} (t_1-r)^{v-1}||s(x,r,\rho(x,r))||dr \\ &\quad -\frac{1}{\Gamma(v)}\int_{t_1}^{t_2} (t_1-r)^{v-1}||s(x,r,\rho(x,r))||dr \\ &\leq \frac{1}{\Gamma(v)}\int_0^{t_2} ||(t_2-r)^{v-1} - (t_1-r)^{v-1}||\ ||s(x,r,\rho(x,r))||dr \\ &\quad +\frac{1}{\Gamma(v)}\int_{t_1}^{t_2} ||(t_1-r)^{v-1}||\ ||s(x,r,\rho(x,r))||dr \\ &\leq \frac{L_1}{\Gamma(v)}\int_0^{t_2} ((t_2-r)^{v-1} - (t_1-r)^{v-1})dr + \frac{L_1}{\Gamma(v)}\int_{t_1}^{t_2} (t_1-r)^{v-1}dr \\ &= \frac{L_1}{\Gamma(v)}\left(\int_0^{t_2} (t_2-r)^{v-1}dr - \int_0^{t_2} (t_1-r)^{v-1}dr + \int_{t_1}^{t_2} (t_1-r)^{v-1}dr\right) \\ &= \frac{L_1}{\Gamma(1+v)}(t_2^v + (t_1-t_2)^v - t_1^v + (t_1-t_2)^v) \\ &\leq \frac{2L_1}{\Gamma(1+v)}(t_1-t_2)^v + \frac{L_1}{\Gamma(1+v)}(t_1-t_1)^v \\ &= \frac{2L_1}{\Gamma(1+v)}(t_1-t_2)^v\end{aligned}$$

$$< \frac{2L_1}{\Gamma(1+v)}\delta^v$$
$$= \epsilon. \tag{1.142}$$

It is clearly seen that, when the same steps are applied to the $a(x,t)$ function, we get same situation. Finally, $|T\rho(x,t_2) - T\rho(x,t_1)| \leq \epsilon$ and $|Ta(x,t_2) - Ta(x,t_1)| \leq \epsilon$ are satisfied, where $\delta = (\epsilon\Gamma(1+v/2L))^{1/v}$. Therefore, $T(M)$ is equicontinuous. So that $\overline{T(M)}$ is compact via the Arzela–Ascoli theorem. □

Theorem 1.6 *Let* $S : [\rho_1, \rho_2] \times [0,\infty) \to [0,\infty)$, *then* $S(x,t)$ *is nondecreasing for each* t *in* $[\rho_1, \rho_2]$. *There exists a positive constants* v_1 *and* v_2 *such that* $B(n)v_1 \leq S(x,t,v_1)$, $B(n)v_2 \geq S(x,t,v_2)$, $0 \leq v_1(x,t) \leq v_2(x,t)$, $\rho_1 \leq t \leq \rho_2$. *This means that the new equation has a positive solution.*

Proof: We only need to consider the fixed point for operator of T. With framework of Lemma 1.2, the considered operator $T : H \to H$ is completely continuous. Let us take two arbitrary ρ_1 and ρ_2,

$$\begin{aligned} T\rho_1(x,t) &= \frac{1}{\Gamma(v)}\int_0^t (t-r)^{v-1}s(x,r,\rho_1(x,r))dr \\ &\leq \frac{1}{\Gamma(v)}\int_0^t (t-r)^{v-1}s(x,r,\rho_2(x,r))dr \\ &= Tp_2(x,t). \end{aligned} \tag{1.143}$$

Hence T is a nondecreasing operator. So that the operator $T : \langle v_1, v_2\rangle \to \langle v_1, v_2\rangle$ is compact and continuous via Lemma 1.2. In that case, H is a normal cone of T. □

1.6.1.2 Uniqueness of Solution

The aim of this section is to prove the uniqueness of solutions to the system (1.136). So the uniqueness of the solution is presented as below,

$$\begin{aligned} ||T\rho_1(x,t) - T\rho_2(x,t)|| &= ||\frac{1}{\Gamma(v)}\int_0^t (t-r)^{v-1}(s(x,r,\rho_1(x,r)) \\ &\quad -s(x,r,\rho_2(x,r)))dr|| \\ &\leq \frac{1}{\Gamma(v)}C_1\int_0^t (t-r)^{v-1}||\rho_1(x,r)-\rho_2(x,r)||dr. \end{aligned} \tag{1.144}$$

So that,

$$||T\rho_1(x,t) - T\rho_2(x,t)|| \leq \left\{\frac{C_1 t^v}{\Gamma(v+1)}\right\} ||\rho_1(x,r) - \rho_2(x,r)||.$$

Similarly,

$$
\begin{aligned}
||Ta_1(x,t) - Ta_2(x,t)|| &= ||\frac{1}{\Gamma(\nu)}\int_0^t (t-r)^{\nu-1}(s(x,r,a_1(x,r)) \\
&\quad -s(x,r,a_2(x,r)))dr|| \\
&\leq \frac{1}{\Gamma(\nu)}C_2\int_0^t (t-r)^{\nu-1}||a_1(x,r)-a_2(x,r)||dr.
\end{aligned} \tag{1.145}
$$

So that,

$$
||Ta_1(x,t) - Ta_2(x,t)|| \leq \left\{\frac{C_2 t^\nu}{\Gamma(\nu+1)}\right\}||a_1(x,r)-a_2(x,r)||.
$$

Therefore, if the following conditions hold,

$$
\left\{\frac{C_1 t^\nu}{\Gamma(\nu+1)}\right\} < 1 \text{ and } \left\{\frac{C_2 t^\nu}{\Gamma(\nu+1)}\right\} < 1.
$$

Then mapping T is a contraction, which implies fixed point, and thus the model has a unique positive solution.

1.6.1.3 Keller–Segel Model with Atangana–Baleanu Derivative in Caputo Sense

We present in this section the existence and uniqueness of solutions of the Keller–Segel model using the Atangana–Baleanu derivative. Let $\Omega = (a,b)$ be an open and bounded subset of R^n. Let $\nu \in (0,1)$ and functions $\rho(x,t), a(x,t) \in H^1(\Omega) \times [0,T]$. Here $\rho(x,t)$ represent the concentration of the chemical substance and the function $a(x,t)$ represent concentration of amoebae. We apply the system (1.135) to the Atangana–Baleanu fractional derivative,

$$
\begin{cases}
{}_0^{ABC}D_t^{\nu}\rho = \sigma_1(x,t,\rho), \\
{}_0^{ABC}D_t^{\nu}a = \sigma_2(x,t,a),
\end{cases} \tag{1.146}
$$

where

$$
\begin{cases}
\sigma_1(x,t,\rho) = D\rho_{xx} - \chi\rho_x a_{xx}, \\
\sigma_2(x,t,a) = D_a a_{xx} + h\rho - ka.
\end{cases} \tag{1.147}
$$

Using the Atangana–Baleanu integral to (1.146), it yields

$$
\begin{cases}
\rho(x,t) = \rho(x,0) + \frac{1-\nu}{B(\nu)}\sigma_1(x,t,\rho(x,t)) \\
\quad +\frac{\nu}{B(\nu)\Gamma(\nu)}\int_0^t \sigma_1(x,r,\rho(x,r))(t-r)^{\nu-1}dr, \\
a(x,t) = a(x,0) + \frac{1-\nu}{B(\nu)}\sigma_2(x,t,a(x,t)) \\
\quad +\frac{\nu}{B(\nu)\Gamma(\nu)}\int_0^t \sigma_2(x,r,a(x,r))(t-r)^{\nu-1}dr,
\end{cases} \tag{1.148}
$$

for all $t \in [0,T]$.

Theorem 1.7 *If the inequality* (1.149) *hold,* σ_1 *and* σ_2 *satisfy Lipschitz condition and contraction.*

$$0 < D\gamma_1^2 + \chi\gamma_2||\frac{\partial^2 a(x,y)}{\partial x^2}|| \leq 1. \tag{1.149}$$

Proof: We would like to start with the kernel σ_1.Let κ_1 and κ_2 are two functions, the following equation is written as:

$$\begin{aligned}&||\sigma_1(x,t,\kappa_1) - \sigma_1(x,t,\kappa_2)||\\&= ||D(\kappa_1(x,t)_{xx} - \kappa_2(x,t)_{xx}) - \chi(\kappa_1(x,t)_x - \kappa_2(x,t)_x)\frac{\partial^2 a(x,t)}{\partial x^2}||.\end{aligned}$$

When we convert the above equation via triangular inequality, we get

$$\begin{aligned}&||\sigma_1(x,t,\kappa_1) - \sigma_1(x,t,\kappa_2)||\\&\leq D||(\kappa_1(x,t)_{xx} - \kappa_2(x,t)_{xx})|| + \chi|| - (\kappa_1(x,t)_x - \kappa_2(x,t)_x)\frac{\partial^2 a(x,t)}{\partial x^2}||.\end{aligned}$$

Using the operator derivative, we can find two constants such as γ_1 and γ_2 :

$$\begin{cases} D||(\kappa_1(x,t)_{xx} - \kappa_2(x,t)_{xx})|| \leq D\gamma_1^2||\kappa_1(x,t) - \kappa_2(x,t)|| \\ \chi|| - (\kappa_1(x,t)_x - \kappa_2(x,t)_x)\frac{\partial^2 a(x,t)}{\partial x^2}|| \leq \chi\gamma_2||\frac{\partial^2 a(x,t)}{\partial x^2}|| \\ \quad ||(\kappa_1(x,t) - \kappa_2(x,t)||. \end{cases} \tag{1.150}$$

When we substitute Eq. (1.150) in below equation, we get:

$$||\sigma_1(x,t,\kappa_1) - \sigma_1(x,t,\kappa_2)|| \leq K||(\kappa_1(x,t) - \kappa_2(x,t)||, \tag{1.151}$$

where

$$K = \left(D\gamma_1^2 + \chi\gamma_2||\frac{\partial^2 a(x,t)}{\partial x^2}||\right).$$

Therefore, σ_1 satisfies the Lipschitz condition. Then we can say that it is a contraction. In the another case, the following inequality can be written because our kernel is linear,

$$\sigma_2(x,t,v_1) - \sigma_2(x,t,v_2) \leq (c\vartheta_1^2 + d)||v_1(x,t) - v_2(x,t)||.$$

Hence, the proof is complete. We can now show that the uniqueness of the solution. □

1.6.1.4 Uniqueness of Solution

The uniqueness solution for system (1.146) is presented as below. Let $\rho_1, \rho_2 \in H^1$ be two solutions of (1.146). Let $\rho = \rho_1 - \rho_2$, the following equation can be written as,

$$\rho = \frac{1-\nu}{B(\nu)}(\sigma_1(x,t,\rho_1(x,t)) - \sigma_1(x,t,\rho_2(x,t))) + \frac{\nu}{B(\nu)\Gamma(\nu)}\int_0^t (\sigma_1(x,r,\rho_1(x,r)) - \sigma_1(x,r,\rho_2(x,r)))dr.$$

If the norms of both sides are taken, by the Gronwall inequality (1.146),

$$\begin{aligned} ||\rho|| \leq & \frac{1-\nu}{B(\nu)}||\sigma_1(x,t,\rho_1(x,t)) - \sigma_1(x,t,\rho_2(x,t))|| \\ & + \frac{\nu}{B(\nu)\Gamma(\nu)}\int_0^t ||\sigma_1(x,r,\rho_1(x,r)) - \sigma_1(x,r,\rho_2(x,r))||dr \\ \leq & K_1 \int_0^t ||\sigma_1(x,t,\rho_1(x,t))||_{H^1} dr. \end{aligned}$$

Similarly, let $a_1, a_2 \in H^1$ be two solutions of (1.146). Let $a = a_1 - a_2$, the following equation can be written as,

$$\begin{aligned} ||a|| \leq & \frac{1-\nu}{B(\nu)}||\sigma_2(x,t,a_1(x,t)) - \sigma_2(x,t,a_2(x,t))|| \\ & + \frac{\nu}{B(\nu)\Gamma(\nu)}\int_0^t ||\sigma_2(x,r,a_1(x,r)) - \sigma_2(x,r,a_2(x,r))||dr \\ \leq & K_2 \int_0^t ||\sigma_2(x,t,a_1(x,t))||_{H^1} dr. \end{aligned} \tag{1.152}$$

Finally, the system (1.146) has a unique solution for the equations ρ and a.

1.6.2 Cancer Treatment Model with Caputo-Fabrizio Fractional Derivative

It is well known that cancer is one of the most common diseases causing deaths in the last century. Emerging in various parts of the human body, this disease becomes unresponsive to the treatment when it is intervened late. There is a great deal of research done for the treatment of cancer, which is the disease of the century. In this chapter, we are going to examine the existence and uniqueness of the cancer treatment model. By developing models for the treatment of cancer, it is aimed to contribute to the studies in this field.

It is assumed that healthy and cancer cells be located in the same area of the organism. Let $\rho(t)$ denotes the concentration of healthy cells, and $a(t)$ denotes the concentration of cancer cells. Then the model is given by [22]:

$$\begin{cases} \dfrac{d\rho(t)}{dt} = \alpha_1 \rho\left(1 - \dfrac{\rho}{S_1}\right) - \beta_1 \rho a - \epsilon D(t)\rho, \\ \dfrac{da(t)}{dt} = \alpha_2 a\left(1 - \dfrac{a}{S_2}\right) - \beta_2 a\rho - D(t)a, \end{cases} \tag{1.153}$$

where $D(t)$ is the strategy of the radiotherapy. It is supposed that $D(t) \equiv \gamma > 0$ when $t \in [nw, nw + L)$(treatment stage) and $D(t) \equiv 0$ when $t \in [nw + L, (n+1)w$ (no treatment stage) for all $n = 0, 1, 2, \ldots$, where w is the radiation treatment time [22]. The system (1.153) with Caputo derivative is given as below,

$$\begin{cases} {}_0^{CF}D_t^{\nu}(\rho(t)) = \alpha_1 \rho \left(1 - \frac{\rho}{S_1}\right) - \beta_1 \rho a - \epsilon D(t)\rho, \\ {}_0^{CF}D_t^{\nu}(a(t)) = \alpha_2 a \left(1 - \frac{a}{S_2}\right) - \beta_2 a\rho - D(t)a. \end{cases} \tag{1.154}$$

1.6.2.1 Existence Solutions

We will give in this section the existence of the solutions for the cancer treatment model by radiotherapy. After that, we also will present the uniqueness of the positive solutions.

Now applying the fractional integral in Eq. (1.153), we obtain the following,

$$\begin{cases} \rho(t) - \rho_0(t) = \frac{2(1-\nu)}{2M(\nu) - \nu M(\nu)} \left[\alpha_1 \rho \left(1 - \frac{\rho}{S_1}\right) - \beta_1 \rho a - \epsilon D(t)\rho\right] \\ \quad + \frac{2\nu}{2M(\nu) - \nu M(\nu)} \int_0^t \left[\alpha_1 \rho \left(1 - \frac{\rho}{S_1}\right) - \beta_1 \rho a - \epsilon D(y)\rho\right] dy, \\ a(t) - a_0(t) = \frac{2(1-\nu)}{2M(\nu) - \nu M(\nu)} \left[\alpha_2 a \left(1 - \frac{a}{S_2}\right) - \beta_2 a\rho - D(t)a\right] \\ \quad + \frac{2\nu}{2M(\nu) - \nu M(\nu)} \int_0^t \left[\alpha_2 a \left(1 - \frac{a}{S_2}\right) - \beta_2 a\rho - D(y)a\right] dy. \end{cases} \tag{1.155}$$

For simplicity, we choose our kernels as $s(t, \rho(t))$ and $s(t, a(t))$ as follows:

$$s(t, \rho(t)) = \alpha_1 \rho \left(1 - \frac{\rho}{S_1}\right) - \beta_1 \rho a - \epsilon D(t)\rho,$$

$$s(t, a(t)) = \alpha_2 a \left(1 - \frac{a}{S_2}\right) - \beta_2 a\rho - D(t)a.$$

First we need to be able to identify an operator. We will then show that this operator is compact. So that the operator which $T : H \to H$. Then we get,

$$\begin{cases} T\rho(t) = \frac{2(1-\nu)}{2M(\nu) - \nu M(\nu)} s(t, \rho(t)) + \frac{2\nu}{2M(\nu) - \nu M(\nu)} \int_0^t s(y, \rho(y))dy, \\ Ta(t) = \frac{2(1-\nu)}{2M(\nu) - \nu M(\nu)} s(a, a(t)) + \frac{2\nu}{2M(\nu) - \nu M(\nu)} \int_0^t s(a, a(y))dy. \end{cases} \tag{1.156}$$

Lemma 1.3 *The mapping $T : H \to H$ is completely continuous.*

Proof: Let $M \subset H$ be bounded. There exists a constants $l, m > 0$ such that $||\rho|| < l$ and $||a|| < m$. Let

$$L_1 = \max_{\substack{0 \le t \le 1 \\ 0 \le \rho \le l}} s(t, \rho(t)) \quad \text{and} \quad L_2 = \max_{\substack{0 \le t \le 1 \\ 0 \le a \le m}} s(t, a(t)),$$

$\forall \rho, a \in M$, we have

$$\begin{aligned}
|T\rho(t)| &= \left| \frac{2-2v}{2M(v) - vM(v)} s(t, \rho(t)) + \frac{2v}{2M(v) - vM(v)} \int_0^t s(y, \rho(y)) dy \right| \\
&\le \left| \frac{2-2v}{2M(v) - vM(v)} |s(t, \rho(t)) + | \frac{2v}{2M(v) - vM(v)} \right| \int_0^t s(y, \rho(y)) dy \Bigg| \\
&\le \left[\frac{2-2v}{2M(v) - vM(v)} + \frac{2v}{2M(v) - vM(v)} c_1 \right] |s(t, \rho(t))| \\
&\le \left[\frac{2-2v}{2M(v) - vM(v)} + \frac{2v}{2M(v) - vM(v)} c_1 \right] |L_1 \\
||T\rho|| &\le \frac{2L_1}{2M(v) - vM(v)} [1 - v + vc_1].
\end{aligned} \tag{1.157}$$

Similarly,

$$\begin{aligned}
|Ta(t)| &= \left| \frac{2-2v}{2M(v) - vM(v)} s(t, a(t)) + \frac{2v}{2M(v) - vM(v)} \int_0^t s(y, a(y)) dy \right| \\
&\le \frac{2-2v}{2M(v) - vM(v)} |s(t, a(t)) + | \frac{2v}{2M(v) - vM(v)} \left| \int_0^t s(y, a(y)) dy \right| \\
&\le \left[\frac{2-2v}{2M(v) - vM(v)} + \frac{2v}{2M(v) - vM(v)} c_2 \right] |s(t, a(t))| \\
&\le \left[\frac{2-2v}{2M(v) - vM(v)} + \frac{2v}{2M(v) - vM(v)} c_2 \right] |L_2 \\
||Ta|| &\le \frac{2L_2}{2M(v) - vM(v)} [1 - v + vc_2].
\end{aligned} \tag{1.158}$$

Hence, $T(M)$ is bounded.

Now in the following part, we will consider $t_1 < t_2$ and $\rho(t), a(t) \in M$ and then for a given $\epsilon > 0$ if $|t_2 - t_1| < \delta$, we have

$$\begin{aligned}
||T\rho(t_2) - T\rho(t_1)|| \le & \left| \frac{2-2v}{2M(v) - vM(v)} (s(t_2, \rho(t_2)) - s(t_1, \rho(t_1))) \right| \\
& + \left| \frac{2v}{2M(v) - vM(v)} \int_0^{t_2} s(y, \rho(y)) dy \right.
\end{aligned}$$

$$
\begin{aligned}
&-\frac{2\nu}{2M(\nu)-\nu M(\nu)}\int_0^{t_1} s(y,\rho(y))dy\Bigg| \\
&\leq \frac{2-2\nu}{2M(\nu)-\nu M(\nu)}|s(t_2,\rho(t_2))-s(t_1,\rho(t_1))| \\
&\quad+\frac{2\nu}{2M(\nu)-\nu M(\nu)}L_1|s(t_2,\rho(t_2))-s(t_1,\rho(t_1))|.
\end{aligned} \tag{1.159}
$$

Now we will investigate the following,

$$
\begin{aligned}
|s(t_2,\rho(t_2))-s(t_1,\rho(t_1))| &\leq \Bigg|\alpha_1(\rho(t_2)-\rho(t_1))\left(1-\frac{\rho(t_2)-\rho(t_1)}{S_1}\right) \\
&\quad -\beta_1(\rho(t_2)-\rho(t_1)a-\epsilon\gamma(\rho(t_2)-\rho(t_1)| \\
&\leq c_3|(\rho(t_2)-\rho(t_1)|-c_4|(\rho(t_2)-\rho(t_1)| \\
&\quad -c_5|(\rho(t_2)-\rho(t_1)| \\
&\leq (c_3-c_4-c_5)|(\rho(t_2)-\rho(t_1)| \\
&\leq C|t_2-t_1|.
\end{aligned} \tag{1.160}
$$

Now putting Eq. (1.159) and the integral part of Eq. (1.158) in Eq. (1.158), we get,

$$
|T\rho(t_2)-T\rho(t_1)| \leq \frac{2\nu}{2M(\nu)-\nu M(\nu)}C|t_2-t_1|+\frac{2(1-\nu)}{2M(\nu)-\nu M(\nu)}L_1|t_2-t_1|, \tag{1.161}
$$

$$
\delta=\frac{\epsilon}{\frac{2\nu}{2M(\nu)-\nu M(\nu)}C+\frac{2(1-\nu)}{2M(\nu)-\nu M(\nu)}L_1}. \tag{1.162}
$$

Such that $|T\rho(t_2)-T\rho(t_1)| \leq \epsilon$.

When we can acquire the following for the function a with same rules, we get

$$
\delta=\frac{\epsilon}{\frac{2\nu}{2M(\nu)-\nu M(\nu)}G+\frac{2(1-\nu)}{2M(\nu)-\nu M(\nu)}L_2}. \tag{1.163}
$$

Such that $|Ta(t_2)-Ta(t_1)| \leq \epsilon$ are satisfied. Therefore $T(M)$ is equicontinuous. So that $\overline{T(M)}$ is compact via the Arzela–Ascoli theorem. □

Theorem 1.8 *Let $N : [\rho_1,\rho_2]\times[0,\infty)\to[0,\infty)$, then $N(t,\cdot)$ is nondecreasing for each t in $[\rho_1,\rho_2]$. There exists positive constants v_1 and v_2. So that $B(n)v_1 \leq S(t,v_1)$, $B(n)v_2 \geq N(t,v_2)$, $0 \leq v_1(t) \leq v_2(t)$, $\rho_1 \leq t \leq \rho_2$. Thus, the equation has a positive solution.*

Proof: We only need to consider the fixed point for operator of T. It is considered that $T : H \to H$ is completely continuous. Let $\rho_1 \leq \rho_2$ and $a_1 \leq a_2$. Here four variables are arbitrary.

$$
\begin{aligned}
T\rho_1(t) &= \frac{2-2v}{2M(v)-vM(v)}|s(t,\rho_1(t))| + \frac{2v}{2M(v)-vM(v)}\int_0^t s(y,\rho_1(y))dy \\
&\leq \frac{2-2v}{2M(v)-vM(v)}|s(t,\rho_2(t))| + \frac{2v}{2M(v)-vM(v)}\int_0^t s(y,\rho_2(y))dy \\
&\leq T\rho_2(x,t)
\end{aligned}
\tag{1.164}
$$

and

$$
\begin{aligned}
Ta_1(t) &= \frac{2-2v}{2M(v)-vM(v)}|s(t,a_1(t))| + \frac{2v}{2M(v)-vM(v)}\int_0^t s(y,a_1(y))dy \\
&\leq \frac{2-2v}{2M(v)-vM(v)}|s(t,a_2(t))| + \frac{2v}{2M(v)-vM(v)}\int_0^t s(y,a_2(y))dy \\
&\leq Ta_2(x,t).
\end{aligned}
\tag{1.165}
$$

Hence, T is a nondecreasing operator. So that the operator $T : \langle v_1, v_2\rangle \rightarrow \langle v_1, v_2\rangle$ is compact and continuous via Lemma 1.2. In that case, H is a normal cone of T.□

1.6.2.2 Uniqueness Solutions

In the previous section, we proved using the fixed point theorem, that the coupled cancer treatment model with Caputo–Fabrizio time fractional derivative has an existing solution. The goal of this section is to show the uniqueness of solutions to the system (1.152) with the initial conditions. Let assume in addition that, we can find two special coupled solutions (ρ_1, ρ_2) and (a_1, a_2). So the uniqueness of the solution is presented as:

$$
\begin{aligned}
|T\rho_1(t) - T\rho_2(t)| &= \left|\frac{2-2v}{2M(v)-vM(v)}\right|(s(t,\rho_1(t)-s(t,\rho_2(t)) \\
&\quad + \frac{2v}{2M(v)-vM(v)}\int_0^t (s(y,\rho_1(y)) - s(y,\rho_2(y)))dy| \\
&\leq \frac{2-2v}{2M(v)-vM(v)}|(s(t,\rho_1(t)-s(t,\rho_2(t))| \\
&\quad + \frac{2v}{2M(v)-vM(v)}\int_0^t |(s(y,\rho_1(y)) - s(y,\rho_2(y)))dy| \\
&\leq \frac{2-2v}{2M(v)-vM(v)}F_1|\rho_1(t)-\rho_2(t)| \\
&\quad + \frac{2v}{2M(v)-vM(v)}F_1|\rho_1(t)-\rho_2(t)|.
\end{aligned}
\tag{1.166}
$$

So that we can write the above Eq. (165),

$$|T\rho_1(t) - T\rho_2(t)| \leq \left\{ \frac{2-2\nu}{2M(\nu) - \nu M(\nu)} F_1 + \frac{2\nu}{2M(\nu) - \nu M(\nu)} F_1 \right\} |\rho_1(t) - \rho_2(t)|.$$

Similarly,

$$\begin{aligned} |Ta_1(t) - Ta_2(t)| &= |\frac{2-2\nu}{2M(\nu) - \nu M(\nu)} |(s(t, \rho_1(t) - s(t, \rho_2(t))) \\ &\quad + \frac{2\nu}{2M(\nu) - \nu M(\nu)} \int_0^t (s(y, \rho_1(y)) - s(y, \rho_2(y)))dy| \\ &\leq \frac{2-2\nu}{2M(\nu) - \nu M(\nu)} F_2 |a_1(t) - a_2(t)| \\ &\quad + \frac{2\nu}{2M(\nu) - \nu M(\nu)} F_2 |a_1(t) - a_2(t)|. \end{aligned} \tag{1.167}$$

So that we can write the above Eq. (166),

$$||Ta_1(t) - Ta_2(t)|| \leq \left\{ \frac{2-2\nu}{2M(\nu) - \nu M(\nu)} F_2 + \frac{2\nu}{2M(\nu) - \nu M(\nu)} F_2 \right\} |a_1(t) - a_2(t)|.$$

Therefore, if the following conditions hold,

$$\left\{ \frac{2-2\nu}{2M(\nu) - \nu M(\nu)} F_1 + \frac{2\nu}{2M(\nu) - \nu M(\nu)} F_1 \right\} < 1 \text{ and}$$

$$\left\{ \frac{2-2\nu}{2M(\nu) - \nu M(\nu)} F_2 + \frac{2\nu}{2M(\nu) - \nu M(\nu)} F_2 \right\} < 1.$$

Then mapping T is a contraction. We can say that the model has a unique positive solution using fixed point theorem.

1.6.2.3 Conclusion

We first integrated this cancer treatment model with the new fractional derivative. After that, we found the existence solution of the cancer treatment model. Finally, we analyzed how the model is the uniqueness positive solution under which conditions. We tried to help the researcher working on cancer education with this work. When the results are examined, it has been shown that the fractional derivative gives important information about the process.

Bibliography

1 Leibniz, G.W. (1853). Leibniz an de L'Hospital (Letter form Hannover, Germany, September 30, 1695). In: *Oeuvres Mathematiques de Leibniz. Correspondance de Leibniz de A. Franck*, pp. 297–302.

2 Leibniz, G.W. (1962). Mathematische Schriften. *Georg. Olms Verlagsbuchhandlung* **5**: 377–382. Vol

3 Letnikov, A.V. (1868). Theory of differentiation of an arbitrary order. *Mat. Sb.* **3**: 1–68 (in Russian).

4 Podlubny, I. (1999). *Fractional Differential Equations, Mathematics in Science and Engineering*, vol. **198**. San Diego, CA: Academic Press.

5 Caputo, M. (1967). Linear models of dissipation whose Q is almost frequency independent, Part II. *Geophys. J. R. Astron. Soc.* **13**: 529–539.

6 Caputo, M. and Fabrizio, M. (2015). A new definition of fractional derivative without singular kernel. *Progr. Fract. Differ. Appl.* **1**: 73–85.

7 Atangana, A. and Baleanu, D. (2016). New fractional derivatives with non-local and non-singular kernel: theory and application to heat transfer model. *Therm. Sci.* **20**: 763–769.

8 Dokuyucu, M.A., Baleanu, D., and Celik, E. (2018). Analysis of Keller-Segel model with Atangana-Baleanu fractional derivative. *Filomat* **32** (16): 5633–5643.

9 Dokuyucu, M.A., Celik, E., Bulut, H., and Baskonus, H.M. (2018). Cancer treatment model with the Caputo–Fabrizio fractional derivative. *Eur. Phys. J. Plus* **133** (3): 92.

10 Samko, S.G., Kilbas, A.A., and Maritchev, O.I. (1987). *Integrals and Derivatives of the Fractional Order and Some of Their Applications.* Minsk: Nauka i Tekhnika (in Russian).

11 Caputo, M. (1969). *Elasticita e Dissipazione.* Bologna: Zanichelli.

12 El-Sayed, A.M.A. (1994). Multivalued fractional differential equations. *Appl. Math. Comput.* **80**: 1–11.

13 El-Sayed, A.M.A. (1995). Fractional order evolution equations. *J. Fract. Calculus* **7**: 89–100.

14 Ochmann, M. and Makarov, S. (1993). Representation of the absorption of nonlinear waves by fractional derivatives. *J. Am. Acoust. Soc.* **94** (6): 3392–3399.

15 Losada, J. and Nieto, J.J. (2015). Properties of a new fractional derivative without singular kernel. *Progr. Fract. Differ. Appl.* **1**: 87–92.

16 Kilbas, A.A., Srivastava, H.M., and Trujillo, J.J. (2006). *Theory and Applications of Fractional Differential Equations*. Amsterdam: Elsevier.

17 Hristov, J. (2015). Diffusion models with weakly singular kernels in the fading memories: how the integral-balance method can be applied? *Therm. Sci.* **19** (3): 947–957.

18 Hristov, J. (2015). Approximate solutions to time-fractional models by integral balance approach. In: *Fractional Dynamics*, Chapter 5 (ed. C. Cattani, H.M. Srivastava, and X.J. Yang), 78–109. De Gruyter Open.

19 Keller, E.F. and sSegel, L.A. (1970). Initiation of slime mold aggregation viewed as an instability. *J. Theor. Biol.* **26**: 399–415.

20 Keller, E.F. and Segel, L.A. (1971). Model for chemotaxis. *J. Theor. Biol.* **30**: 225–234.

21 Oldham, K.B. and Spanier, J. (1974). *The Fractional Calculus*. New York and London: Academic Press.

22 Liu, Z. and Yang, C. (2014). A mathematical model of cancer treatment by radiotherapy. *Comput. Math. Methods Med.* **30** (2): 225–234

2

Generalized Conformable Fractional Operators and Their Applications

Muhammad Adil Khan and Tahir Ullah Khan

Department of Mathematics, University of Peshawar, Peshawar, Khyber Pakhtunkhwa, Pakistan

2.1 Introduction and Preliminaries

Everyone in the field of mathematics is almost aware of a very popular correspondence between Leibnitz and L'hospital which became a source for the emergence of fractional calculus. The correspondence started firstly on 30 September 1695 and extended in a shape of letters in which they discussed the existence of arbitrary order derivative. Before that there was a concept of integer order derivative only. This correspondence attracted a host of researchers toward a new idea and they started drawing useful results in it. The first challenge was the construction of fractional derivative operator, an operator which was the generalization of the classical derivative operator. In this regard, the researchers used two approaches, one is that of the iterations of integral operator and the other is the iteration of derivative operator. In each approach, they iterated the corresponding classical derivative or integral operators and then generalized it to the arbitrary order. In this regard, the contributions of Euler, Laplace, Fourier, Abel, Liouville, Riemann, Grunwald, Letnikov, Hadamard, and in the present century, Weyl, Riesz, Marchaud, Kober, and Caputo are remarkable in this field [1–10]. Most of these researchers initially introduced fractional integrals, on the basis of which the associated fractional derivative and other related results were produced. Some of the most explored and commonly used definitions of fractional integrals are given below.

The right-sided Riemann–Liouville fractional integral operator of order $\beta > 0$ is given by [11]:

$$J^{\beta}_{a^+}\psi(r) = \frac{1}{\Gamma(\beta)}\int_a^r (r-w)^{\beta-1}\psi(w)dw, \quad \text{with } r > a, \tag{2.1}$$

which is based on the iteration of the Riemann integral operator $\int_a^r \psi(w)dw$. The Hadamard fractional integral introduced by J. Hadamard [12], for $\beta > 0$ is given

Fractional Order Analysis: Theory, Methods and Applications, First Edition.
Edited by Hemen Dutta, Ahmet Ocak Akdemir, and Abdon Atangana.

by:

$$H_{a^+}^{\beta}\psi(r) = \frac{1}{\Gamma(\beta)} \int_a^r \left(\log \frac{r}{w}\right)^{\beta-1} \psi(w)\frac{dw}{w}, \quad \text{with } r > a, \tag{2.2}$$

which is based on iterating the integral operator $\int_a^r \psi(w)\frac{dw}{w}$. Udita N. Katugampola has defined a generalized Katugampola integral operator [13], which for $\beta > 0$, $\tau \neq -1$ is given by:

$${}_a^{\tau}I_r^{\beta}\psi(r) = \frac{(\tau+1)^{1-\beta}}{\Gamma(\beta)} \int_a^r (r^{\tau+1} - w^{\tau+1})^{\beta-1}\psi(w)w^{\tau}dw, \quad r > a \tag{2.3}$$

and is the generalization of both the above operators defined in (2.1) and (2.2). The operator in (2.3) is based on the iteration of the integral operator $\int_a^r \psi(w)w^{\tau}dw$. Concurrently, with the operators (2.1), (2.2), and (2.3) their corresponding left-sided versions were also determined. Also by using these fractional integral operators, the associated fractional derivative operators were defined [11–13].

Khalil et al. [6] have discovered novel definitions of fractional derivative and connected integral, which are given below.

Definition 2.1 [6] For $\psi : [0, \infty) \to \mathbb{R}$, the conformable fractional derivative of ψ of order $\alpha \in (0, 1]$, at point $w \in (0, \infty)$ is defined as:

$$T_{\alpha}\psi(w) = \lim_{\epsilon\to 0} \frac{\psi(w + \epsilon w^{1-\alpha}) - \psi(w)}{\epsilon}, \tag{2.4}$$

for $w = 0$, it is defined as:

$$T_{\alpha}\psi(0) = \lim_{w\to 0^+} T_{\alpha}\psi(w).$$

If the conformable fractional derivative of ψ of order α exists, then we say that ψ is α-differentiable.

If ψ is ordinary differentiable, then the connection of conformable fractional derivative with the ordinary derivative for $w > 0$ is given by:

$$T_{\alpha}\psi(w) = w^{1-\alpha}\psi'(w), \tag{2.5}$$

where $\psi'(w)$ denotes the ordinary derivative of ψ at the point w. It is simple to prove that a function could be α-differentiable at a point but not ordinary differentiable, see for detail [6]. This new definition is simple and satisfies almost all basic properties which the ordinary derivative does. These properties are given in the following theorem.

Theorem 2.1 ([6]). *Let $\alpha \in (0,1]$ and ψ_1, ψ_2 be α-differentiable functions at a point $w > 0$. Then for any μ_1, $\mu_2 \in \mathbb{R}$ we have:*

(1) $T_\alpha(\mu_1\psi_1 + \mu_2\psi_2) = \mu_1 T_\alpha(\psi_1) + \mu_2 T_\alpha(\psi_2)$.

(2) $T_\alpha(t^r) = rt^{r-\alpha}, \ \forall\, r \in \mathbb{R}$.

(3) $T_\alpha(c) = 0, \ \forall\, c \in \mathbb{R}$.

(4) $T_\alpha(\psi_1\psi_2) = \psi_2 T_\alpha(\psi_1) + \psi_1 T_\alpha(\psi_2)$.

(5) $T_\alpha\left(\dfrac{\psi_1}{\psi_2}\right) = \dfrac{\psi_2 T_\alpha(\psi_1) - \psi_1 T_\alpha(\psi_2)}{\psi_2^2}$.

The authors in [6] have also defined the conformable fractional integral of order $0 < \alpha \leq 1$ (about which an open problem was posed in [1]). This is defined as under.

Theorem 2.2 ([6]). *Let $\alpha \in (0,1]$, the conformable fractional integral of the continuous function $\psi : [a,b] \subseteq [0,\infty) \to \mathbb{R}$ of order α is defined as:*

$$I_\alpha(\psi)(w) = \int_a^b \psi(w)d_\alpha w := \int_a^b \psi(w)w^{\alpha-1}dw, \tag{2.6}$$

where the integral $\int_a^b dw$, on the right side, represents the classical Riemann integral.

The inverse property is given in the following theorem.

Theorem 2.3 ([6]). *For any continuous function ψ in the domain of I_α, we have $T_\alpha I_\alpha \psi(r) = \psi(r)$.*

In this chapter, to give answer to the open problem given in [1], we define new generalized conformable fractional integral operators (right-sided and left-sided), by iterating conformable integral of order $\alpha \in (0,1]$. We prove semigroup property, linearity, and boundedness property for these operators. After introducing new fractional integral operators, we define the associated right-sided and left-sided generalized conformable fractional derivative operators. The semigroup property linearity and boundedness property are also proved for generalized conformable fractional derivative operator. By making use of these operators, we also define Riemann–Liouville type conformable fractional operators. Our newly defined fractional operators are the generalizations of the Katugampola fractional operators, Riemann–Liouville fractional operators and Hadamard fractional integral operators. We apply our fractional differential operator to a simple function. Also we consider a nonlinear fractional differential equation using this new formulation. We show that this equation is equivalent to a Volterra integral equation and demonstrate the existence and uniqueness of solution to the nonlinear problem. We also extend our results to the area of inequalities. For some more results related to conformable derivative and integral we recommend [14–19].

2.2 Generalized Conformable Fractional Integral Operators

Throughout this chapter, we consider $\alpha \in (0,1]$, $\tau \in \mathbb{R}$ such that $\tau + \alpha \neq 0$. Also we take $0 \leq a < b$, $L_\alpha[a,b] = \{\psi(w) : \int_a^b \psi(w) d_\alpha w < \infty\}$. Define an operator ${}_\alpha^\tau K_{a^+} : L_\alpha[a,b] \to \mathbb{R}$ by:

$$ {}_\alpha^\tau K_{a^+}\psi(r) = \int_a^r \psi(w) w^\tau d_\alpha w, \quad r \in [a,b], $$

and ${}_\alpha^\tau K_{b^-} : L_\alpha[a,b] \to \mathbb{R}$ by:

$$ {}_\alpha^\tau K_{b^-}\psi(r) = \int_r^b \psi(w) w^\tau d_\alpha w, \quad r \in [a,b]. $$

Here $\int d_\alpha w$ represents the conformable fractional integral, which was defined in (2.6). To define new generalized conformable fractional integral operators, we need to prove the following result.

Theorem 2.4 ***(Cauchy integral formula for repeated conformable integrals)*** *Let $\psi \in L_\alpha[a,b]$. Then the n times repeated right-sided and left-sided conformable fractional integrals are given by the single conformable fractional integrals*

$$ \begin{aligned} {}_\alpha^\tau K_{a^+}^n \psi(r) &= \int_a^r r_1^\tau \int_a^{r_1} r_2^\tau \int_a^{r_2} r_3^\tau \ldots \int_a^{r_{n-1}} \psi(r_n) r_n^\tau d_\alpha r_n \ldots d_\alpha r_2 d_\alpha r_1 \\ &= \frac{1}{(n-1)!} \int_a^r \left(\frac{r^{\tau+\alpha} - w^{\tau+\alpha}}{\tau+\alpha} \right)^{n-1} \psi(w) w^\tau d_\alpha w \end{aligned} \tag{2.7} $$

and

$$ \begin{aligned} {}_\alpha^\tau K_{b^-}^n \psi(r) &= \int_r^b r_1^\tau \int_{r_1}^b r_2^\tau \int_{r_2}^b \ldots \int_{r_{n-1}}^b \psi(r_n) r_n^\tau d_\alpha r_n \ldots d_\alpha r_2 d_\alpha r_1 \\ &= \frac{1}{(n-1)!} \int_r^b \left(\frac{w^{\tau+\alpha} - r^{\tau+\alpha}}{\tau+\alpha} \right)^{n-1} \psi(w) w^\tau d_\alpha w, \end{aligned} \tag{2.8} $$

respectively.

Proof: First, we prove (2.7). For $n = 1$, we have

$$ {}^{\tau}_{\alpha}K^{1}_{a^+}\psi(r) = \int_a^r \psi(w)w^{\tau}d_{\alpha}w, $$

which is just the definition of ${}^{\tau}_{\alpha}K_{a^+}$ and hence true.

Now we prove for $n = 2$. Let us define

$$ \varphi(r) = \int_a^r \left(\frac{r^{\tau+\alpha} - w^{\tau+\alpha}}{\tau+\alpha}\right)\psi(w)w^{\tau}d_{\alpha}w, \tag{2.9} $$

which is the right-hand side of (2.7) when $n = 2$. So we want to show that $\varphi(r) = {}^{\tau}_{\alpha}K^{2}_{a^+}\psi(r)$. Since

$$ \varphi(r) = \frac{1}{\tau+\alpha}\left[r^{\tau+\alpha}\int_a^r \psi(w)w^{\tau}d_{\alpha}w - \int_a^r w^{\tau+\alpha}\psi(w)w^{\tau}d_{\alpha}w\right]. \tag{2.10} $$

By taking the conformable derivative of both sides of (2.10) with respect to r, we get

$$ \begin{aligned} T_{\alpha}\varphi(r) &= \frac{1}{\tau+\alpha}\left[r^{\tau+\alpha}T_{\alpha}\int_a^r \psi(w)w^{\tau}d_{\alpha}w\right. \\ &\quad \left. + \int_a^r \psi(w)w^{\tau}d_{\alpha}wT_{\alpha}r^{\tau+\alpha} - T_{\alpha}\int_a^r w^{\tau+\alpha}\psi(w)w^{\tau}d_{\alpha}w\right] \\ &= \frac{1}{\tau+\alpha}\left[r^{\tau+\alpha}\psi(r)r^{\tau} + (\tau+\alpha)r^{\tau}\int_a^r \psi(w)w^{\tau}d_{\alpha}w - r^{\tau+\alpha}\psi(r)r^{\tau}\right] \\ &= r^{\tau}\int_a^r \psi(w)w^{\tau}d_{\alpha}w = r^{\tau}\,{}^{\tau}_{\alpha}K_{a^+}\psi(r). \end{aligned} $$

Since (2.9) implies that $\varphi(a) = 0$, so

$$ \varphi(r) = \varphi(r) - \varphi(a) = \int_a^r T_{\alpha}\varphi(w)d_{\alpha}w = \int_a^r {}^{\tau}_{\alpha}K_{a^+}\psi(w)w^{\tau}d_{\alpha}w = {}^{\tau}_{\alpha}K^{2}_{a^+}\psi(r). $$

Generally for any $n \in \mathbb{N}$, the proof is similar. First, we expand the term $(r^{\tau+\alpha} - w^{\tau+\alpha})^{n-1}$ by thebinomialtheorem, and then write $\varphi(r)$ as written in (2.10) and take all the terms containing $r^{\tau+\alpha}$ outside the integral sign. The process is then similar as above, this shows that (2.7) is true for every positive integer n.

The identity (2.8) can be proved in the similar way by iterating the single integral ${}^{\tau}_{\alpha}K_{b^-}\psi(r) = \int_r^b \psi(w)w^{\tau}d_{\alpha}w$, instead of ${}^{\tau}_{\alpha}K_{a^+}\psi(r) = \int_a^r \psi(w)w^{\tau}d_{\alpha}w$. □

2.2.1 Construction of New Integral Operators

Consider the results obtained in Theorem 2.4.

$$ {}^{\tau}_{\alpha}K^{n}_{a^+}\psi(r) = \frac{1}{(n-1)!}\int_a^r \left(\frac{r^{\tau+\alpha} - w^{\tau+\alpha}}{\tau+\alpha}\right)^{n-1}\psi(w)w^{\tau}d_{\alpha}w, \tag{2.11} $$

$$ {}_{\alpha}^{\tau}K_{b-}^{n}\psi(r) = \frac{1}{(n-1)!}\int_{r}^{b}\left(\frac{w^{\tau+\alpha}-r^{\tau+\alpha}}{\tau+\alpha}\right)^{n-1}\psi(w)w^{\tau}d_{\alpha}w. \tag{2.12} $$

In these cases, we are applying the conformable integral operators ${}_{\alpha}^{\tau}K_{a+}$ or ${}_{\alpha}^{\tau}K_{b-}$, n times, where n is restricted to be a positive integer. We generalize it and use a positive real number instead of positive integer n, which is what the fractional calculus requires, that is, the generalization of integer-order differentiation or n-fold integration. As in the case of previously defined fractional operators (integral or derivative), replacing the positive integer n by a positive real number β and using the gamma function, which is the generalization of the factorial function, we get new right-sided and left-sided generalized conformable fractional integral operators, which are defined below:

Definition 2.2 Let ψ be a conformable integrable function on the interval $[a,b] \subseteq [0,\infty)$. The left-sided and right-sided generalized conformable fractional integrals ${}_{\alpha}^{\tau}K_{a+}^{\beta}$ and ${}_{\alpha}^{\tau}K_{b-}^{\beta}$ of order $\beta > 0$ with $\alpha \in (0,1]$, $\tau \in \mathbb{R}$, $\alpha + \tau \neq 0$ are defined by:

$$ {}_{\alpha}^{\tau}K_{a+}^{\beta}\psi(r) = \frac{1}{\Gamma(\beta)}\int_{a}^{r}\left(\frac{r^{\tau+\alpha}-w^{\tau+\alpha}}{\tau+\alpha}\right)^{\beta-1}\psi(w)w^{\tau}d_{\alpha}w, \quad r > a $$

and

$$ {}_{\alpha}^{\tau}K_{b-}^{\beta}\psi(r) = \frac{1}{\Gamma(\beta)}\int_{r}^{b}\left(\frac{w^{\tau+\alpha}-r^{\tau+\alpha}}{\tau+\alpha}\right)^{\beta-1}\psi(w)w^{\tau}d_{\alpha}w, \quad b > r, $$

respectively, and ${}_{\alpha}^{\tau}K_{a+}^{0}\psi(r) = {}_{\alpha}^{\tau}K_{b-}^{0}\psi(r) = \psi(r)$. Here Γ denotes the well-known gamma function.

Remark 2.1

(1) For $\tau = 0$ in the above Definition 2.2, we get the Riemann–Liouville type conformable fractional integral operators, which are given below:

$$ {}_{\alpha}^{\tau}K_{a+}^{\beta}\psi(r) = \frac{1}{\Gamma(\beta)}\int_{a}^{r}\left(\frac{r^{\alpha}-w^{\alpha}}{\alpha}\right)^{\beta-1}\psi(w)d_{\alpha}w, \quad r > a \tag{2.13} $$

and

$$ {}_{\alpha}^{\tau}K_{b-}^{\beta}\psi(r) = \frac{1}{\Gamma(\beta)}\int_{r}^{b}\left(\frac{w^{\alpha}-r^{\alpha}}{\alpha}\right)^{\beta-1}\psi(w)d_{\alpha}w, \quad b > r, \tag{2.14} $$

respectively, and ${}_{\alpha}^{\tau}K_{a+}^{0}\psi(r) = {}_{\alpha}^{\tau}K_{b-}^{0}\psi(r) = \psi(r)$. Here Γ denotes the well-known gamma function.

Note that the operators in (2.13) and (2.14) can also be obtained by taking the conformable integral operators $\int_{a}^{r}\psi(w)d_{\alpha}w$ and $\int_{r}^{b}\psi(w)d_{\alpha}w$ and iterating in the manner as done above in Theorem 2.4.

(2) Using the definition of conformable integral given in (2.6) and L'Hospital rule, it is straightforward that when $\alpha \to 0$ in (2.13) and (2.14), we get the Hadamard fractional integrals:

$$ {}_{a^+}K_0^{\beta}\psi(r) = \frac{1}{\Gamma(\beta)}\int_a^r \left(\log\frac{r}{w}\right)^{\beta-1}\psi(w)\frac{dw}{w}, \quad r > a, $$

$$ {}_{b^-}K_0^{\beta}\psi(r) = \frac{1}{\Gamma(\beta)}\int_r^b \left(\log\frac{w}{r}\right)^{\beta-1}\psi(w)\frac{dw}{w}, \quad r < b. $$

(3) For $\alpha = 1$, we get the Riemann–Liouville fractional integrals:

$$ {}_{a^+}K_1^{\beta}\psi(r) = \frac{1}{\Gamma(\beta)}\int_a^r (r-w)^{\beta-1}\psi(w)dw, \quad r > a, $$

$$ {}_{b^-}K_1^{\beta}\psi(r) = \frac{1}{\Gamma(\beta)}\int_r^b (w-r)^{\beta-1}\psi(w)dw, \quad r < b. $$

(4) For the case $\beta = 1$ in Definition 2.2, we get the conformable fractional integrals. And when $\alpha = \beta = 1, \tau = 0$, we get classical Riemann integrals.

Now we prove some basic properties for the obtained generalized operators.

Theorem 2.5 *Let $\psi \in L_{\alpha}[a,b]$, where $0 \le a < b$. For $\alpha \in (0,1]$, $\beta > 0$, we have*

$$ \lim_{\beta\to 0} {}_{\alpha}^{\tau}K_{a^+}^{\beta}\psi(r) = {}_{\alpha}^{\tau}K_{a^+}^{0}\psi(r) = \psi(r), \tag{2.15} $$

$$ \lim_{\beta\to 0} {}_{\alpha}^{\tau}K_{b^-}^{\beta}\psi(r) = {}_{\alpha}^{\tau}K_{b^-}^{0}\psi(r) = \psi(r). \tag{2.16} $$

Proof: Applying the relation (2.6), integration by parts and well-known property of beta function:

$$ \begin{aligned} {}_{\alpha}^{\tau}K_{a^+}^{\beta}\psi(r) &= \frac{1}{\Gamma(\beta)}\int_a^r \left(\frac{r^{\tau+\alpha}-w^{\tau+\alpha}}{\tau+\alpha}\right)^{\beta-1}\psi(w)w^{\tau}d_{\alpha}w \\ &= \frac{1}{\beta\Gamma(\beta)}\psi(a)\left(\frac{r^{\tau+\alpha}-p^{\tau+\alpha}}{\tau+\alpha}\right)^{\beta} \\ &\quad + \frac{1}{\beta\Gamma(\beta)}\int_a^r T_{\alpha}\psi(w)\left(\frac{r^{\tau+\alpha}-w^{\tau+\alpha}}{\tau+\alpha}\right)^{\beta}d_{\alpha}w \\ &= \frac{1}{\Gamma(\beta+1)}\psi(a)\left(\frac{r^{\tau+\alpha}-p^{\alpha}}{\tau+\alpha}\right)^{\beta} \\ &\quad + \frac{1}{\Gamma(\beta+1)}\int_a^r T_{\alpha}\psi(w)\left(\frac{r^{\tau+\alpha}-w^{\tau+\alpha}}{\tau+\alpha}\right)^{\beta}d_{\alpha}w. \end{aligned} $$

Taking limit as $\beta \to 0$, we have

$$ \lim_{\beta\to 0} {}_{\alpha}^{\tau}K_{a^+}^{\beta}\psi(r) = {}_{\alpha}^{\tau}K_{a^+}^{0}\psi(r) = \psi(a) + \int_a^r T_{\alpha}\psi(w)w^{\tau}d_{\alpha}w = \psi(r). $$

The proof is similar for (2.16). □

Now we prove semigroup property for this newly defined operator. This property makes possible not only the definition of new integral, but also of new differentiation, by taking enough derivatives of ${}^{\tau}_{\alpha}K^{\beta}_{a^+}\psi(r)$ and ${}^{\tau}_{\alpha}K^{\beta}_{b^-}\psi(r)$, which we will discuss in next section.

Theorem 2.6 ***(Semigroup property)*** *Let* $\psi : [a,b] \subseteq [0,\infty) \to \mathbb{R}$ *be a conformable integrable function. Then for* $\beta_1, \beta_2 > 0$ *and* $\alpha \in (0,1]$, *we have*

$$
\begin{aligned}
{}^{\tau}_{\alpha}K^{\beta_1}_{a^+}\,{}^{\tau}_{\alpha}K^{\beta_2}_{a^+}\psi(r) &= {}^{\tau}_{\alpha}K^{\beta_1+\beta_2}_{a^+}\psi(r) \\
&= \frac{1}{\Gamma(\beta_1+\beta_2)}\int_a^r \left(\frac{r^{\tau+\alpha}-w^{\tau+\alpha}}{\tau+\alpha}\right)^{\beta_1+\beta_2-1}\psi(w)w^{\tau}d_{\alpha}w. \quad (2.17)
\end{aligned}
$$

$$
\begin{aligned}
{}^{\tau}_{\alpha}K^{\beta_1}_{b^-}\,{}^{\tau}_{\alpha}K^{\beta_2}_{b^-}\psi(r) &= {}^{\tau}_{\alpha}K^{\beta_1+\beta_2}_{b^-}\psi(r) \\
&= \frac{1}{\Gamma(\beta_1+\beta_2)}\int_r^b \left(\frac{w^{\tau+\alpha}-r^{\tau+\alpha}}{\tau+\alpha}\right)^{\beta_1+\beta_2-1}\psi(w)w^{\tau}d_{\alpha}w. \quad (2.18)
\end{aligned}
$$

Proof: Consider

$$
\begin{aligned}
&{}^{\tau}_{\alpha}K^{\beta_1}_{a^+}\,{}^{\tau}_{\alpha}K^{\beta_2}_{a^+}\psi(r) \\
&= \frac{1}{\Gamma(\beta_1)}\int_a^r \left(\frac{r^{\tau+\alpha}-w^{\tau+\alpha}}{\tau+\alpha}\right)^{\beta_1-1} {}^{\tau}_{\alpha}K^{\beta_2}_{a^+}\psi(w)w^{\tau}d_{\alpha}w \\
&= \frac{1}{\Gamma(\beta_1)\Gamma(\beta_2)}\int_a^r \left(\frac{r^{\tau+\alpha}-w^{\tau+\alpha}}{\tau+\alpha}\right)^{\beta_1-1} w^{\tau}\int_a^w \left(\frac{w^{\tau+\alpha}-s^{\tau+\alpha}}{\tau+\alpha}\right)^{\beta_2-1} \\
&\quad \psi(s)s^{\tau}d_{\alpha}s d_{\alpha}w \\
&= \frac{(\tau+\alpha)^{2-(\beta_1+\beta_2)}}{\Gamma(\beta_1)\Gamma(\beta_2)}\int_a^r \psi(s)s^{\tau} \\
&\quad \int_s^r (r^{\tau+\alpha}-w^{\tau+\alpha})^{\beta_1-1}(w^{\tau+\alpha}-s^{\tau+\alpha})^{\beta_2-1}w^{\alpha-1+\tau}dw\, d_{\alpha}s,
\end{aligned}
$$

where in the last step, we have exchanged the order of integration using Fubini's theorem and applied the relation (2.6). Changing variables to l defined by $w^{\tau+\alpha} = s^{\tau+\alpha} + (r^{\tau+\alpha} - s^{\tau+\alpha})l$ in the inner integral

$$
\begin{aligned}
&{}^{\tau}_{\alpha}K^{\beta_1}_{a^+}\,{}^{\tau}_{\alpha}K^{\beta_2}_{a^+}\psi(r) \\
&= \frac{(\tau+\alpha)^{1-(\beta_1+\beta_2)}}{\Gamma(\beta_1)\Gamma(\beta_2)}\int_a^r (r^{\tau+\alpha}-s^{\tau+\alpha})^{\beta_1+\beta_2-1}\psi(s)s^{\tau}\int_0^1 l^{\beta_2-1}(1-l)^{\beta_1-1}dl\, d_{\alpha}s \\
&= \frac{1}{\Gamma(\beta_1+\beta_2)}\int_a^r \left(\frac{r^{\tau+\alpha}-s^{\tau+\alpha}}{\tau+\alpha}\right)^{\beta_1+\beta_2-1}\psi(s)s^{\tau}d_{\alpha}s \\
&= {}^{\tau}_{\alpha}K^{\beta_1+\beta_2}_{a^+}\psi(r),
\end{aligned}
$$

where

$$\int_0^1 l^{\beta_2-1}(1-l)^{\beta_1-1}dl = \frac{\Gamma(\beta_1)\Gamma(\beta_2)}{\Gamma(\beta_1+\beta_2)}$$

is the well-known Euler beta function.

Equation (2.18) can be proved in the same way. This completes the proof. □

If we consider a bounded interval $[a, b]$, such that $p \geq 0$. The operators ${}_{\alpha}^{\tau}K_{a^+}^{\beta}$ and ${}_{\alpha}^{\tau}K_{b^-}^{\beta}$ associate the function ${}_{\alpha}^{\tau}K_{a^+}^{\beta}\psi(r)$ and ${}_{\alpha}^{\tau}K_{b^-}^{\beta}\psi(r)$ to each conformable integrable function ψ on $[a, b]$. Thus, these are linear operators, which is proved in the following theorem.

Theorem 2.7 ***(Linearity)*** *The operators* ${}_{\alpha}^{\tau}K_{a^+}^{\beta}$ *and* ${}_{\alpha}^{\tau}K_{b^-}^{\beta}$ *are linear operators on* $L_{\alpha}[a, b]$. *That is, define*

$${}_{\alpha}^{\tau}K_{a^+}^{\beta},\ {}_{\alpha}^{\tau}K_{b^-}^{\beta} : L_{\alpha}[a, b] \to L_{\alpha}[a, b],$$

then

$$\begin{aligned} {}_{\alpha}^{\tau}K_{a^+}^{\beta}(\mu_1\psi_1 + \mu_2\psi_2) &= \mu_1\, {}_{\alpha}^{\tau}K_{a^+}^{\beta}\psi_1 + \mu_2\, {}_{\alpha}^{\tau}K_{a^+}^{\beta}\psi_2, \\ {}_{\alpha}^{\tau}K_{b^-}^{\beta}(\mu_1\psi_1 + \mu_2\psi_2) &= \mu_1\, {}_{\alpha}^{\tau}K_{b^-}^{\beta}\psi_1 + \mu_2\, {}_{\alpha}^{\tau}K_{b^-}^{\beta}\psi_2. \end{aligned}$$

For all $\psi_1,\ \psi_2 \in L_{\alpha}[a, b]$ *and* $\mu_1,\ \mu_2 \in \mathbb{R}$.

Proof: The proof is simple, consider

$$\begin{aligned} {}_{\alpha}^{\tau}K_{a^+}^{\beta}(\mu_1\psi_1 + \mu_2\psi_2)(r) &= \frac{1}{\Gamma(\beta)}\int_a^r \left(\frac{r^{\tau+\alpha} - w^{\tau+\alpha}}{\tau+\alpha}\right)^{\beta-1} (\mu_1\psi_1 + \mu_2\psi_2)(w)w^{\tau}d_{\alpha}w \\ &= \frac{\mu_1}{\Gamma(\beta)}\int_a^r \left(\frac{r^{\tau+\alpha} - w^{\tau+\alpha}}{\tau+\alpha}\right)^{\beta-1} \psi_1(w)w^{\tau}d_{\alpha}w \\ &\quad + \frac{\mu_2}{\Gamma(\beta)}\int_a^r \left(\frac{r^{\tau+\alpha} - w^{\tau+\alpha}}{\tau+\alpha}\right)^{\beta-1} \psi_2(w)w^{\tau}d_{\alpha}w \\ &= \mu_1\, {}_{\alpha}^{\tau}K_{a^+}^{\beta}\psi_1(r) + \mu_2\, {}_{\alpha}^{\tau}K_{a^+}^{\beta}\psi_2(r). \end{aligned}$$

Similarly,

$${}_{\alpha}^{\tau}K_{b^-}^{\beta}(\mu_1\psi_1 + \mu_2\psi_2)(r) = \mu_1\, {}_{\alpha}^{\tau}K_{b^-}^{\beta}\psi_1(r) + \mu_2\, {}_{\alpha}^{\tau}K_{b^-}^{\beta}\psi_2(r).$$

□

In the following theorem, we prove that the operators ${}_{\alpha}^{\tau}K_{a^+}^{\beta}$ and ${}_{\alpha}^{\tau}K_{b^-}^{\beta}$ are bounded on the space $L_{\alpha}[a, b]$.

Theorem 2.8 ***(Boundedness)*** *The operators* ${}_{\alpha}^{\tau}K_{a^+}^{\beta}$ *and* ${}_{\alpha}^{\tau}K_{b^-}^{\beta}$ *are bounded operators on* $L_{\alpha}[a, b]$. *That is, define*

$${}_{\alpha}^{\tau}K_{a^+}^{\beta},\ {}_{\alpha}^{\tau}K_{b^-}^{\beta} : L_{\alpha}[a, b] \to L_{\alpha}[a, b],$$

then

$$\| {}_{\alpha}^{\tau}K_{a^+}^{\beta}\psi\| \leq M\|\psi\|_C, \quad \| {}_{\alpha}^{\tau}K_{b^-}^{\beta}\psi\| \leq M\|\psi\|_C, \tag{2.19}$$

where $\|\psi\|_C = \max_{r\in[a,b]} |\psi(r)|$, $M = \frac{|(\tau+\alpha)^{-\beta}|}{\beta+1}(b^{\tau+\alpha} - a^{\tau+\alpha})^{\beta}$.

Proof: The proof is simple, we consider

$$\begin{aligned}
\| {}_{\alpha}^{\tau}K_{a^+}^{\beta}\psi(r)\| &= \left\| \frac{1}{\Gamma(\beta)} \int_a^r \left(\frac{r^{\tau+\alpha} - w^{\tau+\alpha}}{\tau+\alpha} \right)^{\beta-1} \psi(w) w^{\tau} d_{\alpha}w \right\| \\
&\leq \frac{|(\tau+\alpha)^{1-\beta}|}{\Gamma(\beta)} \|\psi\|_C \int_a^r (r^{\tau+\alpha} - w^{\tau+\alpha})^{\beta-1} w^{\tau+\alpha-1} dw \\
&\leq \frac{|(\tau+\alpha)^{1-\beta}|}{\Gamma(\beta+1)} \|\psi\|_C (b^{\tau+\alpha} - a^{\tau+\alpha})^{\beta}.
\end{aligned} \tag{2.20}$$

In the case of the right generalized conformable fractional integral operator ${}_{\alpha}^{\tau}K_{b^-}^{\beta}$, the proof is similar. □

2.3 Generalized Conformable Fractional Derivative

Because the Riemann–Liouville approach to the generalized conformable fractional integrals began with an expression involving repeated conformable integration of a function. One can adopt the Grunwald–Letnikov approach to construct a fractional derivative operator firstly, on the basis of which the fractional integral operator can be defined. However, it is also possible to frame a definition for the generalized conformable fractional derivative using the definition already obtained above for the related integral. Now keeping the above integral operators under consideration, we define the right- and left-sided generalized conformable fractional derivative operators as follows.

Definition 2.3 Let ψ be a conformable integrable function on the interval $[a, b]$. The right- and left-sided generalized conformable fractional derivative operators ${}_{\alpha}^{\tau}T_{a^+}^{\beta}$ and ${}_{\alpha}^{\tau}T_{b^-}^{\beta}$ of order $0 < \beta < 1$, $\alpha \in (0, 1]$ with $p \geq 0$ are defined by:

$${}_{\alpha}^{\tau}T_{a^+}^{\beta}\psi(r) = \frac{r^{-\tau}}{\Gamma(1-\beta)} T_{\alpha} \int_a^r \left(\frac{r^{\tau+\alpha} - w^{\tau+\alpha}}{\tau+\alpha} \right)^{-\beta} \psi(w) w^{\tau} d_{\alpha}w, \quad r > a,$$

$${}_{\alpha}^{\tau}T_{b^-}^{\beta}\psi(r) = \frac{r^{-\tau}}{\Gamma(1-\beta)} T_{\alpha} \int_r^b \left(\frac{r^{\tau+\alpha} - w^{\tau+\alpha}}{\tau+\alpha} \right)^{-\beta} \psi(w) w^{\tau} d_{\alpha}w, \quad b > r,$$

respectively, and ${}_{\alpha}^{\tau}T_{a^+}^{0}\psi(r) = {}_{\alpha}^{\tau}T_{b^-}^{0}\psi(r) = \psi(r)$. Here Γ denotes the gamma function and T_{α} denotes the conformable derivative of order α.

To proceed further, we need to prove the following theorem which shows the relation between the fractional integral and derivative operators.

Theorem 2.9 *Let $\psi : [a,b] \subseteq [0,\infty) \to \mathbb{R}$ be a conformable integrable function. Then for $0 < \beta < 1$ and $\alpha \in (0,1]$, we have*

$$ {}^{\tau}_{\alpha}T^{\beta}_{a^+}\psi(r) = \frac{\psi(a)(\tau+\alpha)^{\beta}(r^{\tau+\alpha}-a^{\tau+\alpha})^{-\beta}}{\Gamma(1-\beta)} + {}^{\tau}_{\alpha}K^{1-\beta}_{a^+}\left(r^{1-(\tau+\alpha)}\frac{d}{dr}\psi(r)\right). \tag{2.21}$$

Proof: Let $u'(w) = w^{\tau+\alpha-1}(r^{\tau+\alpha}-w^{\tau+\alpha})^{-\beta}$, $v(w) = \psi(w) - \psi(a)$. Consider

$$\int_a^r u'(w)v(w)dw = \int_a^r w^{\tau+\alpha-1}(r^{\tau+\alpha}-w^{\tau+\alpha})^{-\beta}(\psi(w)-\psi(a))dw$$

$$\Longrightarrow \frac{d}{dr}\int_a^r u'(w)v(w)dw = \frac{d}{dr}\int_a^r w^{\tau+\alpha-1}(r^{\tau+\alpha}-w^{\tau+\alpha})^{-\beta}(\psi(w)-\psi(a))dw$$

multiplying both sides by $\frac{(\tau+\alpha)^{\beta}r^{1-(\tau+\alpha)}}{\Gamma(1-\beta)}$, we get

$$\begin{aligned}
&\frac{(\tau+\alpha)^{\beta}r^{1-(\tau+\alpha)}}{\Gamma(1-\beta)}\frac{d}{dr}\int_a^r u'(w)v(w)dw \\
&= \frac{(\tau+\alpha)^{\beta}r^{1-(\tau+\alpha)}}{\Gamma(1-\beta)}\frac{d}{dr}\int_a^r w^{\tau+\alpha-1}(r^{\tau+\alpha}-w^{\tau+\alpha})^{-\beta}(\psi(w)-\psi(a))dw \\
&= \mathbf{I}_1 + \mathbf{I}_2,
\end{aligned} \tag{2.22}$$

where

$$\begin{aligned}
\mathbf{I}_1 &= \frac{(\tau+\alpha)^{\beta}r^{1-(\tau+\alpha)}}{\Gamma(1-\beta)}\frac{d}{dr}\int_a^r w^{\tau}w^{\alpha-1}(r^{\tau+\alpha}-w^{\tau+\alpha})^{-\beta}\psi(w)dw \\
&= \frac{r^{1-(\tau+\alpha)}}{\Gamma(1-\beta)}\frac{d}{dr}\int_a^r w^{\tau}\left(\frac{r^{\tau+\alpha}-w^{\tau+\alpha}}{\tau+\alpha}\right)^{-\beta}\psi(w)d_{\alpha}w \\
&= \frac{r^{1-(\tau+\alpha)}.r^{\alpha-1}}{\Gamma(1-\beta)}T_{\alpha}\int_a^r w^{\tau}\left(\frac{r^{\tau+\alpha}-w^{\tau+\alpha}}{\tau+\alpha}\right)^{-\beta}\psi(w)d_{\alpha}w = {}^{\tau}_{\alpha}T^{\beta}_{a^+}\psi(r).
\end{aligned} \tag{2.23}$$

Also,

$$\begin{aligned}
\mathbf{I}_2 &= \frac{(\tau+\alpha)^{\beta}r^{1-(\tau+\alpha)}}{\Gamma(1-\beta)}\frac{d}{dr}\int_a^r w^{\tau+\alpha-1}(r^{\tau+\alpha}-w^{\tau+\alpha})^{-\beta}\psi(a)dw \\
&= \frac{r^{1-(\tau+\alpha)}(\tau+\alpha)^{\beta-1}}{\Gamma(1-\beta)}\frac{d}{dr}\int_a^r (-(\tau+\alpha)w^{\tau+\alpha-1})(r^{\tau+\alpha}-w^{\tau+\alpha})^{-\beta}\psi(a)dw \\
&= \frac{r^{1-(\tau+\alpha)}(\tau+\alpha)^{\beta-1}\psi(a)}{(\beta-1)\Gamma(1-\beta)}\frac{d}{dr}(r^{\tau+\alpha}-a^{\tau+\alpha})^{-\beta+1} \\
&= \frac{-\psi(a)}{\Gamma(1-\beta)}\left(\frac{r^{\tau+\alpha}-a^{\tau+\alpha}}{\tau+\alpha}\right)^{-\beta},
\end{aligned} \tag{2.24}$$

putting values in (2.22), we get

$$\frac{(\tau+\alpha)^{\beta}r^{1-(\tau+\alpha)}}{\Gamma(1-\beta)}\frac{d}{dr}\int_a^r u'(w)v(w)dw = {}_{\alpha}^{\tau}T_{a^+}^{\beta}\psi(r) - \frac{\psi(a)}{\Gamma(1-\beta)}\left(\frac{r^{\tau+\alpha}-a^{\tau+\alpha}}{\tau+\alpha}\right)^{-\beta}. \tag{2.25}$$

Now considering the left side of (2.25) and differentiating the integral with respect to variable r, we get

$$\begin{aligned}\frac{(\tau+\alpha)^{\beta}r^{1-(\tau+\alpha)}}{\Gamma(1-\beta)}\frac{d}{dr}\int_a^r u'(w)v(w)dw &= \frac{(\tau+\alpha)^{\beta}}{\Gamma(1-\beta)}\int_a^r (r^{\tau+\alpha}-w^{\tau+\alpha})^{-\beta}\psi'(w)dw \\ &= {}_{\alpha}^{\tau}K_{a^+}^{1-\beta}\left(r^{1-(\tau+\alpha)}\frac{d}{dr}\psi(r)\right).\end{aligned} \tag{2.26}$$

From (2.25) and (2.26), we get the required result. □

We prove the inverse property for the defined generalized operators in the following theorem.

Theorem 2.10 ***(Inverse property)*** *For any continuous function ψ in the domain of ${}_{\alpha}^{\tau}K_{a^+}^{\beta}$, ${}_{\alpha}^{\tau}K_{b^-}^{\beta}$, ${}_{\alpha}^{\tau}T_{a^+}^{\beta}$, and ${}_{\alpha}^{\tau}T_{b^-}^{\beta}$, we have*

$${}_{\alpha}^{\tau}T_{a^+}^{\beta}\,{}_{\alpha}^{\tau}K_{a^+}^{\beta}\psi(r) = \psi(r), \quad {}_{\alpha}^{\tau}T_{b^-}^{\beta}\,{}_{\alpha}^{\tau}K_{b^-}^{\beta}\psi(r) = \psi(r). \tag{2.27}$$

Similarly,

$${}_{\alpha}^{\tau}K_{a^+}^{\beta}\,{}_{\alpha}^{\tau}T_{a^+}^{\beta}\psi(r) = \psi(r), \quad {}_{\alpha}^{\tau}K_{b^-}^{\beta}\,{}_{\alpha}^{\tau}T_{b^-}^{\beta}\psi(r) = \psi(r). \tag{2.28}$$

Proof: Consider

$$\begin{aligned}&{}_{\alpha}^{\tau}T_{a^+}^{\beta}\,{}_{\alpha}^{\tau}K_{a^+}^{\beta}\psi(r)\\ &= \frac{r^{-\tau}}{\Gamma(1-\beta)}T_{\alpha}\int_a^r\left(\frac{r^{\tau+\alpha}-w^{\tau+\alpha}}{\tau+\alpha}\right)^{-\beta}{}_{\alpha}^{\tau}K_{a^+}^{\beta}\psi(w)w^{\tau}d_{\alpha}w\\ &= \frac{r^{-\tau}}{\Gamma(1-\beta)\Gamma(\beta)}T_{\alpha}\int_a^r\left(\frac{r^{\tau+\alpha}-w^{\tau+\alpha}}{\tau+\alpha}\right)^{-\beta}w^{\tau}\\ &\quad\int_a^w\left(\frac{w^{\tau+\alpha}-s^{\tau+\alpha}}{\tau+\alpha}\right)^{\beta-1}\psi(s)s^{\tau}d_{\alpha}s\,d_{\alpha}w\\ &= \frac{r^{-\tau}(\tau+\alpha)}{\Gamma(1-\beta)\Gamma(\beta)}T_{\alpha}\int_a^r\int_a^w(r^{\tau+\alpha}-w^{\tau+\alpha})^{-\beta}(w^{\tau+\alpha}-s^{\tau+\alpha})^{\beta-1}w^{\tau}\psi(s)s^{\tau}d_{\alpha}s\,d_{\alpha}w.\end{aligned}$$

Switching the order of integration and changing variables to u, define $w^{\tau+\alpha} = s^{\tau+\alpha} + (r^{\tau+\alpha} - s^{\tau+\alpha})u$,

$$
\begin{aligned}
{}^{\tau}_{\alpha}T^{\beta}_{a^+}\,{}^{\tau}_{\alpha}K^{\beta}_{a^+}\psi(r) &= \frac{r^{-\tau}}{\Gamma(1-\beta)\Gamma(\beta)} T_{\alpha} \int_a^r \int_0^1 u^{\beta-1}(1-u)^{-\beta}\psi(s)s^{\tau}du\, d_{\alpha}s \\
&= r^{-\tau}T_{\alpha}\int_a^r \psi(s)s^{\tau}d_{\alpha}s \\
&= \psi(r),
\end{aligned}
$$

where

$$\int_0^1 u^{\beta-1}(1-u)^{-\beta}du = B(\beta,\, 1-\beta) = \Gamma(\beta)\Gamma(1-\beta).$$

Similarly, we can prove that

$${}^{\tau}_{\alpha}T^{\beta}_{b^-}\,{}^{\tau}_{\alpha}K^{\beta}_{b^-}\psi(r) = \psi(r).$$

To prove (2.28), we proceed as under:

Applying the operator ${}^{\tau}_{\alpha}K^{\beta}_{a^+}$ to both sides of (2.21) and using the relations (2.5) and (2.17), we get

$$
\begin{aligned}
&{}^{\tau}_{\alpha}K^{\beta}_{a^+}\,{}^{\tau}_{\alpha}T^{\beta}_{a^+}\psi(r) \\
&= \frac{\psi(a)(\tau+\alpha)^{\beta}}{\Gamma(1-\beta)}\,{}^{\tau}_{\alpha}K^{\beta}_{a^+}(r^{\tau+\alpha}-a^{\tau+\alpha})^{-\beta} + {}^{\tau}_{\alpha}K^{\beta}_{a^+}({}^{\tau}_{\alpha}K^{1-\beta}_{a^+}(r^{-\tau}T_{\alpha}\psi(r))) \\
&= \frac{\psi(a)(\tau+\alpha)^{\beta}}{\Gamma(1-\beta)}\frac{1}{\Gamma(\beta)}\int_a^r \left(\frac{r^{\tau+\alpha}-w^{\tau+\alpha}}{\tau+\alpha}\right)^{\beta-1}(w^{\tau+\alpha}-a^{\tau+\alpha})^{-\beta}w^{\tau}d_{\alpha}w \\
&\quad + {}^{\tau}_{\alpha}K_{a^+}(r^{-\tau}T_{\alpha}\psi(r)) \\
&= \frac{\psi(a)(\tau+\alpha)}{\Gamma(\beta)\Gamma(1-\beta)}\int_a^r (r^{\tau+\alpha}-w^{\tau+\alpha})^{\beta-1}(w^{\tau+\alpha}-a^{\tau+\alpha})^{-\beta}w^{\tau+\alpha-1}dw + \psi(r)-\psi(a) \\
&= \frac{\psi(a)(\tau+\alpha)}{\Gamma(\beta)\Gamma(1-\beta)}\int_a^r (r^{\tau+\alpha}-w^{\tau+\alpha})^{\beta-1}(w^{\tau+\alpha}-a^{\tau+\alpha})^{-\beta}w^{\tau+\alpha-1}dw + \psi(r)-\psi(a).
\end{aligned}
\tag{2.29}
$$

Changing variables to u defined by, $w^{\tau+\alpha} = a^{\tau+\alpha} + (r^{\tau+\alpha} - a^{\tau+\alpha})u$, in the inner integral, we get from (2.29):

$$
\begin{aligned}
{}^{\tau}_{\alpha}K^{\beta}_{a^+}\,{}^{\tau}_{\alpha}T^{\beta}_{a^+}\psi(r) &= \frac{\psi(a)}{\Gamma(\beta)\Gamma(1-\beta)}\int_0^1 u^{-\beta}(1-u)^{\beta-1}du + \psi(r)-\psi(a) \\
&= \frac{\psi(a)}{\Gamma(\beta)\Gamma(1-\beta)}\Gamma(\beta)\Gamma(1-\beta) + \psi(r)-\psi(a) \\
&= \psi(r),
\end{aligned}
$$

which is the required proof. □

Theorem 2.11 ***(Linearity)*** *The generalized conformable fractional derivative operators are linear on its domain, that is:*

$$
\begin{aligned}
{}^{\tau}_{\alpha}T^{\beta}_{a^+}(\mu_1\psi_1+\mu_2\psi_2) &= \mu_1\,{}^{\tau}_{\alpha}T^{\beta}_{a^+}\psi_1+\mu_2\,{}^{\tau}_{\alpha}T^{\beta}_{a^+}\psi_2,\\
{}^{\tau}_{\alpha}T^{\beta}_{b^-}(\mu_1\psi_1+\mu_2\psi_2) &= \mu_1\,{}^{\tau}_{\alpha}T^{\beta}_{b^-}\psi_1+\mu_2\,{}^{\tau}_{\alpha}T^{\beta}_{b^-}\psi_2,
\end{aligned}
$$

for all $\psi_1,\ \psi_2 \in L_\alpha[a,b]$ *and* $\mu_1,\ \mu_2 \in \mathbb{R}$.

Proof: The proof is similar to the proof of Theorem 2.7. □

Theorem 2.12 ***(Semigroup property)*** *Let* $\psi : [a,b] \subseteq [0,\infty) \to \mathbb{R}$ *be a conformable integrable function. Then for* $0<\beta_1<1$, $0<\beta_2<1$ *and* $\alpha \in (0,1]$, *we have*

$$
\begin{aligned}
{}^{\tau}_{\alpha}T^{\beta_1}_{a^+}\,{}^{\tau}_{\alpha}T^{\beta_2}_{a^+}\psi(r) &= {}^{\tau}_{\alpha}T^{\beta_1+\beta_2}_{a^+}\psi(r),\\
{}^{\tau}_{\alpha}T^{\beta_1}_{b^-}\,{}^{\tau}_{\alpha}T^{\beta_2}_{b^-}\psi(r) &= {}^{\tau}_{\alpha}T^{\beta_1+\beta_2}_{b^-}\psi(r).
\end{aligned}
$$

Proof: The proof is similar to the proof of Theorem 2.6. □

We consider an example to illustrate the results. We shall find the generalized conformable fractional derivative of the power function and explore the response for different values of α, β, λ, and τ.

Example 2.1 Consider the function $\psi(r)=r^{\lambda}$, $\lambda \in \mathbb{R}$, $r \geq 0$. Then for $0<\beta<1$, $\alpha \in (0,1]$, we have

$$
{}^{\tau}_{\alpha}T^{\beta}_{0^+}\psi(r) = \frac{\Gamma(1+\frac{\lambda}{\tau+\alpha})(\tau+\alpha)^{\beta-1}}{\Gamma(1-\beta+\frac{\lambda}{\tau+\alpha})}\,r^{-\beta(\tau+\alpha)+\lambda}.
$$

Proof: The proof is simple by taking $p=0$ in the above Definition 2.3, we get

$$
\begin{aligned}
{}^{\tau}_{\alpha}T^{\beta}_{0^+}r^{\lambda} &= \frac{r^{-\tau}}{\Gamma(1-\beta)}T_\alpha\int_0^r\left(\frac{r^{\tau+\alpha}-w^{\tau+\alpha}}{\tau+\alpha}\right)^{-\beta} w^{\lambda}w^{\tau}d_\alpha w\\
&= \frac{r^{-\tau}(\tau+\alpha)^{\beta}}{\Gamma(1-\beta)}T_\alpha\int_0^r (r^{\tau+\alpha}-w^{\tau+\alpha})^{-\beta}w^{\lambda}w^{\tau+\alpha-1}dw. \qquad (2.30)
\end{aligned}
$$

Making the substitution $w^{\tau+\alpha}=ur^{\tau+\alpha}$ in (2.30) and using the relations

$$
\int_0^1 u^{a-1}(1-u)^{b-1}du = B(a,\ b) = \frac{\Gamma(a)\Gamma(b)}{\Gamma(a+b)} \quad \text{and} \quad \Gamma(t+1)=t\Gamma(t),
$$

we get the required result. □

Remark 2.2 It is interesting to note that for $\alpha = 1,\ \beta = 1$ and $\tau = 0$, we obtain ${}^{0}_{1}T^{1}_{0^+} r^{\lambda} = \lambda r^{\lambda-1}$, as one would expect the ordinary derivative.

To understand the behavior of different parameters involved in the newly established fractional operators, we use graphical representation. In Figure 2.1, the effect of β is presented while other parameters are kept fix. In Figure 2.2, the effect of α is shown for different fix values of other parameters. The same thing is done for λ and τ in Figures 2.3 and 2.4 respectively. The effect of the parameter λ is given in the following figures:

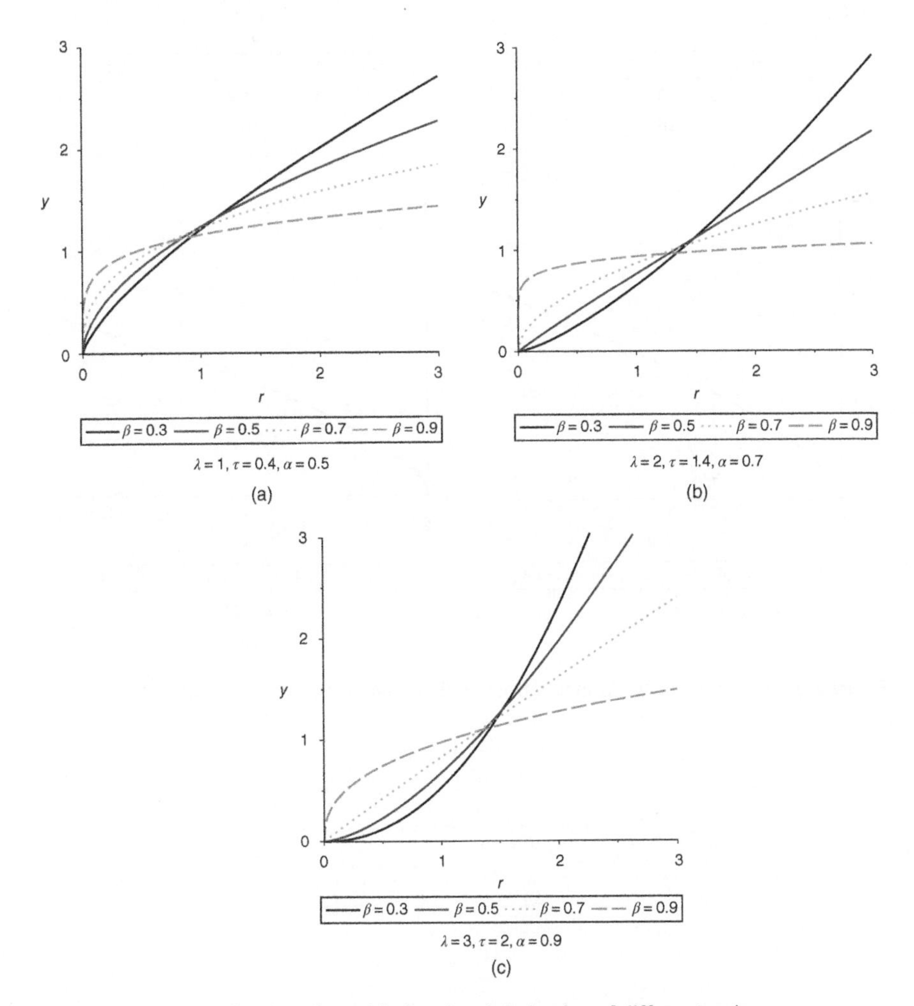

Figure 2.1 Generalized conformable fractional derivative of different orders.

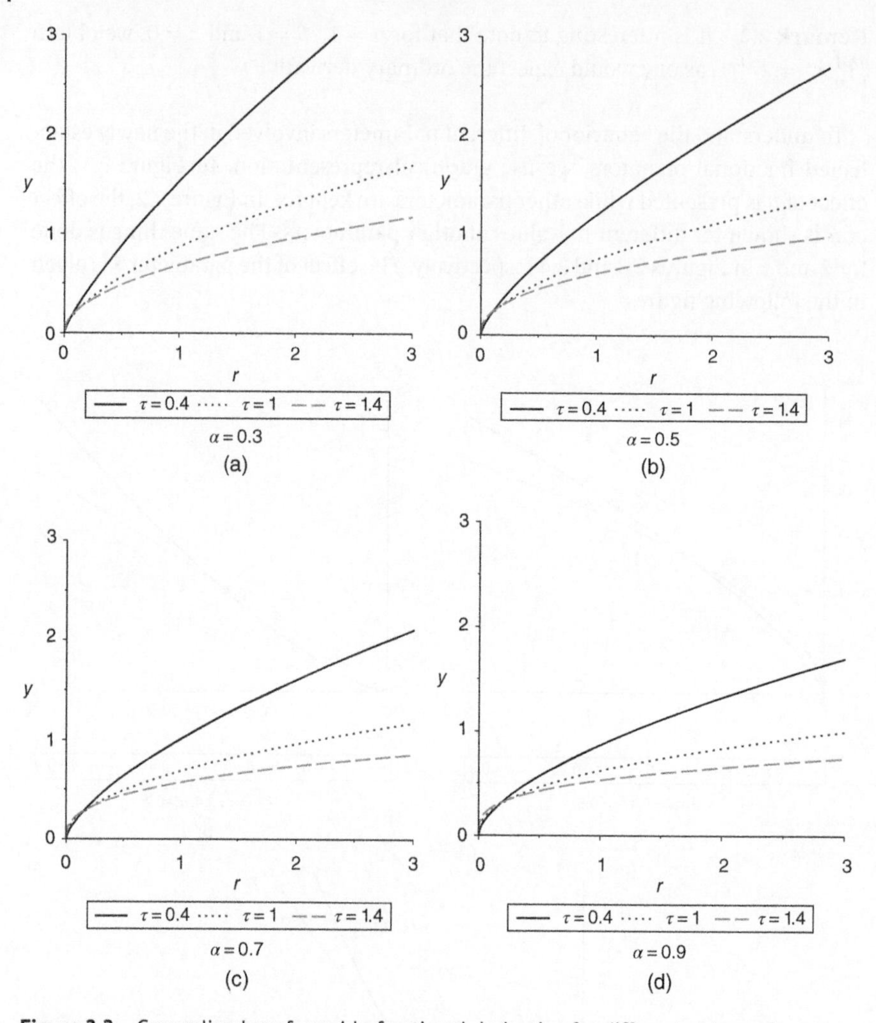

Figure 2.2 Generalized conformable fractional derivative for different values of α.

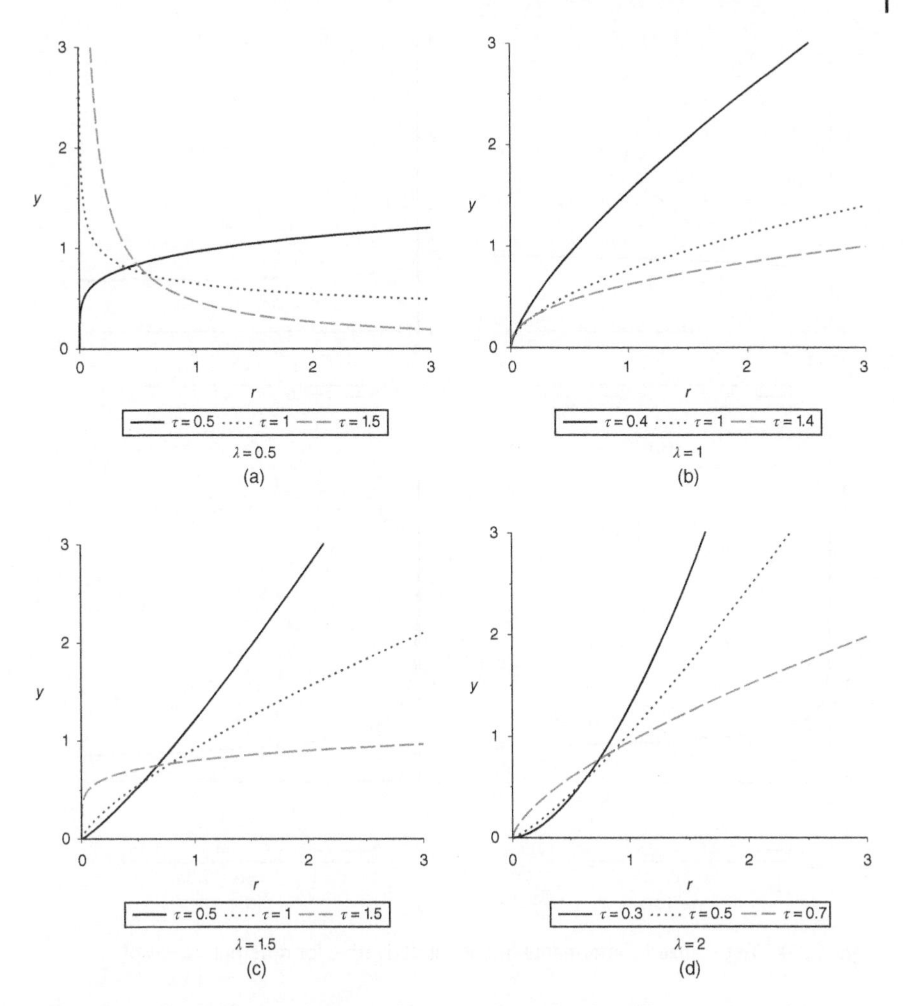

Figure 2.3 Generalized conformable fractional derivative for different values of λ.

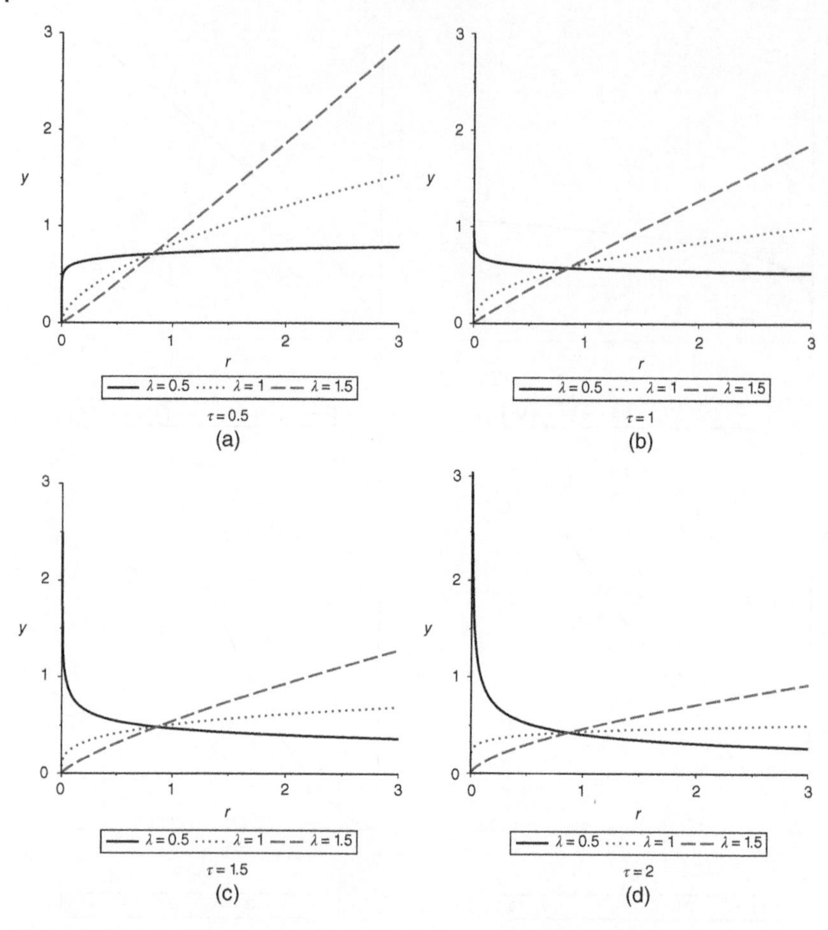

Figure 2.4 Generalized conformable fractional derivative for different values of τ.

2.4 Applications to Integral Equations and Fractional Differential Equations

As mentioned above that our newly obtained fractional operators generalize R–L operators, Hadamard operators, Katugampola operators, which have remarkable applications in various fields [2, 3, 5, 9, 10, 20]. One of the possible applications has been given below (after this section) in the form of open problem related to image denoising. Moreover in the following results, we apply our operators to the field of integral equations and fractional differential equations. We observe that our integral operator is a kind of Volterra integral operator, and this can be used as a solution of the nonlinear problem given below. Further we prove the existence and uniqueness of that solution.

2.4.1 Equivalence Between the Generalized Nonlinear Problem and the Volterra Integral Equation

Consider the nonlinear fractional differential equation of order $\beta \in (0,1)$.

$$ {}_{\alpha}^{\tau}T_{a^+}^{\beta}\psi(r) = f(r,\psi(r)), \quad r \in [a,b], \tag{2.31} $$

where $f : [a,b] \times \mathbb{R} \to \mathbb{R}$ is a continuous function with respect to all its arguments. We seek condition that guarantee the existence and uniqueness of solution to the problem (2.31) in the set of functions

$$ A = \{\psi \in C([a,b]) : {}_{\alpha}^{\tau}T_{a^+}^{\beta}\psi(r) \in C([a,b])\}. $$

First, let us observe that, for $\psi \in C([a,b])$, the problem (2.31) is equivalent to the problem of finding solution to the following Volterra integral equation

$$ \psi(r) = \frac{1}{\Gamma(\beta)} \int_a^r \left(\frac{r^{\tau+\alpha} - w^{\tau+\alpha}}{\tau+\alpha}\right)^{\beta-1} f(w,\psi(w)) w^{\tau} d_{\alpha}w. \tag{2.32} $$

Indeed, if $\psi \in C([a,b])$ satisfies (2.31), then applying operator ${}_{\alpha}^{\tau}K_{a^+}^{\beta}$ to the both sides of (2.31), and applying relation (2.28), we obtain Eq. (2.32). Conversely, taking $r \to a^+$ and applying operator ${}_{\alpha}^{\tau}T_{a^+}^{\beta}$ to both sides of (2.32), using the relation (2.27), we arrive to problem (2.31).

2.4.2 Existence and Uniqueness of Solution for the Nonlinear Problem

In the following theorem, the existence and uniqueness of solution to the problem (2.31) are proved.

Theorem 2.13 *Let $f : [a,b] \times \mathbb{R} \to \mathbb{R}$ be a continuous function and Lipschitz with respect to thesecond variable, i.e.*

$$ |f(r,x_1) - f(r,x_2)| < L|x_1 - x_2|, \tag{2.33} $$

for all $r \in [a,b]$ and for all $x_1, x_2 \in \mathbb{R}$, $L > 0$. Then, the problem (2.31) *possesses a unique solution.*

Proof: We start by showing that for the problem (2.31), there exists a unique solution $\psi \in C([a,b])$. Let us recall that the problem (2.31) is equivalent to the problem of finding solutions to the Volterra integral equation (2.32). This allows us to use the well-known method for nonlinear Volterra integral equations, where first we prove existence and uniqueness of solutions on a subinterval of $[a,b]$.

Let us choose $p < r_1 < q$ to be such that the following condition is satisfied

$$ 0 < L\frac{(\tau+\alpha)^{-\beta}}{\Gamma(\beta+1)}(r_1^{\tau+\alpha} - a^{\tau+\alpha})^{\beta} < 1. \tag{2.34} $$

We shall prove the existence of a unique solution ψ to (2.32) on the subinterval $[a, r_1] \subseteq [a, b]$. Let us define the following integral operator, $S : C[a, b] \to C[a, b]$, by:

$$S\psi(r) = \frac{1}{\Gamma(\beta)} \int_a^r \left(\frac{r^{\tau+\alpha} - w^{\tau+\alpha}}{\tau + \alpha} \right)^{\beta-1} f(w, \psi(w)) w^\tau d_\alpha w. \tag{2.35}$$

Note that S is well defined and is a bounded operator as proved in Theorem 2.8. Using Theorem 2.8 and condition (2.33), we get

$$\begin{aligned}
&\|S\psi_1 - S\psi_2\|_{C([a,r_1])} \\
&= \left\| \frac{1}{\Gamma(\beta)} \int_a^r \left(\frac{r^{\tau+\alpha} - w^{\tau+\alpha}}{\tau + \alpha} \right)^{\beta-1} (f(w, \psi_1(w)) - f(w, \psi_2(w))) w^\tau d_\alpha w \right\|_{C([a,r_1])} \\
&= \| {}_\alpha^\tau K_{a^+}^\beta (f(w, \psi_1(w)) - f(w, \psi_2(w))) \|_{C([a,r_1])} \\
&\leq L \| {}_\alpha^\tau K_{a^+}^\beta (\psi_1(w) - \psi_2(w)) \|_{C([a,r_1])} \\
&\leq L \frac{|(\tau + \alpha)^{-\beta}|}{\beta + 1} (b^{\tau+\alpha} - a^{\tau+\alpha})^\beta \|\psi_1(w) - \psi_2(w)\|_{C([a,r_1])},
\end{aligned}$$

and because condition (2.34) is satisfied, by the Banach fixed point theorem, there exists a uniquesolution $\psi^{*1} \in C([a, r_1])$ to Eq. (2.32) on the interval $[a, r_1]$. Moreover, if we define the sequence $\psi_m^1(r) := S^m \psi_0(r)$, for $m = 1, 2, 3, \ldots,$

$$S^m \psi_0(r) = \frac{1}{\Gamma(\beta)} \int_a^r \left(\frac{r^{\tau+\alpha} - w^{\tau+\alpha}}{\tau + \alpha} \right)^{\beta-1} f(w, S^{m-1} \psi_0(w)) w^\tau d_\alpha w \tag{2.36}$$

then, again, by the Banach fixed point theorem, we obtain the solution ψ^{*1} as a limit of the sequence ψ_m^1, i.e.

$$\lim_{m \to \infty} \|\psi_m^1 - \psi^{*1}\|_{C([a,r_1])} = 0. \tag{2.37}$$

Now, let us choose $r_2 = r_1 + h_1$, with $h_1 > 0$ such that $r_2 < b$ and

$$0 < L \frac{(\tau + \alpha)^{-\beta}}{\Gamma(\beta + 1)} (r_2^{\tau+\alpha} - r_1^{\tau+\alpha})^\beta < 1. \tag{2.38}$$

Consider the interval $[r_1, r_2]$ and write Eq. (2.32) in the form of:

$$\begin{aligned}
\psi(r) = &\frac{1}{\Gamma(\beta)} \int_{r_1}^r \left(\frac{r^{\tau+\alpha} - w^{\tau+\alpha}}{\tau + \alpha} \right)^{\beta-1} f(w, \psi(w)) w^\tau d_\alpha w \\
&+ \frac{1}{\Gamma(\beta)} \int_a^{r_1} \left(\frac{r^{\tau+\alpha} - w^{\tau+\alpha}}{\tau + \alpha} \right)^{\beta-1} f(w, \psi(w)) w^\tau d_\alpha w.
\end{aligned} \tag{2.39}$$

Because on the interval $[a, r_1]$, Eq. (2.39) possesses a unique solution, we can rewrite (2.39) as:

$$\psi(r) = \psi_{01}(r) + \frac{1}{\Gamma(\beta)} \int_{r_1}^r \left(\frac{r^{\tau+\alpha} - w^{\tau+\alpha}}{\tau + \alpha} \right)^{\beta-1} f(w, \psi(w)) w^\tau d_\alpha w, \tag{2.40}$$

where

$$\psi_{01}(r) = \frac{1}{\Gamma(\beta)} \int_a^{r_1} \left(\frac{r^{\tau+\alpha} - w^{\tau+\alpha}}{\tau+\alpha} \right)^{\beta-1} f(w, \psi(w)) w^{\tau} d_{\alpha} w. \tag{2.41}$$

By the same argument as before, we prove that there exists a uniquesolution $\psi^{*2} \in C([r_1, r_2])$ to Eq. (2.32) on $[r_1, r_2]$. Repeating the process as above, choosing $r_k = r_{k-1} + h_{k-1}$ such that $h_{k-1} > 0, t_k < b$,

$$0 < L \frac{(\tau+\alpha)^{-\beta}}{\Gamma(\beta+1)} (r_k^{\tau+\alpha} - r_{k-1}^{\tau+\alpha})^{\beta} < 1. \tag{2.42}$$

We see that Eq. (2.32) possesses a solution $\psi^{*k} \in C([\psi_{k-1}, \psi_k])$ on each interval $[\psi_{k-1}, \psi_k]$, $(k = 1, \dots, l)$, where $p = \psi_0 < \psi_1 < \cdots < \psi_l = q$, and we conclude that for problem (2.31), thereexists a unique solution $\psi \in C([a, b])$. It remains to prove that $\psi \in A$, i.e. we need to show that ${}_{\alpha}^{\tau}T_{0^+}^{\beta} \psi(r) \in C([a, b])$. Recall that our solution ψ can be approximated by the sequence $\psi_m(r) = S_m \psi_0(r)$, i.e.

$$\lim_{m \to \infty} \|\psi_m - \psi\|_{C([a,b])} = 0 \tag{2.43}$$

with the choice of certain ψ_m on each interval $[a, r_1], \dots, [r_{l-1}, b]$. Using (2.31) and the Lipschitz type condition (2.33), we have

$$\| {}_{\alpha}^{\tau}T_{a^+}^{\beta} \psi_m - {}_{\alpha}^{\tau}T_{a^+}^{\beta} \psi \|_{C([a,b])} = \| f(r, \psi_m) - f(r, \psi) \|_{C([a,b])} \le L \|\psi_m - \psi\|_{C([a,b])}.$$

Then taking $m \to \infty$, we get

$$\lim_{m \to \infty} \| {}_{\alpha}^{\tau}T_{a^+}^{\beta} \psi_m - {}_{\alpha}^{\tau}T_{a^+}^{\beta} \psi \|_{C([a,b])} = 0.$$

Since ${}_{\alpha}^{\tau}T_{a^+}^{\beta} \psi_m(r) = f(r, \psi_m(r))$ is continuous on $[a, b]$, we have that ${}_{\alpha}^{\tau}T_{a^+}^{\beta}$ belongs to the space $C([a, b])$. This completes the proof. □

2.5 Applications to the Field of Inequalities

In this section, an identity for conformable fractional integrals is proved. Then, by using this identity some integral inequalities are presented. These inequalities are connected with the left hand side of the Hermite–Hadamard inequality and contain conformable fractional integrals. At the end, applications to some special means and error estimates for the midpoint formula are provided. The resultspresented in this section have been given in [21, 22].

To proceed further, we present some basic concepts related to this area, which will help us in proving our main results.

The class of convex functions is well known in the literature and is usually defined in the following way: a function $\psi : \mathtt{I} \to \mathbb{R}$, $\mathtt{I} \subseteq \mathbb{R}$ is said to be convex on $\mathtt{I}$ if the inequality

$$\psi(\lambda x + (1-\lambda)y) \le \lambda \psi(x) + (1-\lambda)\psi(y) \tag{2.44}$$

holds for all $x, y \in \mathtt{I}$ and $\lambda \in [0,1]$. Also, we say that ψ is concave, if the inequality (2.44) is reversed. Many important inequalities have been obtained for this class of functions but here we will present only one of them in following:

If $\psi : \mathtt{I} \to \mathbb{R}$ is a convex function on the interval $\mathtt{I}$, then for any $a, b \in \mathtt{I}$ with $a \neq b$, we have the following double inequality:

$$\psi\left(\frac{a+b}{2}\right) \leq \frac{1}{b-a}\int_a^b \psi(t)dt \leq \frac{\psi(a)+\psi(b)}{2}. \tag{2.45}$$

Both the inequalities hold in reversed direction if the function ψ is concave on the interval $\mathtt{I}$. This remarkable result was given in ([23], 1893) and is well known in the literature as Hermite–Hadamard inequality. Since, its discovery this inequality has become the center of interest for many prolific researchers and received a considerable attention. Also, a number of extensions, generalizations, and variants of (2.45) have been provided in the theory of mathematical inequalities. For example, see [24–38] and the references cited therein. In [39], Dragomir and Agarwal proved the following results connected with the right hand part of Hermite–Hadamard inequality.

Lemma 2.1 [39] *Let $\psi : \mathcal{I}^\circ \subseteq \mathbb{R} \to \mathbb{R}$ be a differentiable mapping on $\mathcal{I}^\circ$ and $a, b \in \mathcal{I}^\circ$ with $a < b$. If $\psi' \in \mathtt{L}[a,b]$, then the following identity holds*

$$\frac{\psi(a)+\psi(b)}{2} - \frac{1}{b-a}\int_a^b \psi(x)dx = \frac{b-a}{2}\int_0^1 (1-2t)\psi'(ta+(1-t)b)dt. \tag{2.46}$$

Theorem 2.14 [39] *Let $\psi : \mathcal{I}^\circ \subseteq \mathbb{R} \to \mathbb{R}$ be a differentiable mapping on $\mathcal{I}^\circ$ and $a, b \in \mathcal{I}^\circ$ with $a < b$. If $\psi' \in \mathtt{L}[a,b]$ and $|\psi'|$ is convex on $[a,b]$, then we have the following inequality:*

$$\left|\frac{\psi(a)+\psi(b)}{2} - \frac{1}{b-a}\int_a^b \psi(x)dx\right| \leq \frac{(b-a)}{4}\left(\frac{|\psi'(a)|+|\psi'(b)|}{2}\right). \tag{2.47}$$

In [40], U.S. Kirmaci gave the following results.

Lemma 2.2 [40] *Let $\psi : \mathcal{I}^\circ \subseteq \mathbb{R} \to \mathbb{R}$ be a differentiable mapping on $\mathcal{I}^\circ$, $a, b \in \mathcal{I}^\circ$ with $a < b$. If $\psi' \in \mathtt{L}[a,b]$, then the following equality holds*

$$\frac{1}{b-a}\int_a^b \psi(x)dx - \psi\left(\frac{a+b}{2}\right)$$
$$= (b-a)\left[\int_0^{\frac{1}{2}} t\psi'(ta+(1-t)b)dt + \int_{\frac{1}{2}}^1 (t-1)\psi'(ta+(1-t)b)dt\right]. \tag{2.48}$$

Theorem 2.15 [40] *Let $\psi : I^\circ \subseteq \mathbb{R} \to \mathbb{R}$ be a differentiable mapping on I° and $a, b \in I^\circ$ with $a < b$. If $\psi' \in L[a,b]$ and $|\psi'|$ is convex on $[a,b]$, then we have the following inequality:*

$$\left| \frac{1}{b-a} \int_a^b \psi(x)dx - \psi\left(\frac{a+b}{2}\right) \right| \leq \frac{(b-a)(|\psi'(a)| + |\psi'(b)|)}{8}. \tag{2.49}$$

Recently, Anderson [41] investigated the following conformable integral version of Hermite–Hadamard inequality:

Theorem 2.16 [41] *Let $\alpha \in (0,1]$ and $\psi : [a,b] \to \mathbb{R}$ be an α-differentiable function with $0 < a < b$, such that $D_\alpha f$ is increasing, then we have the following inequality*

$$\frac{\alpha}{b^\alpha - a^\alpha} \int_a^b \psi(t) d_\alpha t \leq \frac{\psi(a) + \psi(b)}{2}. \tag{2.50}$$

Moreover, if the function ψ is decreasing on $[a,b]$, then we have

$$\psi\left(\frac{a+b}{2}\right) \leq \frac{\alpha}{b^\alpha - a^\alpha} \int_a^b \psi(t) d_\alpha t. \tag{2.51}$$

Remark 2.3 It is obvious that, if we choose $\alpha = 1$, then the inequalities (2.50) and (2.51) reduce to inequality (2.45).

2.5.1 Inequalities Related to the Left Side of Hermite–Hadamard Inequality

We begin, this section with the following lemma associated with inequality (2.51), which is needed in the derivation of our main results.

Lemma 2.3 *Let $a, b \in \mathbb{R}$ with $0 \leq a < b$, and let $\psi : [a,b] \to \mathbb{R}$ be a differentiable function on (a,b) for $\alpha \in (0,1]$. If $D_\alpha(\psi) \in L^1_\alpha([a,b])$, then the following identity holds*

$$\begin{aligned}
&\psi\left(\frac{a+b}{2}\right) - \frac{\alpha}{b^\alpha - a^\alpha} \int_a^b \psi(s) d_\alpha s \\
&= \frac{(b-a)}{b^\alpha - a^\alpha} \left[\int_0^{\frac{1}{2}} (((1-t)a + tb)^{2\alpha-1} - a^\alpha((1-t)a + tb)^{\alpha-1}) \right. \\
&\times D_\alpha(\psi)((1-t)a + tb) t^{1-\alpha} d_\alpha t + \int_{\frac{1}{2}}^1 (((1-t)a + tb)^{2\alpha-1} - b^\alpha((1-t)a + tb)^{\alpha-1}) \\
&\left. \times D_\alpha(\psi)((1-t)a + tb) t^{1-\alpha} d_\alpha t \right].
\end{aligned} \tag{2.52}$$

Proof: Integrating by parts, we have

$$I = \int_0^{\frac{1}{2}} (((1-t)a+tb)^{2\alpha-1} - a^\alpha((1-t)a+tb)^{\alpha-1})\mathrm{D}_\alpha(\psi)((1-t)a+tb)dt$$

$$+\int_{\frac{1}{2}}^{1} (((1-t)a+tb)^{2\alpha-1} - b^\alpha((1-t)a+tb)^{\alpha-1})\mathrm{D}_\alpha(\psi)((1-t)a+tb)dt$$

$$= \int_0^{\frac{1}{2}} (((1-t)a+tb)^\alpha - a^\alpha)\psi'((1-t)a+tb)dt$$

$$+\int_{\frac{1}{2}}^{1} (((1-t)a+tb)^\alpha - b^\alpha)\psi'((1-t)a+tb)dt$$

$$= (((1-t)a+tb)^\alpha - a^\alpha)\frac{\psi((1-t)a+tb)}{b-a}\Big|_0^{\frac{1}{2}}$$

$$-\int_0^{\frac{1}{2}} \alpha((1-t)a+tb)^{\alpha-1}(b-a)\frac{\psi((1-t)a+tb)}{b-a}dt$$

$$+(((1-t)a+tb)^\alpha - b^\alpha)\frac{\psi((1-t)a+tb)}{b-a}\Big|_{\frac{1}{2}}^{1}$$

$$-\int_{\frac{1}{2}}^{1} \alpha((1-t)a+tb)^{\alpha-1}(b-a)\frac{\psi((1-t)a+tb)}{b-a}dt$$

$$= \frac{1}{b-a}\left[\left(\left(\frac{a+b}{2}\right)^\alpha - a^\alpha\right)\psi\left(\frac{a+b}{2}\right) - \alpha\int_a^{\frac{a+b}{2}} \psi(s)d_\alpha s\right]$$

$$+\frac{1}{b-a}\left[\left(b^\alpha - \left(\frac{a+b}{2}\right)^\alpha\right)\psi\left(\frac{a+b}{2}\right) - \alpha\int_{\frac{a+b}{2}}^{b} \psi(s)d_\alpha s\right]$$

$$= \frac{b^\alpha - a^\alpha}{b-a}\psi\left(\frac{a+b}{2}\right) - \frac{\alpha}{b-a}\int_a^b \psi(s)d_\alpha s,$$

where, we have used the change of variable $s = (1-t)a + tb$. Then multiplying both sides by $\frac{b-a}{b^\alpha - a^\alpha}$ to get the desired result in (2.52). □

Remark 2.4 By setting $\alpha = 1$, the identity in (2.52) reduces to (2.48).

Theorem 2.17 *Let $a, b \in \mathbb{R}$ with $0 \le a < b$, and let $\psi : [a,b] \to \mathbb{R}$ be a differentiable function on (a,b) for $\alpha \in (0,1]$. If $\mathrm{D}_\alpha(\psi) \in L_\alpha^1([a,b])$ and $|\psi'|$ is convex on $[a,b]$, then we have the following inequality:*

$$\left|\psi\left(\frac{a+b}{2}\right) - \frac{\alpha}{b^\alpha - a^\alpha}\int_a^b \psi(s)d_\alpha s\right| \tag{2.53}$$

$$\le \frac{b-a}{b^\alpha - a^\alpha}\left[\frac{|\psi'(a)|}{192}[13b^\alpha - 35a^\alpha]\right.$$

$$\left. + \frac{|\psi'(b)|}{192}[19b^\alpha - 29a^\alpha] + (ab^{\alpha-1} + a^{\alpha-1}b)\left[\frac{11|\psi'(a)| + 5|\psi'(b)|}{192}\right]\right].$$

Proof: First of all, we consider Lemma 2.3 and then using the convexity of $x^{\alpha-1}$ and $-x^{\alpha}$ $(x > 0)$ for $\alpha \in (0, 1]$. Also, since the function $|\psi'|$ is convex, therefore we have

$$\left|\psi\left(\frac{a+b}{2}\right) - \frac{\alpha}{b^\alpha - a^\alpha}\int_a^b \psi(s)d_\alpha s\right|$$
$$\leq \frac{b-a}{b^\alpha - a^\alpha}\left[\int_0^{\frac{1}{2}} (((1-t)a+tb)^\alpha - a^\alpha)|\psi'((1-t)a+tb)|dt\right.$$
$$\left.+\int_{\frac{1}{2}}^1 (b^\alpha - ((1-t)a+tb)^\alpha)|\psi'((1-t)a+tb)|dt\right]$$
$$= \frac{b-a}{b^\alpha - a^\alpha}\left[\int_0^{\frac{1}{2}} (((1-t)a+tb)^{\alpha-1+1} - a^\alpha)|\psi'((1-t)a+tb)|dt\right.$$
$$\left.+\int_{\frac{1}{2}}^1 (b^\alpha - ((1-t)a+tb)^\alpha)|\psi'((1-t)a+tb)|dt\right]$$
$$\leq \frac{b-a}{b^\alpha - a^\alpha}\left[\int_0^{\frac{1}{2}} (((1-t)a+tb)^{\alpha-1}((1-t)a+tb) - a^\alpha)|\psi'((1-t)a+tb)|dt\right.$$
$$\left.+\int_{\frac{1}{2}}^1 (b^\alpha - ((1-t)a^\alpha+tb^\alpha))|\psi'((1-t)a+tb)|dt\right]$$
$$\leq \frac{b-a}{b^\alpha - a^\alpha}\left[\int_0^{\frac{1}{2}} (((1-t)a^{\alpha-1}+tb^{\alpha-1})((1-t)a+tb) - a^\alpha)|\psi'((1-t)a+tb)|dt\right.$$
$$\left.+\int_{\frac{1}{2}}^1 (b^\alpha - ((1-t)a^\alpha+tb^\alpha))|\psi'((1-t)a+tb)|dt\right]$$
$$\leq \frac{b-a}{b^\alpha - a^\alpha}\left[\int_0^{\frac{1}{2}} (((1-t)a^{\alpha-1}+tb^{\alpha-1})((1-t)a+tb) - a^\alpha)[(1-t)|\psi'(a)|]\right.$$
$$+t|\psi'(b))|dt$$
$$\left.+\int_{\frac{1}{2}}^1 (b^\alpha - ((1-t)a^\alpha+tb^\alpha))[(1-t)|\psi'(a)| + t|\psi'(b))|]dt\right].$$

Evaluating all the above integrals, we have the following

$$\frac{b-a}{b^\alpha - a^\alpha}\left[\int_0^{\frac{1}{2}} (((1-t)a^{\alpha-1}+tb^{\alpha-1})((1-t)a+tb) - a^\alpha)\right.$$
$$[(1-t)|\psi'(a)| + t|\psi'(b))|]dt$$
$$\left.+\int_{\frac{1}{2}}^1 (b^\alpha - ((1-t)a^\alpha+tb^\alpha))[(1-t)|\psi'(a)| + t|\psi'(b))|]dt\right]$$

$$= \frac{b-a}{b^\alpha - a^\alpha}\left[\frac{15}{64}a^\alpha|\psi'(a)| + \frac{11}{192}a^\alpha|\psi'(b)| + \frac{11}{192}ab^{\alpha-1}|\psi'(a)| + \frac{5}{192}ab^{\alpha-1}|\psi'(b)| + \frac{11}{192}a^{\alpha-1}b|\psi'(a)| + \frac{5}{192}a^{\alpha-1}b|\psi'(b)| + \frac{5}{192}b^\alpha|\psi'(a)| + \frac{1}{64}b^\alpha|\psi'(b)| - \frac{1}{24}a^\alpha|\psi'(a)| - \frac{1}{12}a^\alpha|\psi'(b)| - \frac{1}{12}b^\alpha|\psi'(a)| - \frac{7}{24}b^\alpha|\psi'(b)| - \frac{3}{8}a^\alpha|\psi'(a)| - \frac{1}{8}a^\alpha|\psi'(b)| + \frac{1}{8}b^\alpha|\psi'(a)| + \frac{3}{8}b^\alpha|\psi'(b)|\right]$$

$$= \frac{b-a}{b^\alpha - a^\alpha}\left[\frac{|\psi'(a)|}{192}[13b^\alpha - 35a^\alpha] + \frac{|\psi'(b)|}{192}[19b^\alpha - 29a^\alpha] + (ab^{\alpha-1} + a^{\alpha-1}b) \times \left[\frac{11|\psi'(a)| + 5|\psi'(b)|}{192}\right]\right].$$

□

Remark 2.5 By putting $\alpha = 1$ in (2.53), we get the inequality in (2.49).

Theorem 2.18 *Let $a, b \in \mathbb{R}$ with $0 \le a < b$, and let $\psi : [a, b] \to \mathbb{R}$ be a differentiable function on (a, b) for $\alpha \in (0, 1]$. If $D_\alpha(\psi) \in L^1_\alpha([a, b])$ and $|\psi'|^b$ is convex on $[a, b]$ for $s > 1$, then the following inequality holds*

$$\left|\psi\left(\frac{a+b}{2}\right) - \frac{\alpha}{b^\alpha - a^\alpha}\int_a^b \psi(s)d_\alpha s\right| \le \frac{(b-a)}{b^\alpha - a^\alpha}\left[(A_1(\alpha))^{1-\frac{1}{b}}\{A_2(\alpha)|\psi'(a)|^b + A_3(\alpha)|\psi'(b)|^b\}^{\frac{1}{b}} + (B_1(\alpha))^{1-\frac{1}{b}}\{B_2(\alpha)|\psi'(a)|^b + B_3(\alpha)|\psi'(b)|^b\}^{\frac{1}{b}}\right], \tag{2.54}$$

where

$$A_1(\alpha) = \left[\frac{(a+b)^{\alpha+1} - (2a)^{\alpha+1}}{2^{\alpha+1}(\alpha+1)(b-a)}\right] - \frac{a^\alpha}{2}, \quad B_1(\alpha) = \frac{b^\alpha}{2} - \left[\frac{(2b)^{\alpha+1} - (a+b)^{\alpha+1}}{2^{\alpha+1}(\alpha+1)(b-a)}\right],$$

$$A_2(\alpha) = \frac{(a+b)^{\alpha+1}}{2^{\alpha+1}(\alpha+1)(b-a)}\left[\frac{(b-a)(\alpha+2) - (a+b)}{2(b-a)(\alpha+2)}\right] - \frac{a^{\alpha+1}}{(b-a)(\alpha+2)}\left[\frac{(a+b)^{\alpha+1} - (2a)^{\alpha+1}}{2^{\alpha+1}(\alpha+1)(b-a)}\right] - \frac{3a^\alpha}{8},$$

$$B_2(\alpha) = \frac{(a+b)^{\alpha+1}}{2^{\alpha+1}(\alpha+1)(b-a)}\left[\frac{2(b-a)(\alpha+2) - (a+b)}{2(b-a)(\alpha+2)}\right] - \frac{b^{\alpha+1}}{(b-a)(\alpha+1)(\alpha+2)} - \frac{3b^\alpha}{8},$$

$$A_3(\alpha) = \frac{(a+b)^{\alpha+1}}{2^{\alpha+1}(\alpha+1)(b-a)}\left[\frac{(b-a)(\alpha+2) - (a+b)}{2(b-a)(\alpha+2)}\right] - \frac{a^{\alpha+2}}{(b-a)^2(\alpha+1)(\alpha+2)} - \frac{a^\alpha}{2},$$

$$B_3(\alpha) = \frac{(a+b)^{\alpha+1}}{2^{\alpha+1}(\alpha+1)(b-a)}\left[\frac{2(b-a)(\alpha+2)+(a+b)}{2(b-a)(\alpha+2)}\right]$$
$$-\frac{b^{\alpha+1}}{(b-a)(\alpha+1)}\left[\frac{(a+b)^{\alpha+1}-b}{(\alpha+2)(b-a)}\right]+\frac{b^{\alpha}}{2}.$$

Proof: Using Lemma 2.3, it follows that

$$\left|\psi\left(\frac{a+b}{2}\right)-\frac{\alpha}{b^{\alpha}-a^{\alpha}}\int_a^b \psi(s)d_{\alpha}s\right|$$
$$=\left|\frac{(b-a)}{b^{\alpha}-a^{\alpha}}\left[\int_0^{\frac{1}{2}}(((1-t)a+tb)^{2\alpha-1}-a^{\alpha}((1-t)a+tb)^{\alpha-1})\right.\right.$$
$$\mathrm{D}_{\alpha}(\psi)((1-t)a+tb)dt$$
$$\left.\left.+\int_{\frac{1}{2}}^{1}(((1-t)a+tb)^{2\alpha-1}-b^{\alpha}((1-t)a+tb)^{\alpha-1})\mathrm{D}_{\alpha}(\psi)((1-t)a+tb)dt\right]\right|$$
$$\leq\frac{(b-a)}{b^{\alpha}-a^{\alpha}}\left[\int_0^{\frac{1}{2}}(((1-t)a+tb)^{\alpha}-a^{\alpha})|\psi'((1-t)a+tb)|dt\right.$$
$$\left.+\int_{\frac{1}{2}}^{1}(b^{\alpha}-((1-t)a+tb)^{\alpha})|\psi'((1-t)a+tb)|dt\right].$$

Now by the power-mean inequality,

$$\int_0^{\frac{1}{2}}(((1-t)a+tb)^{\alpha}-a^{\alpha})|\psi'((1-t)a+tb)|dt$$
$$\leq\left(\int_0^{\frac{1}{2}}(((1-t)a+tb)^{\alpha}-a^{\alpha})dt\right)^{1-\frac{1}{b}}$$
$$\times\left(\int_0^{\frac{1}{2}}(((1-t)a+tb)^{\alpha}-a^{\alpha})|\psi'((1-t)a+tb)|^{b}dt\right)^{\frac{1}{b}},$$

and similarly, we have

$$\int_{\frac{1}{2}}^{1}(b^{\alpha}-((1-t)a+tb)^{\alpha})|\psi'((1-t)a+tb)|dt$$
$$\leq\left(\int_{\frac{1}{2}}^{1}(b^{\alpha}-((1-t)a+tb)^{\alpha})dt\right)^{1-\frac{1}{b}}$$
$$\times\left(\int_{\frac{1}{2}}^{1}(b^{\alpha}-((1-t)a+tb)^{\alpha})|\psi'((1-t)a+tb)|^{b}dt\right)^{\frac{1}{b}}.$$

Now by the convexity $|\psi'|^b$ from above, we have

$$
\begin{aligned}
&\int_0^{\frac{1}{2}} (((1-t)a+tb)^\alpha - a^\alpha)|\psi'((1-t)a+tb)|^b dt \\
&\leq \int_0^{\frac{1}{2}} (((1-t)a+tb)^\alpha - a^\alpha)[(1-t)|\psi'(a)|^b + t|\psi'(b)|^b]dt \\
&= |\psi'(a)|^b \int_0^{\frac{1}{2}} (((1-t)a+tb)^\alpha - a^\alpha)(1-t)dt + |\psi'(b)|^b \\
&\quad \int_0^{\frac{1}{2}} (((1-t)a+tb)^\alpha - a^\alpha)t\, dt \\
&= |\psi'(a)|^b \left(\frac{(a+b)^{\alpha+1}}{2^{\alpha+1}(\alpha+1)(b-a)} \left[\frac{(b-a)(\alpha+2)-(a+b)}{2(b-a)(\alpha+2)} \right] \right. \\
&\quad \left. - \frac{a^{\alpha+1}}{(b-a)(\alpha+2)} \left[\frac{(a+b)^{\alpha+1}-(2a)^{\alpha+1}}{2^{\alpha+1}(\alpha+1)(b-a)} \right] - \frac{3a^\alpha}{8} \right) \\
&\quad + |\psi'(b)|^b \left(\frac{(a+b)^{\alpha+1}}{2^{\alpha+1}(\alpha+1)(b-a)} \left[\frac{(b-a)(\alpha+2)-(a+b)}{2(b-a)(\alpha+2)} \right] \right. \\
&\quad \left. - \frac{a^{\alpha+2}}{(b-a)^2(\alpha+1)(\alpha+2)} - \frac{a^\alpha}{2} \right)
\end{aligned}
$$

and

$$
\begin{aligned}
&\int_{\frac{1}{2}}^1 (b^\alpha - ((1-t)a+tb)^\alpha)|\psi'((1-t)a+tb)|^b dt \\
&\leq \int_{\frac{1}{2}}^1 (b^\alpha - ((1-t)a+tb)^\alpha)[(1-t)|\psi'(a)|^b + t|\psi'(b)|^b]dt \\
&= |\psi'(a)|^b \int_{\frac{1}{2}}^1 (b^\alpha - ((1-t)a+tb)^\alpha)(1-t)dt + |\psi'(b)|^b \\
&\quad \int_{\frac{1}{2}}^1 (b^\alpha - ((1-t)a+tb)^\alpha)tdt \\
&= |\psi'(a)|^b \left(\frac{(a+b)^{\alpha+1}}{2^{\alpha+1}(\alpha+1)(b-a)} \left[\frac{2(b-a)(\alpha+2)-(a+b)}{2(b-a)(\alpha+2)} \right] \right. \\
&\quad \left. - \frac{b^{\alpha+1}}{(b-a)(\alpha+1)(\alpha+2)} - \frac{3b^\alpha}{8} \right) + |\psi'(b)|^b \left(\frac{(a+b)^{\alpha+1}}{2^{\alpha+1}(\alpha+1)(b-a)} \right. \\
&\quad \left. \left[\frac{2(b-a)(\alpha+2)+(a+b)}{2(b-a)(\alpha+2)} \right] - \frac{b^{\alpha+1}}{(b-a)(\alpha+1)} \left[\frac{(a+b)^{\alpha+1}-b}{(\alpha+2)(b-a)} \right] + \frac{b^\alpha}{2} \right),
\end{aligned}
$$

where, we have also used the facts that

$$
\int_0^{\frac{1}{2}} (((1-t)a+tb)^\alpha - a^\alpha)dt = \left[\frac{(a+b)^{\alpha+1}-(2a)^{\alpha+1}}{2^{\alpha+1}(\alpha+1)(b-a)} \right] - \frac{a^\alpha}{2}
$$

$$\int_{\frac{1}{2}}^{1} (b^{\alpha} - ((1-t)a + tb)^{\alpha})dt = \frac{b^{\alpha}}{2} - \left[\frac{(2b)^{\alpha+1} - (a+b)^{\alpha+1}}{2^{\alpha+1}(\alpha+1)(b-a)}\right].$$

Hence, we have the result in (2.54). □

Remark 2.6 By setting $\alpha = 1$ in (2.54), we get the following inequality:

$$\begin{aligned}
&\left|\psi\left(\frac{a+b}{2}\right) - \frac{1}{b-a}\int_a^b \psi(s)ds\right| \\
&\leq \left(\frac{b-a}{8}\right)^{1-\frac{1}{b}} \left[\{A_2(1)|\psi'(a)|^b + A_3(1)|\psi'(b)|^b\}^{\frac{1}{b}}\right. \\
&\left. + \{B_2(1)|\psi'(a)|^b + B_3(1)|\psi'(b)|^b\}^{\frac{1}{b}}\right],
\end{aligned} \tag{2.55}$$

where

$$\begin{aligned}
A_2(1) &= \frac{(a+b)^2(b-2a) - a^2(b^2-3a^2+2ab) - 3a(b-a)^2}{24(b-a)^2}, \\
B_2(1) &= \frac{(a+b)^2(5b-7a) - 8b^2(b-a) - 18b(b-a)}{48(b-a)^2}, \\
A_3(1) &= \frac{(a+b)^2(b-2a) - 4a^3 - 12a(b-a)^2}{24(b-a)^2}, \\
B_3(1) &= \frac{(a+b)^2(7b-5a) - 8b^2(a+b)^2 + 8b^3 - 24b(b-a)^2}{48(b-a)^2}.
\end{aligned}$$

Theorem 2.19 *Let $a, b \in \mathbb{R}$ with $0 \leq a < b$, and let $\psi : [a,b] \to \mathbb{R}$ be a differentiable function on (a,b) for $\alpha \in (0,1]$. If $D_{\alpha}(\psi) \in L_{\alpha}^{1}([a,b])$ and $|\psi'|^b$ is concave on $[a,b]$ for $s > 1$, then the following inequality holds*

$$\begin{aligned}
&\left|\psi\left(\frac{a+b}{2}\right) - \frac{\alpha}{b^{\alpha}-a^{\alpha}}\int_a^b \psi(s)d_{\alpha}s\right| \\
&\leq \frac{(b-a)}{b^{\alpha}-a^{\alpha}}\left[A_1(\alpha)\psi'\left(\frac{C_1(\alpha)}{A_1(\alpha)}\right) + B_1(\alpha)\psi'\left(\frac{C_2(\alpha)}{B_1(\alpha)}\right)\right],
\end{aligned} \tag{2.56}$$

where

$$A_1(\alpha) = \left[\frac{(a+b)^{\alpha+1} - (2a)^{\alpha+1}}{2^{\alpha+1}(\alpha+1)(b-a)}\right] - \frac{a^{\alpha}}{2}, \quad B_1(\alpha) = \frac{b^{\alpha}}{2} - \left[\frac{(2b)^{\alpha+1} - (a+b)^{\alpha+1}}{2^{\alpha+1}(\alpha+1)(b-a)}\right],$$

$$C_1(\alpha) = \frac{(a+b)^2}{4(b-a)}\left[\frac{(a+b)^{\alpha} - 2^{\alpha-1}a^{\alpha}(\alpha+2)}{2^{\alpha}(\alpha+2)}\right] - \frac{\alpha a^{\alpha+2}}{2(\alpha+2)(b-a)},$$

$$C_2(\alpha) = \frac{(a+b)^2}{4(b-a)}\left[\frac{(a+b)^{\alpha} - 2^{\alpha-1}b^{\alpha}(\alpha+2)}{2^{\alpha}(\alpha+2)}\right] - \frac{\alpha b^{\alpha+2}}{2(\alpha+2)(b-a)}.$$

Proof: By power mean inequality, we have

$$(t|\psi'(a)| + (1-t)|\psi'(b)|)^b \leq t|\psi'(a)|^b + (1-t)|\psi'(b)|^b$$
$$\leq |\psi'(ta + (1-t)b)|^b \text{ (by concavity of } |\psi'|^b),$$

and therefore

$$|\psi'(ta + (1-t)b)| \geq t|\psi'(a)| + (1-t)|\psi'(b)|,$$

which shows that $|\psi'|$ is also concave. Now taking into consideration Lemma 2.3, it follows that

$$\left|\psi\left(\frac{a+b}{2}\right) - \frac{\alpha}{b^\alpha - a^\alpha}\int_a^b \psi(s)d_\alpha s\right|$$
$$= \left|\frac{(b-a)}{b^\alpha - a^\alpha}\left[\int_0^{\frac{1}{2}} (((1-t)a+tb)^{2\alpha-1} - a^\alpha((1-t)a+tb)^{\alpha-1})\mathrm{D}_\alpha(\psi)((1-t)a+tb)dt\right.\right.$$
$$\left.\left.+ \int_{\frac{1}{2}}^1 (((1-t)a+tb)^{2\alpha-1} - b^\alpha((1-t)a+tb)^{\alpha-1})\mathrm{D}_\alpha(\psi)((1-t)a+tb)dt\right]\right|$$
$$\leq \frac{(b-a)}{b^\alpha - a^\alpha}\left[\int_0^{\frac{1}{2}} (((1-t)a+tb)^\alpha - a^\alpha)|\psi'((1-t)a+tb)|dt\right.$$
$$\left.+ \int_{\frac{1}{2}}^1 (b^\alpha - ((1-t)a+tb)^\alpha)|\psi'((1-t)a+tb)|dt\right]$$

and applying Jensen's integral inequality, we have

$$\int_0^{\frac{1}{2}} (((1-t)a+tb)^\alpha - a^\alpha)|\psi'((1-t)a+tb)|dt$$
$$\leq \left(\int_0^{\frac{1}{2}} (((1-t)a+tb)^\alpha - a^\alpha)\right)\psi'$$
$$\left(\frac{\int_0^{\frac{1}{2}} (((1-t)a+tb)^\alpha - a^\alpha)((1-t)a+tb)dt}{\int_0^{\frac{1}{2}} (((1-t)a+tb)^\alpha - a^\alpha)}\right)$$
$$= \mathrm{A}_1(\alpha)\psi'\left(\frac{\mathrm{C}_1(\alpha)}{\mathrm{A}_1(\alpha)}\right).$$

Equivalently, we have

$$\int_{\frac{1}{2}}^1 (b^\alpha - ((1-t)a+tb)^\alpha)|\psi'((1-t)a+tb)|dt$$
$$\leq \left(b^\alpha - \int_0^{\frac{1}{2}} (((1-t)a+tb)^\alpha)\right)\psi'$$

$$\left(\frac{\int_{\frac{1}{2}}^{1}(b^{\alpha}-((1-t)a+tb)^{\alpha})((1-t)a+tb)dt}{\int_{\frac{1}{2}}^{1}(b^{\alpha}-((1-t)a+tb)^{\alpha})dt}\right)$$

$$=\mathrm{B}_1(\alpha)\psi'\left(\frac{\mathrm{C}_2(\alpha)}{\mathrm{B}_1(\alpha)}\right),$$

where, we have also used the following facts that

$$\int_{0}^{\frac{1}{2}}(((1-t)a+tb)^{\alpha}-a^{\alpha})dt=\mathrm{A}_1(\alpha)=\left[\frac{(a+b)^{\alpha+1}-(2a)^{\alpha+1}}{2^{\alpha+1}(\alpha+1)(b-a)}\right]-\frac{a^{\alpha}}{2},$$

$$\int_{\frac{1}{2}}^{1}(b^{\alpha}-((1-t)a+tb)^{\alpha})dt=\mathrm{B}_1(\alpha)=\frac{b^{\alpha}}{2}-\left[\frac{(2b)^{\alpha+1}-(a+b)^{\alpha+1}}{2^{\alpha+1}(\alpha+1)(b-a)}\right],$$

$$\int_{0}^{\frac{1}{2}}(((1-t)a+tb)^{\alpha}-a^{\alpha})((1-t)a+tb)dt$$

$$=\mathrm{C}_1(\alpha)=\frac{(a+b)^2}{4(b-a)}\left[\frac{(a+b)^{\alpha}-2^{\alpha-1}a^{\alpha}(\alpha+2)}{2^{\alpha}(\alpha+2)}\right]-\frac{\alpha a^{\alpha+2}}{2(\alpha+2)(b-a)}$$

and

$$\int_{\frac{1}{2}}^{1}(b^{\alpha}-((1-t)a+tb)^{\alpha})((1-t)a+tb)dt$$

$$=\mathrm{C}_2(\alpha)=\frac{(a+b)^2}{4(b-a)}\left[\frac{(a+b)^{\alpha}-2^{\alpha-1}b^{\alpha}(\alpha+2)}{2^{\alpha}(\alpha+2)}\right]-\frac{\alpha b^{\alpha+2}}{2(\alpha+2)(b-a)}.$$

□

Remark 2.7 If we set $\alpha = 1$ in (2.56), then we have the following

$$\left|\psi\left(\frac{a+b}{2}\right)-\frac{1}{b-a}\int_{a}^{b}\psi(s)ds\right|$$
$$\le\frac{(b-a)}{8}\left[\psi'\left(\frac{(a+b)^4-3a(a+b)^2-4a^3}{3(b-a)^2}\right)\right.$$
$$\left.+\psi'\left(\frac{(a+b)^4-3b(a+b)^2-4b^3}{3(b-a)^2}\right)\right]. \tag{2.57}$$

Remark 2.8 Several important variants of Hermite–Hadamard inequality have been provided in the literature, such as the versions established by Anderson [41], Sarikaya et al. [42], and Set et al. [43]. Recently, Set et al. [44] presented some Hermite–Hadamard type inequalities for conformable fractional integrals. They have obtainedresults for the version of Hermite–Hadamard type inequalities given in [43], while our resultsare devoted to the version obtained by Anderson [41].

2.5.1.1 Applications to Special Means of Real Numbers

We begin, this section by considering some particular means for arbitrary positive real numbers $a, b \in \mathbb{R}^+$ such that $a \neq b$. So, for this purpose we recall the following well-known definitions in the literature:

(1) *The arithmetic mean:*

$$\mathbb{A} = \mathbb{A}(a, b) := \frac{a+b}{2}, \quad a, b \in \mathbb{R}^+.$$

(2) *The logarithmic mean:*

$$\mathbb{L} = \mathbb{L}(a, b) := \frac{b-a}{\ln b - \ln a}, \quad a \neq b,\ a, b \in \mathbb{R}^+.$$

(3) *The generalized logarithmic* (α, r)*th mean:*

$$\mathbb{L}_{(\alpha,r)} = \mathbb{L}_{(\alpha,r)}(a, b) = \left[\frac{\alpha(b^{r+\alpha} - a^{r+\alpha})}{(b^\alpha - a^\alpha)(r+\alpha)}\right]^{\frac{1}{r}}, \ a \neq b,\ r \neq 0, -\alpha,\ \alpha \in (0, 1],\ r \in \mathbb{R}.$$

Now, by making use of the results obtained in Section 2.2, we give some applications to special means of real numbers.

Proposition 2.1 *Let $a, b \in \mathbb{R}$ with $0 < a < b, r > 1$ and $\alpha \in (0, 1]$, then the following holds:*

$$\begin{aligned}
&|\mathbb{A}^r(a, b) - \mathbb{L}^r_{(\alpha,r)}(a, b)| \\
&\leq \frac{(r-1)(b-a)}{b^\alpha - a^\alpha}\left[\frac{|a|^{(r-1)}}{192}[13b^\alpha - 35a^\alpha] + \frac{|b|^{(r-1)}}{192}[19b^\alpha - 29a^\alpha]\right. \\
&\left.+ (ab^{\alpha-1} + a^{\alpha-1}b)\left\{\frac{11|a|^{(r-1)} + 5|b|^{(r-1)}}{192}\right\}\right].
\end{aligned}$$

Proof: The result follows from Theorem 2.17 for the convex function $\psi(x) = x^r, x > 0$. □

Proposition 2.2 *Let $a, b \in \mathbb{R}$ with $0 < a < b$ and $r > 1$. Then for $s > 1$ and $\alpha \in (0, 1]$, we have the following inequality:*

$$\begin{aligned}
&|\mathbb{A}^r(a, b) - \mathbb{L}^r_{(\alpha,r)}(a, b))| \leq \frac{(r-1)(b-a)}{b^\alpha - a^\alpha} \\
&[(A_1(\alpha))^{1-\frac{1}{b}}\{A_2(\alpha)|a|^{(r-1)q} + A_3(\alpha)|b|^{(r-1)q}\}^{\frac{1}{b}} \\
&+ (B_1(\alpha))^{1-\frac{1}{b}}\{B_2(\alpha)|a|^{(r-1)q} + B_3(\alpha)|b|^{(r-1)q}\}^{\frac{1}{b}}\Big].
\end{aligned}$$

Proof: One can obtain the result from Theorem 2.18 by using the convex function $\psi(x) = x^r, x > 0$. □

Proposition 2.3 *Let $a, b \in \mathbb{R}$ with $0 < a < b$ and $\alpha \in (0, 1]$, then we have*

$$|A^{-1}(a,b) - L^{-1}_{(\alpha,-1)}(a,b)|$$
$$\leq \frac{(b-a)}{b^\alpha - a^\alpha}\left[\frac{|a|^{-2}}{192}[13b^\alpha - 35a^\alpha] + \frac{|b|^{-2}}{192}[19b^\alpha - 29a^\alpha]\right.$$
$$\left. + (ab^{\alpha-1} + a^{\alpha-1}b)\left[\frac{11|a|^{-2} + 5|b|^{-2}}{192}\right]\right].$$

Proof: The statement of results follows from Theorem 2.17 for the convex function $\psi(x) = \frac{1}{x}\ x > 0$. □

Proposition 2.4 *Let $a, b \in \mathbb{R}$ with $0 < a < b$. Then for $s > 1$ and $\alpha \in (0, 1]$, we have*

$$|A^{-1}(a,b) - L^{-1}_{(\alpha,-1)}(a,b)| \leq \frac{(b-a)}{b^\alpha - a^\alpha}\left[(A_1(\alpha))^{1-\frac{1}{b}}\{A_2(\alpha)|a|^{-2q} + A_3(\alpha)|b|^{-2q}\}^{\frac{1}{b}}\right.$$
$$\left. + (B_1(\alpha))^{1-\frac{1}{b}}\{B_2(\alpha)|a|^{-2q} + B_3(\alpha)|b|^{-2q}\}^{\frac{1}{b}}\right].$$

Proof: We can get the inequality from Theorem 2.18 by using the convex function $\psi(x) = \frac{1}{x}\ x > 0$. □

2.5.1.2 Applications to the Midpoint Formula

Let P be the partition of the points $a = x_0 < x_1 < \cdots < x_{n-1} < x_n = b$ of the interval $[a, b]$ and consider the quadrature formula

$$\int_a^b \psi(x) d_\alpha x = \mathrm{T}_\alpha(\psi, \mathrm{P}) + \mathrm{E}_\alpha(\psi, \mathrm{P}), \tag{2.58}$$

where

$$\mathrm{T}_\alpha(\psi, \mathrm{P}) = \sum_{i=0}^{n-1} \psi\left(\frac{x_i + x_{i+1}}{2}\right)\frac{(x^\alpha_{i+1} - x^\alpha_i)}{\alpha} \tag{2.59}$$

is the midpoint version and $\mathrm{E}_\alpha(\psi, \mathrm{P})$ denotes the associated approximation error. Here, we are going to derive some new estimates for the midpoint formula.

Proposition 2.5 *Let $a, b \in \mathbb{R}$ with $0 \leq a < b$, and let $\psi : [a, b] \to \mathbb{R}$ be a differentiable function on (a, b) for $\alpha \in (0, 1]$. If $D_\alpha(\psi) \in L^1_\alpha([a, b])$ and $|\psi'|$ is convex on $[a, b]$, then we have*

$$|E_\alpha(\psi, \mathrm{P})|$$
$$\leq \sum_{i=0}^{n-1} \frac{(x^\alpha_{i+1} - x^\alpha_i)}{\alpha}\left[\frac{|\psi'(x_i)|}{192}[13x^\alpha_{i+1} - 35x^\alpha_i] + \frac{|\psi'(x_{i+1})|}{192}[19x^\alpha_{i+1} - 29x^\alpha_i]\right.$$
$$\left. + (x_i x^{\alpha-1}_{i+1} + x^{\alpha-1}_i x_{i+1})\left[\frac{11|\psi'(x_i)| + 5|\psi'(x_{i+1})|}{192}\right]\right].$$

Proof: Applying Theorem 2.17 on the subintervals $[x_i, x_{i+1}]$ $(i = 0, 1, \dots, n-1)$ of the partition P, we have

$$\left|\psi\left(\frac{x_i + x_{i+1}}{2}\right)\frac{(x_{i+1}^\alpha - x_i^\alpha)}{\alpha} - \int_{x_i}^{x_{i+1}} \psi(x) d_\alpha x\right|$$
$$\leq \frac{(x_{i+1}^\alpha - x_i^\alpha)}{\alpha}\left[\frac{|\psi'(x_i)|}{192}[13x_{i+1}^\alpha - 35x_i^\alpha] + \frac{|\psi'(x_{i+1})|}{192}[19x_{i+1}^\alpha - 29x_i^\alpha]\right.$$
$$\left. + (x_i x_{i+1}^{\alpha-1} + x_i^{\alpha-1}x_{i+1})\left[\frac{11|\psi'(x_i)| + 5|\psi'(x_{i+1})|}{192}\right]\right].$$

Hence, from above

$$\left|\int_a^b \psi(x) d_\alpha x - \mathrm{T}_\alpha(\psi, \mathrm{P})\right|$$
$$= \left|\sum_{i=0}^{n-1}\left\{\int_{x_i}^{x_{i+1}} \psi(x) d_\alpha x - \psi\left(\frac{x_i + x_{i+1}}{2}\right)\frac{(x_{i+1}^\alpha - x_i^\alpha)}{\alpha}\right\}\right|$$
$$\leq \sum_{i=0}^{n-1}\left|\left\{\int_{x_i}^{x_{i+1}} \psi(x) d_\alpha x - \psi\left(\frac{x_i + x_{i+1}}{2}\right)\frac{(x_{i+1}^\alpha - x_i^\alpha)}{\alpha}\right\}\right|$$
$$\leq \sum_{i=0}^{n-1}\frac{(x_{i+1}^\alpha - x_i^\alpha)}{\alpha}\left[\frac{|\psi'(x_i)|}{192}[13x_{i+1}^\alpha - 35x_i^\alpha] + \frac{|\psi'(x_{i+1})|}{192}[19x_{i+1}^\alpha - 29x_i^\alpha]\right.$$
$$\left. + (x_i x_{i+1}^{\alpha-1} + x_i^{\alpha-1}x_{i+1})\left[\frac{11|\psi'(x_i)| + 5|\psi'(x_{i+1})|}{192}\right]\right].$$

□

Proposition 2.6 *Let $a, b \in \mathbb{R}$ with $0 \leq a < b$, and let $\psi : [a, b] \to \mathbb{R}$ be a differentiable function on (a, b) for $\alpha \in (0, 1]$. If $D_\alpha(\psi) \in L_\alpha^1([a, b])$ and $|\psi'|^b$ is convex on $[a, b]$ with $s > 1$, then we have*

$$|E_\alpha(\psi, \mathrm{P})| \leq \sum_{i=0}^{n-1}\frac{(x_{i+1} - x_i)}{\alpha}\left[(A_1(\alpha))^{1-\frac{1}{b}}\{A_2(\alpha)|\psi'(x_i)|^b + A_3(\alpha)|\psi'(x_{i+1})|^b\}^{\frac{1}{b}}\right.$$
$$\left. + (B_1(\alpha))^{1-\frac{1}{b}}\{B_2(\alpha)|\psi'(x_i)|^b + B_3(\alpha)|\psi'(x_{i+1})|^b\}^{\frac{1}{b}}\right].$$

Proof: The proof is analogous to that of Proposition 2.5 only by using Theorem 2.18. □

2.5.2 Inequalities Related to the Right Side of Hermite–Hadamard Inequality

In this section, first an identity for conformable fractional integrals is established. Second, by using this identity some Hermite–Hadamard type inequalities for conformable fractional integrals are obtained. At the end applications to some special means of real numbers are given and then a couple of new error estimates for the trapezoidal formula are addressed. In order to prove our main results we need a lemma, which we present in this section.

Lemma 2.4 *Let $\alpha \in (0,1]$, $a, b \in \mathbb{R}$ with $0 \le a < b$ and $\psi : [a,b] \to \mathbb{R}$ be an α-fractional differentiable function on (a,b). Then, the identity*

$$\frac{\psi(a)+\psi(b)}{2} - \frac{\alpha}{b^\alpha - a^\alpha}\int_a^b \psi(w)d_\alpha w$$
$$= \frac{(b-a)}{2(b^\alpha - a^\alpha)}\left[\int_0^1 ((ta+(1-t)b)^{2\alpha-1} - b^\alpha(ta+(1-t)b)^{\alpha-1})\right.$$
$$\times D_\alpha(\psi)(ta+(1-t)b)t^{1-\alpha}d_\alpha t + \int_0^1 ((ta+(1-t)b)^{2\alpha-1}$$
$$\left. - a^\alpha(ta+(1-t)b)^{\alpha-1}) \times D_\alpha(\psi)(ta+(1-t)b)t^{1-\alpha}d_\alpha t\right]$$

holds if $D_\alpha(\psi) \in L^1_\alpha([a,b])$.

Proof: Let $w = ta + (1-t)b$. Then making use of integration by parts, we get

$$\int_0^1 ((ta+(1-t)b)^{2\alpha-1} - b^\alpha(ta+(1-t)b)^{\alpha-1})D_\alpha(\psi)(ta+(1-t)b)dt$$
$$+ \int_0^1 ((ta+(1-t)b)^{2\alpha-1} - a^\alpha(ta+(1-t)b)^{\alpha-1})D_\alpha(\psi)(ta+(1-t)b)dt$$
$$= \int_0^1 ((ta+(1-t)b)^\alpha - b^\alpha)\psi'(ta+(1-t)b)dt$$
$$+ \int_0^1 ((ta+(1-t)b)^\alpha - a^\alpha)\psi'(ta+(1-t)b)dt$$
$$=((ta+(1-t)b)^\alpha - b^\alpha)\frac{\psi(ta+(1-t)b)}{a-b}\Big|_0^1$$
$$- \int_0^1 \alpha(ta+(1-t)b)^{\alpha-1}(a-b)\frac{\psi(ta+(1-t)b)}{a-b}dt$$
$$+ ((ta+(1-t)b)^\alpha - a^\alpha)\frac{\psi(ta+(1-t)b)}{a-b}\Big|_0^1$$
$$- \int_0^1 \alpha(ta+(1-t)b)^{\alpha-1}(a-b)\frac{\psi(ta+(1-t)b)}{a-b}dt$$
$$=\frac{1}{b-a}\left[(b^\alpha - a^\alpha)\psi(a) - \alpha\int_a^b \psi(w)d_\alpha w\right]$$
$$+ \frac{1}{b-a}\left[(b^\alpha - a^\alpha)\psi(b) - \alpha\int_a^b \psi(w)d_\alpha w\right]$$
$$=\frac{b^\alpha - a^\alpha}{b-a}(\psi(a)+\psi(b)) - \frac{2\alpha}{b-a}\int_a^b \psi(w)d_\alpha w. \tag{2.60}$$

□

Remark 2.9 We clearly see that the identity given in Lemma 2.4 reduces to the identity given in Lemma 2.1 if $\alpha = 1$.

Theorem 2.20 *Let $\alpha \in (0,1]$, $a, b \in \mathbb{R}$ with $0 \le a < b$ and $\psi : [a,b] \to \mathbb{R}$ be an α-differentiable function. Then, the inequality*

$$\left|\frac{\psi(a)+\psi(b)}{2} - \frac{\alpha}{b^\alpha - a^\alpha}\int_a^b \psi(w)d_\alpha w\right|$$
$$\le \frac{b-a}{2(b^\alpha - a^\alpha)}\left[\frac{(|\psi'(a)|+|\psi'(b)|)(5b^\alpha - 7a^\alpha + ab^{\alpha-1} + a^{\alpha-1}b)}{12}\right] \tag{2.61}$$

takes place if $D_\alpha(\psi) \in L^1_\alpha([a,b])$ and $|\psi'|$ is convex on $[a,b]$.

Proof: It follows from Lemma 2.4 and the convexities of the functions $w \to w^{\alpha-1}$ and $w \to -w^\alpha$ on $(0,\infty)$ together with the convexity of $|\psi'|$ on $[a,b]$ that

$$\left|\frac{\psi(a)+\psi(b)}{2} - \frac{\alpha}{b^\alpha - a^\alpha}\int_a^b \psi(w)d_\alpha w\right|$$
$$\le \frac{b-a}{2(b^\alpha - a^\alpha)}\left[\int_0^1 ((ta+(1-t)b)^\alpha - a^\alpha)|\psi'(ta+(1-t)b)|dt\right.$$
$$\left.+\int_0^1 (b^\alpha - (ta+(1-t)b)^\alpha)|\psi'(ta+(1-t)b)|dt\right]$$
$$\le \frac{b-a}{2(b^\alpha - a^\alpha)}\left[\int_0^1 ((ta+(1-t)b)^{\alpha-1}(ta+(1-t)b)-a^\alpha)|\psi'(ta+(1-t)b)|dt\right.$$
$$\left.+\int_0^1 (b^\alpha - ((1-t)a^\alpha + tb^\alpha))|\psi'(ta+(1-t)b)|dt\right]$$
$$\le \frac{b-a}{2(b^\alpha - a^\alpha)}\left[\int_0^1 (((1-t)a^{\alpha-1} + tb^{\alpha-1})(ta+(1-t)b) - a^\alpha)\right.$$
$$\left.|\psi'(ta+(1-t)b)|dt + \int_0^1 (b^\alpha - ((1-t)a^\alpha + tb^\alpha))|\psi'(ta+(1-t)b)|dt\right]$$
$$\le \frac{b-a}{2(b^\alpha - a^\alpha)}\left[\int_0^1 (((1-t)a^{\alpha-1}+tb^{\alpha-1})(ta+(1-t)b)-a^\alpha)[(1-t)|\psi'(a)|]\right.$$
$$\left.+t|\psi'(b))|]dt + \int_0^1 (b^\alpha - ((1-t)a^\alpha + tb^\alpha))[(1-t)|\psi'(a)| + t|\psi'(b))|]dt\right]$$
$$= \frac{b-a}{b^\alpha - a^\alpha}\left[\frac{1}{4}a^\alpha|\psi'(a)| + \frac{1}{12}a^\alpha|\psi'(b)| + \frac{1}{12}ab^{\alpha-1}|\psi'(a)| + \frac{1}{12}ab^{\alpha-1}|\psi'(b)\right.$$
$$| + \frac{1}{12}a^{\alpha-1}b|\psi'(a)|$$
$$+\frac{1}{12}a^{\alpha-1}b|\psi'(b)| + \frac{1}{12}b^\alpha|\psi'(a)| + \frac{1}{4}b^\alpha|\psi'(b)| - \frac{1}{2}a^\alpha|\psi'(a)|$$
$$\left.-\frac{1}{2}a^\alpha|\psi'(b)| + \frac{1}{2}b^\alpha|\psi'(a)| + \frac{1}{2}b^\alpha|\psi'(b)| - \frac{1}{3}a^\alpha|\psi'(a)|\right]$$
$$-\frac{1}{6}a^\alpha|\psi'(b)| - \frac{1}{6}b^\alpha|\psi'(a)| - \frac{1}{3}b^\alpha|\psi'(b)|$$

$$=\frac{b-a}{2(b^\alpha-a^\alpha)}\left[\frac{(|\psi'(a)|+|\psi'(b)|)(5b^\alpha-7a^\alpha+ab^{\alpha-1}+a^{\alpha-1}b)}{12}\right].$$ □

Remark 2.10 Let $\alpha = 1$. Then, the above inequality in Theorem 2.20 becomes

$$\left|\frac{\psi(a)+\psi(b)}{2}-\frac{1}{b-a}\int_a^b\psi(w)dw\right|\leq\frac{b-a}{4}[|\psi'(a)|+|\psi'(b)|].$$

Theorem 2.21 *Let $\alpha\in(0,1]$, $a,b\in\mathbb{R}$ with $0\leq a<b$ and $\psi:[a,b]\to\mathbb{R}$ be an α-differentiable function. Then, the inequality*

$$\left|\frac{\psi(a)+\psi(b)}{2}-\frac{\alpha}{b^\alpha-a^\alpha}\int_a^b\psi(w)d_\alpha w\right|$$
$$\leq\frac{b-a}{2(b^\alpha-a^\alpha)}\left[\frac{(|\psi'(a)|+|\psi'(b)|)(5b^\alpha-7a^\alpha+ab^{\alpha-1}+a^{\alpha-1}b)}{12}\right]\tag{2.62}$$

takes place if $D_\alpha(\psi)\in L^1_\alpha([a,b])$ and $|\psi'|$ is convex on $[a,b]$.

Proof: It follows from Lemma 2.4 and the convexities of the functions $w\to w^{\alpha-1}$ and $w\to -w^\alpha$ on $(0,\infty)$ together with the convexity of $|\psi'|$ on $[a,b]$ that

$$\left|\frac{\psi(a)+\psi(b)}{2}-\frac{\alpha}{b^\alpha-a^\alpha}\int_a^b\psi(w)d_\alpha w\right|$$
$$\leq\frac{b-a}{2(b^\alpha-a^\alpha)}\left[\int_0^1((ta+(1-t)b)^\alpha-a^\alpha)|\psi'(ta+(1-t)b)|dt\right.$$
$$\left.+\int_0^1(b^\alpha-(ta+(1-t)b)^\alpha)|\psi'(ta+(1-t)b)|dt\right]$$
$$\leq\frac{b-a}{2(b^\alpha-a^\alpha)}\left[\int_0^1((ta+(1-t)b)^{\alpha-1}(ta+(1-t)b)-a^\alpha)|\psi'(ta+(1-t)b)|dt\right.$$
$$\left.+\int_0^1(b^\alpha-((1-t)a^\alpha+tb^\alpha))|\psi'(ta+(1-t)b)|dt\right]$$
$$\leq\frac{b-a}{2(b^\alpha-a^\alpha)}\left[\int_0^1(((1-t)a^{\alpha-1}+tb^{\alpha-1})(ta+(1-t)b)-a^\alpha)\right.$$
$$\left.|\psi'(ta+(1-t)b)|dt+\int_0^1(b^\alpha-((1-t)a^\alpha+tb^\alpha))|\psi'(ta+(1-t)b)|dt\right]$$
$$\leq\frac{b-a}{2(b^\alpha-a^\alpha)}\left[\int_0^1(((1-t)a^{\alpha-1}+tb^{\alpha-1})(ta+(1-t)b)-a^\alpha)\right.$$
$$[(1-t)|\psi'(a)|+t|\psi'(b))|]dt$$
$$\left.+\int_0^1(b^\alpha-((1-t)a^\alpha+tb^\alpha))[(1-t)|\psi'(a)|+t|\psi'(b))|]dt\right]$$

$$=\frac{b-a}{b^\alpha-a^\alpha}\left[\frac{1}{4}a^\alpha|\psi'(a)|+\frac{1}{12}a^\alpha|\psi'(b)|+\frac{1}{12}ab^{\alpha-1}|\psi'(a)|+\frac{1}{12}ab^{\alpha-1}|\psi'(b)|\right.$$
$$+\frac{1}{12}a^{\alpha-1}b|\psi'(a)|+\frac{1}{12}a^{\alpha-1}b|\psi'(b)|+\frac{1}{12}b^\alpha|\psi'(a)|+\frac{1}{4}b^\alpha|\psi'(b)|$$
$$-\frac{1}{2}a^\alpha|\psi'(a)|-\frac{1}{2}a^\alpha|\psi'(b)|+\frac{1}{2}b^\alpha|\psi'(a)|+\frac{1}{2}b^\alpha|\psi'(b)|-\frac{1}{3}a^\alpha|\psi'(a)|$$
$$\left.-\frac{1}{6}a^\alpha|\psi'(b)|-\frac{1}{6}b^\alpha|\psi'(a)|-\frac{1}{3}b^\alpha|\psi'(b)|\right]$$
$$=\frac{b-a}{2(b^\alpha-a^\alpha)}\left[\frac{(|\psi'(a)|+|\psi'(b)|)(5b^\alpha-7a^\alpha+ab^{\alpha-1}+a^{\alpha-1}b)}{12}\right].$$
□

Remark 2.11 Let $\alpha = 1$. Then, inequality (2.62) becomes

$$\left|\frac{\psi(a)+\psi(b)}{2}-\frac{1}{b-a}\int_a^b\psi(w)dw\right|\leq\frac{b-a}{4}[|\psi'(a)|+|\psi'(b)|].$$

Theorem 2.22 *Let $\alpha \in (0, 1]$, $q > 1$, $a, b \in \mathbb{R}$ with $0 \leq a < b$ and $\psi : [a, b] \to \mathbb{R}$ be an α-differentiable function on (a, b). Then, the inequality*

$$\left|\frac{\psi(a)+\psi(b)}{2}-\frac{\alpha}{b^\alpha-a^\alpha}\int_a^b\psi(w)d_\alpha w\right|$$
$$\leq\frac{b-a}{2(b^\alpha-a^\alpha)}[(A_1(\alpha))^{1-1/q}\{A_2(\alpha)|\psi'(a)|^q+A_3(\alpha)|\psi'(b)|^q\}^{1/q}$$
$$+(B_1(\alpha))^{1-1/q}\{B_2(\alpha)|\psi'(a)|^q+B_3(\alpha)|\psi'(b)|^q\}^{1/q}] \tag{2.63}$$

is valid if $D_\alpha(\psi) \in L^1_\alpha([a, b])$ and $|\psi'|^q$ is convex on $[a, b]$, where

$$A_1(\alpha)=\left[\frac{a^{\alpha+1}-b^{\alpha+1}}{(\alpha+1)(a-b)}\right]-a^\alpha,\quad B_1(\alpha)=b^\alpha-\left[\frac{a^{\alpha+1}-b^{\alpha+1}}{(\alpha+1)(a-b)}\right],$$
$$A_2(\alpha)=\left[\frac{-b^{\alpha+1}}{(\alpha+1)(a-b)}+\frac{a^{\alpha+2}-b^{\alpha+2}}{(\alpha+1)(\alpha+2)(a-b)^2}-\frac{a^\alpha}{2}\right],$$
$$B_2(\alpha)=\left[\frac{b^\alpha}{2}+\frac{b^{\alpha+1}}{(\alpha+1)(a-b)}+\frac{a^{\alpha+2}-b^{\alpha+2}}{(\alpha+1)(\alpha+2)(a-b)^2}\right],$$
$$A_3(\alpha)=\left[\frac{a^{\alpha+1}}{(\alpha+1)(a-b)}-\frac{a^{\alpha+2}-b^{\alpha+2}}{(\alpha+1)(\alpha+2)(a-b)^2}-\frac{a^\alpha}{2}\right],$$
$$B_3(\alpha)=\left[\frac{b^\alpha}{2}-\frac{a^{\alpha+1}}{(\alpha+1)(a-b)}+\frac{a^{\alpha+2}-b^{\alpha+2}}{(\alpha+1)(\alpha+2)(a-b)^2}\right].$$

Proof: From Lemma 2.4 and the well-known Hölder mean inequality together with the convexity of $|\psi'|^q$ on the interval $[a, b]$, we clearly see that

$$\left|\frac{\psi(a)+\psi(b)}{2}-\frac{\alpha}{b^\alpha-a^\alpha}\int_a^b\psi(w)d_\alpha w\right|$$

$$
\begin{aligned}
&= \left| \frac{b-a}{2(b^\alpha - a^\alpha)} \left[\int_0^1 ((ta+(1-t)b)^{2\alpha-1} - a^\alpha (ta+(1-t)b)^{\alpha-1}) \right. \right. \\
&\quad \mathrm{D}_\alpha(\psi)(ta+(1-t)b)dt \\
&\quad \left. \left. + \int_0^1 ((ta+(1-t)b)^{2\alpha-1} - b^\alpha (ta+(1-t)b)^{\alpha-1}) \mathrm{D}_\alpha(\psi)(ta+(1-t)b)dt \right] \right| \\
&\le \frac{b-a}{2(b^\alpha - a^\alpha)} \left[\int_0^1 ((ta+(1-t)b)^\alpha - a^\alpha)|\psi'(ta+(1-t)b)|dt \right. \\
&\quad \left. + \int_0^1 (b^\alpha - (ta+(1-t)b)^\alpha)|\psi'(ta+(1-t)b)|dt \right],
\end{aligned} \tag{2.64}
$$

$$
\begin{aligned}
&\int_0^1 ((ta+(1-t)b)^\alpha - a^\alpha)|\psi'(ta+(1-t)b)|dt \\
&\le \left(\int_0^1 ((ta+(1-t)b)^\alpha - a^\alpha)dt \right)^{1-1/q} \\
&\quad \times \left(\int_0^1 ((ta+(1-t)b)^\alpha - a^\alpha)|\psi'(ta+(1-t)b)|^q dt \right)^{1/q},
\end{aligned} \tag{2.65}
$$

$$
\begin{aligned}
&\int_0^1 (b^\alpha - (ta+(1-t)b)^\alpha)|\psi'(ta+(1-t)b)|dt \\
&\le \left(\int_0^1 (b^\alpha - (ta+(1-t)b)^\alpha)dt \right)^{1-1/q} \\
&\quad \times \left(\int_0^1 (b^\alpha - (ta+(1-t)b)^\alpha)|\psi'(ta+(1-t)b)|^q dt \right)^{1/q},
\end{aligned} \tag{2.66}
$$

$$
\begin{aligned}
&\int_0^1 ((ta+(1-t)b)^\alpha - a^\alpha)|\psi'(ta+(1-t)b)|^q dt \\
&\le \int_0^1 ((ta+(1-t)b)^\alpha - a^\alpha)[(1-t)|\psi'(a)|^q + t|\psi'(b)|^q]dt \\
&= |\psi'(a)|^q \int_0^1 ((ta+(1-t)b)^\alpha - a^\alpha)(1-t)dt \\
&\quad + |\psi'(b)|^q \int_0^1 ((ta+(1-t)b)^\alpha - a^\alpha)tdt \\
&= |\psi'(a)|^q \left[\frac{-b^{\alpha+1}}{(\alpha+1)(a-b)} + \frac{a^{\alpha+2} - b^{\alpha+2}}{(\alpha+1)(\alpha+2)(a-b)^2} - \frac{a^\alpha}{2} \right] \\
&\quad + |\psi'(b)|^q \left[\frac{\alpha+1}{(\alpha+1)(a-b)} - \frac{a^{\alpha+2} - b^{\alpha+2}}{(\alpha+1)(\alpha+2)(a-b)^2} - \frac{a^\alpha}{2} \right],
\end{aligned} \tag{2.67}
$$

$$\begin{aligned}
&\int_0^1 (b^\alpha - (ta+(1-t)b)^\alpha)|\psi'(ta+(1-t)b)|^q dt \\
\leq &\int_0^1 (b^\alpha - (ta+(1-t)b)^\alpha)[(1-t)|\psi'(a)|^q + t|\psi'(b)|^q]dt \\
= &|\psi'(a)|^q \int_0^1 (b^\alpha - (ta+(1-t)b)^\alpha)(1-t)dt \\
&+ |\psi'(b)|^q \int_0^1 (b^\alpha - (ta+(1-t)b)^\alpha)tdt \\
= &|\psi'(a)|^q \left[\frac{b^\alpha}{2} + \frac{b^{\alpha+1}}{(\alpha+1)(a-b)} + \frac{a^{\alpha+2}-b^{\alpha+2}}{(\alpha+1)(\alpha+2)(a-b)^2}\right] \\
&+ |\psi'(b)|^q \left[\frac{b^\alpha}{2} - \frac{a^{\alpha+1}}{(\alpha+1)(a-b)} + \frac{a^{\alpha+2}-b^{\alpha+2}}{(\alpha+1)(\alpha+2)(a-b)^2}\right]. \qquad (2.68)
\end{aligned}$$

Therefore, setting these in above, we get the required result. □

Remark 2.12 Let $\alpha = 1$. Then, inequality (2.63) becomes

$$\begin{aligned}
&\left|\frac{\psi(a)+\psi(b)}{2} - \frac{1}{b-a}\int_a^b \psi(w)dw\right| \\
\leq &\frac{1}{2}\left(\frac{b-a}{2}\right)^{1-1/q} [\{A_2(1)|\psi'(a)|^q + A_3(1)|\psi'(b)|^q\}^{1/q} \\
&+ \{B_2(1)|\psi'(a)|^q + B_3(1)|\psi'(b)|^q\}^{1/q}]
\end{aligned}$$

with

$$\begin{aligned}
A_2(1) &= \frac{b-a}{3}, \quad B_2(1) = \frac{(a+b)^2+2ab}{6(a-b)}, \\
A_3(1) &= \frac{b-a}{6}, \quad B_3(1) = \frac{b-a}{3}.
\end{aligned}$$

Theorem 2.23 *Let $\alpha \in (0,1]$, $q > 1$, $a, b \in \mathbb{R}$ with $0 \leq a < b$ and $\psi : [a,b] \to \mathbb{R}$ be an α-differentiable function on (a,b). Then, the inequality*

$$\begin{aligned}
&\left|\frac{\psi(a)+\psi(b)}{2} - \frac{\alpha}{b^\alpha - a^\alpha}\int_a^b \psi(w)d_\alpha w\right| \\
\leq &\frac{b-a}{2(b^\alpha - a^\alpha)}\left[A_1(\alpha)\psi'\left(\frac{C_1(\alpha)}{A_1(\alpha)}\right) + B_1(\alpha)\psi'\left(\frac{C_2(\alpha)}{B_1(\alpha)}\right)\right] \qquad (2.69)
\end{aligned}$$

holds if $D_\alpha(\psi) \in L_\alpha^1([a,b])$ and $|\psi'|^q$ is concave on $[a,b]$, where $A_1(\alpha)$ and $B_1(\alpha)$ are defined as in Theorem 2.5, and $C_1(\alpha)$ and $C_2(\alpha)$ are defined by

$$C_1(\alpha) = \left[\frac{a^{\alpha+2}-b^{\alpha+2}}{(\alpha+2)(a-b)} - \frac{a^\alpha(a-b)}{2}\right], \quad C_2(\alpha) = \left[\frac{b^\alpha(a+b)}{2} - \frac{a^{\alpha+2}-b^{\alpha+2}}{(\alpha+2)(a-b)}\right].$$

Proof: It follows the concavity of $|\psi'|^q$ and the Hölder mean inequality that

$$(t|\psi'(a)|+(1-t)|\psi'(b)|)^q \leq t|\psi'(a)|^q+(1-t)|\psi'(b)|^q \leq |\psi'(ta+(1-t)b)|^q,$$
$$|\psi'(ta+(1-t)b)| \geq t|\psi'(a)|+(1-t)|\psi'(b)|,$$

which implies that $|\psi'|$ is also concave. Making use of Lemma 2.4 and the Jensen integral inequality, we have

$$\begin{aligned}
&\left|\frac{\psi(a)+\psi(b)}{2}-\frac{\alpha}{b^\alpha-a^\alpha}\int_a^b \psi(w)d_\alpha w\right| \\
=&\left|\frac{b-a}{2(b^\alpha-a^\alpha)}\left[\int_0^1 ((ta+(1-t)b)^{2\alpha-1}-a^\alpha(ta+(1-t)b)^{\alpha-1})\right.\right. \\
&\mathrm{D}_\alpha(\psi)(ta+(1-t)b)dt \\
&\left.\left.+\int_0^1 ((ta+(1-t)b)^{2\alpha-1}-b^\alpha(ta+(1-t)b)^{\alpha-1})\mathrm{D}_\alpha(\psi)(ta+(1-t)b)dt\right]\right| \\
\leq&\frac{b-a}{2(b^\alpha-a^\alpha)}\left[\int_0^1 ((ta+(1-t)b)^\alpha-a^\alpha)|\psi'(ta+(1-t)b)|dt\right. \\
&\left.+\int_0^1 (b^\alpha-(ta+(1-t)b)^\alpha)|\psi'(ta+(1-t)b)|dt\right],
\end{aligned} \tag{2.70}$$

$$\begin{aligned}
&\int_0^1 ((ta+(1-t)b)^\alpha-a^\alpha)|\psi'(ta+(1-t)b)|dt \\
\leq&\left(\int_0^1 ((ta+(1-t)b)^\alpha-a^\alpha)dt\right) \\
&\times\psi'\left(\frac{\int_0^1((ta+(1-t)b)^\alpha-a^\alpha)(ta+(1-t)b)dt}{\int_0^1((ta+(1-t)b)^\alpha-a^\alpha)dt}\right) \\
=&\mathrm{A}_1(\alpha)\psi'\left(\frac{\mathrm{C}_1(\alpha)}{\mathrm{A}_1(\alpha)}\right),
\end{aligned} \tag{2.71}$$

$$\begin{aligned}
&\int_0^1 (b^\alpha-(ta+(1-t)b)^\alpha)|\psi'(ta+(1-t)b)|dt \\
\leq&\left(b^\alpha-\int_0^1 ((ta+(1-t)b)^\alpha)dt\right) \\
&\times\psi'\left(\frac{\int_0^1(b^\alpha-(ta+(1-t)b)^\alpha)(ta+(1-t)b)dt}{\int_0^1(b^\alpha-(ta+(1-t)b)^\alpha)dt}\right) \\
=&\mathrm{B}_1(\alpha)\psi'\left(\frac{\mathrm{C}_2(\alpha)}{\mathrm{B}_1(\alpha)}\right).
\end{aligned} \tag{2.72}$$

Therefore, inequality (2.69) follows easily by substituting the last two inequalities in above. This completes the proof. □

Remark 2.13 Let $\alpha = 1$. Then, inequality (2.63) leads to

$$\left|\frac{\psi(a)+\psi(b)}{2} - \frac{1}{b-a}\int_a^b \psi(w)dw\right|$$
$$\leq \frac{b-a}{4}\left[\psi'\left(\frac{2b^2-a^2+5ab}{3(b-a)}\right) + \psi'\left(\frac{b^2-2a^2+ab}{3(b-a)}\right)\right].$$

2.5.2.1 Applications to Special Means of Real Numbers

We get several new inequalities for the arithmetic, logarithmic, and generalized logarithmic means as follows.

Theorem 2.24 *Let $a, b \in \mathbb{R}$ with $0 < a < b$, $r > 1$, $q > 1$ and $\alpha \in (0, 1]$. Then we have*

$$|A(a^r, b^r) - L^r_{(\alpha,r)}(a, b)|$$
$$\leq \frac{r(b-a)(5b^\alpha - 7a^\alpha + ab^{\alpha-1} + a^{\alpha-1}b)}{12(b^\alpha - a^\alpha)} A(|a|^{r-1}, |b|^{r-1}),$$

$$|A(a^r, b^r) - L^r_{(\alpha,r)}(a, b))|$$
$$\leq \frac{r(b-a)}{2(b^\alpha - a^\alpha)}[(A_1(\alpha))^{1-1/q}\{A_2(\alpha)|a|^{(r-1)q} + A_3(\alpha)|b|^{(r-1)q}\}^{1/q}$$
$$+ (B_1(\alpha))^{1-1/q}\{B_2(\alpha)|a|^{(r-1)q} + B_3(\alpha)|b|^{(r-1)q}\}^{1/q}],$$

$$|A(a^{-1}, b^{-1}) - L^{-1}_{(\alpha,-1)}(a, b)| \leq \frac{(b-a)(5b^\alpha - 7a^\alpha + ab^{\alpha-1} + a^{\alpha-1}b)}{12(b^\alpha - a^\alpha)} A(a^{-2}, b^{-2}),$$

$$|A(a^{-1}, b^{-1}) - L^{-1}_{(\alpha,-1)}(a, b)| \leq \frac{b-a}{2(b^\alpha - a^\alpha)}[(A_1(\alpha))^{1-1/q}\{A_2(\alpha)|a|^{-2q}$$
$$+A_3(\alpha)|b|^{-2q}\}^{1/q} + (B_1(\alpha))^{1-1/q}\{B_2(\alpha)|a|^{-2q} + B_3(\alpha)|b|^{-2q}\}^{\frac{1}{q}}\Big],$$

where $A_1(\alpha)$, $A_2(\alpha)$, $A_3(\alpha)$, $B_1(\alpha)$, $B_2(\alpha)$ and $B_3(\alpha)$ are defined as in Theorem 2.22.

Proof: From Theorems 2.21 and 2.22 together with the convexities of the functions $w \to w^r$ and $w \to 1/w$ on the interval $(0, \infty)$, the results directly follows. □

2.5.2.2 Applications to the Trapezoidal Formula

Let Δ be a division $a = w_0 < w_1 < \cdots < w_{n-1} < w_n = b$ of the interval $[a, b]$ and consider the quadrature formula

$$\int_a^b \psi(w)d_\alpha w = \mathrm{T}_\alpha(\psi, \Delta) + \mathrm{E}_\alpha(\psi, \Delta),$$

where

$$\mathrm{T}_\alpha(\psi, \Delta) = \sum_{i=0}^{n-1} \frac{\psi(w_i) + \psi(w_{i+1})}{2} \frac{(w_{i+1}^\alpha - w_i^\alpha)}{\alpha}$$

is the trapezoidal version and $\mathrm{E}_\alpha(\psi, \Delta)$ denotes the associated approximation error. In this section, we are going to derive several new error estimations for the trapezoidal formula.

Theorem 2.25 *Let $\alpha \in (0, 1]$, $a, b \in \mathbb{R}$ with $0 \le a < b$, $\psi : [a, b] \to \mathbb{R}$ be an α-differentiable function on (a, b) and Δ be a division $a = w_0 < w_1 < \cdots < w_{n-1} < w_n = b$ of the interval $[a, b]$. Then, the inequality*

$$|E_\alpha(\psi, \Delta)| \le \frac{1}{12\alpha} \max\{|\psi'(a)|, |\psi'(b)|\}$$
$$\sum_{i=0}^{n-1} (w_{i+1} - w_i)(5w_{i+1}^\alpha - 7w_i^\alpha + w_i w_{i+1}^{\alpha-1} + w_i^{\alpha-1} w_{i+1})$$

holds if $D_\alpha(\psi) \in L_\alpha^1([a, b])$ and $|\psi'|$ is convex on $[a, b]$.

Proof: Applying Theorem 2.21 on the subinterval $[w_i, w_{i+1}]$ $(i = 0, 1, \ldots, n-1)$ of the division Δ, we have

$$\left| \frac{\psi(w_i) + \psi(w_{i+1})}{2} \frac{(w_{i+1}^\alpha - w_i^\alpha)}{\alpha} - \int_{w_i}^{w_{i+1}} \psi(w) d_\alpha w \right|$$
$$\le \frac{(w_{i+1} - w_i)}{2\alpha} \left[\frac{(|\psi'(w_i)| + |\psi'(w_{i+1})|)(5w_{i+1}^\alpha - 7w_i^\alpha + w_i w_{i+1}^{\alpha-1} + w_i^{\alpha-1} w_{i+1})}{12} \right]. \tag{2.73}$$

From the convexity of $|\psi'(w)|$ on the interval $[a, b]$, we have

$$|\mathrm{E}_\alpha(\psi, \Delta)| = \left| \mathrm{T}_\alpha(\psi, \Delta) - \int_a^b \psi(w) d_\alpha w \right|$$
$$= \left| \sum_{i=0}^{n-1} \left[\frac{\psi(w_i) + \psi(w_{i+1})}{2} \frac{(w_{i+1}^\alpha - w_i^\alpha)}{\alpha} - \int_{w_i}^{w_{i+1}} \psi(w) d_\alpha w \right] \right|$$
$$\le \sum_{i=0}^{n-1} \left| \frac{\psi(w_i) + \psi(w_{i+1})}{2} \frac{(w_{i+1}^\alpha - w_i^\alpha)}{\alpha} - \int_{w_i}^{w_{i+1}} \psi(w) d_\alpha w \right|$$
$$\le \frac{1}{2\alpha} \sum_{i=0}^{n-1} (w_{i+1} - w_i)$$
$$\left[\frac{(|\psi'(w_i)| + |\psi'(w_{i+1})|)(5w_{i+1}^\alpha - 7w_i^\alpha + w_i w_{i+1}^{\alpha-1} + w_i^{\alpha-1} w_{i+1})}{12} \right]$$

$$
\begin{aligned}
&= \frac{1}{12\alpha} \sum_{i=0}^{n-1} (w_{i+1} - w_i) \\
&\left[\frac{(|\psi'(w_i)| + |\psi'(w_{i+1})|)(5w_{i+1}^{\alpha} - 7w_i^{\alpha} + w_i w_{i+1}^{\alpha-1} + w_i^{\alpha-1} w_{i+1})}{2} \right] \\
&\leq \frac{1}{12\alpha} \sum_{i=0}^{n-1} (w_{i+1} - w_i)(5w_{i+1}^{\alpha} - 7w_i^{\alpha} + w_i w_{i+1}^{\alpha-1} + w_i^{\alpha-1} w_{i+1}) \\
&\quad \max\{|\psi'(w_i)|, |\psi'(w_{i+1})|\} \\
&\leq \frac{1}{12\alpha} \max\{|\psi'(a)|, |\psi'(b)|\} \\
&\quad \sum_{i=0}^{n-1} (w_{i+1} - w_i)(5w_{i+1}^{\alpha} - 7w_i^{\alpha} + w_i w_{i+1}^{\alpha-1} + w_i^{\alpha-1} w_{i+1}).
\end{aligned}
$$

□

Making use of the analogous arguments as in the proof of Theorem 2.25, we get Theorem 2.26 immediately.

Theorem 2.26 *Let $\alpha \in (0, 1]$, $q > 1$, $a, b \in \mathbb{R}$ with $0 \leq a < b$, $\psi : [a, b] \to \mathbb{R}$ be an α-differentiable function on (a, b) and Δ be a division $a = w_0 < w_1 < \cdots < w_{n-1} < w_n = b$ of the interval $[a, b]$. Then, the inequality*

$$
\begin{aligned}
|E_\alpha(\psi, \Delta| \leq \sum_{i=0}^{n-1} \frac{(w_{i+1} - w_i)}{2\alpha} &\left[(A_1(\alpha))^{1-1/q} \{A_2(\alpha)|\psi'(w_i)|^q \right. \\
&+ A_3(\alpha)|\psi'(w_{i+1})|^q\}^{1/q} \\
&\left. + (B_1(\alpha))^{1-1/q} \{B_2(\alpha)|\psi'(w_i)|^q + B_3(\alpha)|\psi'(w_{i+1})|^q\}^{1/q} \right]
\end{aligned}
$$

takes place if $D_\alpha(\psi) \in L_\alpha^1([a, b])$ and $|\psi'|^q$ is convex on $[a, b]$, where $A_1(\alpha)$, $A_2(\alpha)$, $A_3(\alpha)$, $B_1(\alpha)$, $B_2(\alpha)$, and $B_3(\alpha)$ are defined as in Theorem 2.22.

Bibliography

1 Abdeljawad, T. (2015). On conformable fractional calculus. *J. Comput. Appl. Math.* **279**: 57–66.

2 Abdeljawad, T.and Atici, F. (2012). On the definitions of Nabla fractional operators. *Abstr. Appl. Anal.* **2012**: Article ID 406757, 13 pages.

3 Abdeljawad, T. (2013). Dual identities in fractional difference calculus within Riemann. *Adv. Differ. Equ.* **2013**: 1–16.

4 Atici, F.F. and Eloe, P.W. (2009). Initial value problems in discrete fractional calculus. *Proc. Am. Math. Soc.* **137**: 981–989.

5 Gray, H.L. and Zhang, N.F.(1988). On a new definition of the fractional difference.*Math. Comput.* **50**(182): 513–529.

6 Khalil, R., Alhorani, M., Yousef, A., and Sababheh, M. (2014). A new definition of fractional derivative. *J. Comput. Appl. Math.* **264**: 65–70.

7 Munkhammar, J.D. (2004). *Riemann-Liouville fractional derivatives and the Taylor-Riemann Series, Examensarbete i matematik, 10 poäng Handledare och examinator*. Andreas Strömbergsson.

8 Miller, K.S. and Ross, B. (1993). *An Introduction to the Fractional Calculus and Fractional Differential Equations*. New York: Wiley.

9 Podlubny, I. (1999). *Fractional Differential Equations*. San Diego, CA: Academic Press.

10 Samko, S.G., Kilbas, A.A.,and Marichev, O.I. (1993). *Fractional Integrals and Derivatives, Theory and Applications*. Yverdon: Gordon and Breach.

11 Riemann, B. (1953). *Versuch Einer Allgemeinen Auffassung der Integration und Differentiation. Gesammelte Mathematische Werke und Wissenschaftlicher Nachlass*. Teubner, Leipzig, 1876. New York: Dover.

12 Hadamard, J. (1892). Essai sur l'etude des fonctions donnees par leur development de Taylor. *J. Pure Appl. Math.* **4** (8): 101–186.

13 Katugampola, U.N.(2011). New approach to a generalized fractional integral.*Appl. Math. Comput.* **218**(3): 860–865.

14 Anderson, D.R. (2017). Second-order self-adjoint differential equations using a proportional-derivative controller. *Commun. Appl. Nonlinear Anal.* **24** (1): 17–48.

15 Anderson, D.R. and Ulness, D.J. (2015). Newly defined conformable derivatives. *Adv. Dyn. Syst. Appl.* **58**: 109–137.

16 Atangana, A., Baleanu, D., and Alsaedi, A. (2015). New properties of conformable derivative. *Open Math.* **13**: 1–10.

17 Hammad, M.A. and Khalil, R. (2014). Conformable fractional heat differential equations. *Int. J. Pure Appl. Mth.* **94** (2): 215–221.

18 Hammad, M.A. and Khalil, R. (2014). Abel's formula and Wronskian for conformable fractional differential equations. *Int. J. Differ. Equ. Appl.* **13**: 177–183.

19 Iyiola, O.S. and Nwaeze, E.R. (2016). Some new results on the new conformable fractional calculus with application using D. Alambert approach. *Progr. Fract. Differ. Appl.* **2**: 115–122.

20 Hamid, A.J. and Rabha, W.I. (2012). Denoising algorithm based on generalized fractional integral operator with two parameters. *Discrete Dyn. Nat.Soc.* **2012**: Article ID 529849, 14 pages.

21 Khan, M.A., Ali, T., Dragomir, S.S., and Sarikaya, M.Z. (2018). Hermite-Hadamard type inequalities for conformable fractional integrals. *Rev. R. Acad. Cienc. Exactas Fis. Nat. Ser. A Mat.* **112** (4): 1033–1048.

22 Chu, Y.M., Khan, M.A., Ali, T., and Dragomir, S.S. (2017). Inequalities for a-fractional differentiable functions. *J. Inequal. Appl.* **93**: 12 pages.

23 Hadamard, J.(1893). Étude sur les propriétés des fonctions entières et en particulier d́une fonction considérée par Riemann.*J. Math. Pures Appl.* **58**: 171–215.

24 Khan, M.A., Khurshid, Y., Ali, T., and Rehman, N. (2016). Inequalities for three times differentiable functions. *Punjab Univ. J. Math.* **48** (2): 35–48.

25 Bai, R.F., Qi, F., and Xi, B.Y. (2013). HermiteHadamard type inequalities for the m-and (α, m)-logarithmically convex functions. *Filomat* **27** (1): 1–7.

26 Dragomir, S.S. (1992). Two mappings in connection to Hadamard's inequality. *J. Math. Anal. Appl.* **167**: 49–56.

27 Dragomir, S.S. and Fitzpatrick, S. (1999). The Hadamard inequalities for s-convex functions in the second sense. *Demonstr. Math.* **32** (4): 687–696.

28 Dragomir, S.S. and Mcandrew, A. (2005). Refinment of the HermiteHadamard inequality for convex functions. *J. Inequal. Pure Appl. Math.* **6**: 1–6.

29 Dragomir, S.S.and Pearce, C.E.M.(2000). *Selected Topics on Hermite-Hadamard Inequalities and Applications*. Victoria University.

30 Dragomir, S.S., Pecaric, J., and Persson, L.E. (1995). Some inequalities of Hadamard type. *Soochow J. Math.* **21**: 335–341.

31 Kirmaci, U.S., Bakula, M.K., Ozdemir, M.E., and Pecaric, J. (2007). Hadamard-type inequalities for s-convex functions. *Appl. Math. Comput.* **193**: 26–35.

32 Matloka, M.(2013). On some Hadamard-type inequalities for (h_1, h_2)-preinvex functions on the co-ordinates.*J. Inequal. Appl.* **2013**: 227.

33 Noor, M.A. (2007). On HermiteHadamard integral inequalities for involving two log-preinvex functions. *J. Inequal. Pure Appl. Math.* **3**: 75–81.

34 Noor, M.A., Noor, K.I., and Awan, M.U. (2014). HermiteHadamard inequalities for relative semi-convex functions and applications. *Filomat* **28** (2): 221–230.

35 Pachpatte, B.G. (2003). On some inequalities[SAQ1] for convex functions. *RGMIA Res. Rep. Coll.* **6(E)**: 1–9.

36 Pachpatte, B.G. (2005). *Mathematical Inequalities*, vol. **67**. Amsterdam, Holland: North-Holland Library, Elsevier Science.

37 Sarikaya, M.Z., Saglam, A., and Yildrim, H. (2008). On some Hadamard-type inequalities for h-convex functions. *J. Math. Inequal.* **2**: 335–341.

38 Wu, Y., Qi, F., and Niu, D.W. (2015). Integral inequalities of HermiteHadamard type for the product of strongly logarithmically convex and other convex functions. *Maejo Int. J. Sci. Technol.* **9** (3): 394–402.

39 Dragomir, S.S. and Agarwal, R.P. (1998). Two inequalities for differentiable mappings and applications to special means of real numbers and trapezoidal formula. *Appl. Math. Lett.* **11** (5): 91–95.

40 Kirmaci, U.S. (2004). Inequalities for differentiable mappings and applications to special means of real numbers and to midpoint formula. *Appl. Math. Comput.* **147**: 137–146.

41 Anderson, D.R.(2016). Taylor's formula and integral inequalities for conformable fractional derivatives. In: *Contributions in Mathematics and Engineering* (ed. P. Pardalos and T.M. Rassias), 25–44. Springer.

42 Sarikaya, M.Z., Set, E., Yaldiz, H., and Başak, N. (2013). Hermite-Hadamard's inequalities for fractional integrals and related fractional inequalities. *Math. Comput.Modell.* **57**: 2403–2407.

43 Set, E., Akdemir, A.O., and Mumcu, I. (2018)Hermite-Hadamard's inequality and its extensions for conformable fractional integrals of any orderα>0. *Creat. Math. Inform.* **2**: 197–206.

44 Set, E., Gözpinar, A., and Ekinci, A. (2017). HermiteHadamard type inequalities via conformable fractional integrals. *Acta Math. Univ. Comenian* **LXXXVI**: 309–320-.

3

Analysis of New Trends of Fractional Differential Equations

Abdon Atangana[1,2] *and Ali Akgül*[3]

[1]*Faculty of Natural and Agricultural Science, Institute for Groundwater Studies, University of Free State, Bloemfontein, South Africa*
[2]*Department of Medical Research, China Medical University Hospital, China Medical University, Taichung, Taiwan*
[3]*Department of Mathematics, Art and Science Faculty, Siirt University, Siirt, Turkey*

3.1 Introduction

Differential equations have been recognized as useful mathematical tools to model real world problems arising in all the fields of science, engineering, and technology. Nevertheless, one needs to recall that they are obtained as a combination of differential or integral operators. These operators are used to construct ordinary, partial differential equations, with integer, noninteger, and variable orders. Each class of differential or integral operator can be used to specific purpose, for instance, the classical differential and integral operators can be used to model simple classical problems that follow non-Markovian processes, meaning processes with no memory. As example, one can quote movement of underground water in homogeneous aquifers many others can be mentioned, but will not be listed here. Nevertheless when, dealing with nonclassical problem, differential operator based on rate of change cannot be used to its limitation to capture memory and other complex heterogeneities, suitable operators can be used. Now the model will be constructed due to observed fact, for instance, if the process follow strictly decays law process with sign of crossover, the Caputo–Fabrizio differential and integral operators can be used to depict such process. Other interesting processes that can be depicted using these fractional operators, such as the diffusion with crossovers without explicitly taking into account inertial effects, reaction terms, and external forces. More precisely, these operators can be used in case of obstacles or obstacles following the process known as Ornstein–Uhlenbeck process. However, when dealing with time-dependent nonlocality, such differential or integral operators as they unable to depict such physical problems following such processes, therefore

Fractional Order Analysis: Theory, Methods and Applications, First Edition.
Edited by Hemen Dutta, Ahmet Ocak Akdemir, and Abdon Atangana.

other differential and integral operators are needed. For example, when dealing with processed displaying tent-shaped distribution, with tails longer for example longer the expected Gaussian distribution, Caputo–Fabrizio fractional differential and integral operators are not suitable candidate for modeling such behavior, in this case, differential and integral fractional differential based on power can be used efficiently. Specific example of such problem can be the flow of pollution within a long fracture, where the distribution can be compared to that of power law with long tail. Many other processes follow such behavior and can be depicted using the Caputo or Riemann–Liouville power fractional differential and integral operators. Nevertheless, while the power is able to capture some non-Markovian processes, it is worth noting that, there are many physical problem that display more than one behavior, for instance, the subsurface water flowing in dual medial, from the matrix soil to fracture if the matrix soil is permeable, this gives the flow to different processes, exponential decay law and long tailed, such behavior cannot be depicted by either Caputo–Riemann–Liouville power fractional operators or Caputo–Fabrizio exponential decay law fractional integral operators [1–5]. For more details see Refs [6–10].

Thus in this case, the Atangana–Baleanu fractional differential and integral operators are suitable operators to depict such behavior. Although these four differential and integral operators have opened doors to mankind to depict some problems they faced in their daily journeys, while many physical observed facts have been understood due to the applicability of these mathematical tools, one needs to understand that, there are still many problems in nature that cannot be described efficiently using these mathematical tools, thus very recently Atangana introduced new concept where the differential operator is equipped with two orders, the first order account for fractional order while the second account for the fractal dimension, his study was extended to the concept of variable fractal dimension, the concept was recently applied to capture new chaotic behavior with great success. Nevertheless, the differential operator was considered in Riemann–Liouville sense. In this chapter, we consider the operator to be in Caputo sense [11–14].

3.2 Theory

In this section, we present very important properties of the used differential operators. We assume that the used function is differentiable. We consider the following problem:

$$ {}_{0}^{FFPC}D_t^{\alpha,\beta} f(t) = \frac{1}{\Gamma(1-\alpha)} \int_0^t \frac{df(\tau)}{d\tau^\beta} (t-\tau)^{-\alpha} d\tau. \tag{3.1} $$

Assuming that f is differentiable. Then, we get

$$ {}_{0}^{FFPC}D_t^{\alpha,\beta} f(t) = \frac{1}{\Gamma(1-\alpha)} \int_0^t \frac{df(\tau)}{d\tau} \frac{\tau^{1-\beta}}{\beta} (t-\tau)^{-\alpha} d\tau. \tag{3.2} $$

Thus, we obtain

$$ \begin{aligned} \left| {}_{0}^{FFPC}D_t^{\alpha,\beta} f(t) \right| &= \frac{1}{\Gamma(1-\alpha)} \left| \int_0^t \frac{df(\tau)}{d\tau} \frac{\tau^{1-\beta}}{\beta} (t-\tau)^{-\alpha} d\tau \right| \\ &\leq \frac{1}{\Gamma(1-\alpha)} \int_0^t \left| \frac{df(\tau)}{d\tau} \right| \frac{\tau^{1-\beta}}{\beta} (t-\tau)^{-\alpha} d\tau \\ &< \frac{1}{\Gamma(1-\alpha)} \int_0^t \sup_{\tau\in[0,t]} \left| \frac{df(\tau)}{d\tau} \right| \frac{\tau^{1-\beta}}{\beta} (t-\tau)^{-\alpha} d\tau \\ &< \frac{\left\| \frac{df}{dt} \right\|_\infty}{\beta\Gamma(1-\alpha)} \int_0^t \tau^{1-\beta} (t-\tau)^{-\alpha} d\tau. \end{aligned} $$

Let $\tau = th$. Then, we acquire

$$ \begin{aligned} \left| {}_{0}^{FFPC}D_t^{\alpha,\beta} f(t) \right| &< \frac{\left\| \frac{df}{dt} \right\|_\infty}{\beta\Gamma(1-\alpha)} \int_0^1 (th)^{1-\beta} (t-th)^{-\alpha} t\, dh \\ &< \frac{\left\| \frac{df}{dt} \right\|_\infty t^{-\beta-\alpha}}{\beta\Gamma(1-\alpha)} \int_0^1 (h)^{1-\beta} (1-h)^{-\alpha} dh \\ &< \frac{\left\| \frac{df}{dt} \right\|_\infty t^{-\beta-\alpha}}{\beta\Gamma(1-\alpha)} \int_0^1 (h)^{2-\beta-1} (1-h)^{1-\alpha-1} dh \\ &< \frac{\left\| \frac{df}{dt} \right\|_\infty t^{-\beta-\alpha}}{\beta\Gamma(1-\alpha)} B(2-\beta, 1-\alpha), \end{aligned} $$

where $\alpha < 1$ and $\beta < 2$.

We take into consideration the following problem:

$$ {}_{0}^{FFMC}D_t^{\alpha,\beta} f(t) = \frac{AB(\alpha)}{1-\alpha} \int_0^t \frac{df(\tau)}{d\tau^\beta} E_\alpha \left(\frac{-\alpha}{1-\alpha} (t-\tau)^\alpha \right) d\tau. \tag{3.3} $$

Assuming that f is differentiable. Then, we get

$$ {}_{0}^{FFMC}D_t^{\alpha,\beta} f(t) = \frac{AB(\alpha)}{1-\alpha} \int_0^t \frac{df(\tau)}{d\tau} \frac{\tau^{1-\beta}}{\beta} E_\alpha \left(\frac{-\alpha}{1-\alpha} (t-\tau)^\alpha \right) d\tau. \tag{3.4} $$

Then, we get

$$ \begin{aligned} \left| {}_{0}^{FFMC}D_t^{\alpha,\beta} f(t) \right| &= \frac{AB(\alpha)}{1-\alpha} \left| \int_0^t \frac{df(\tau)}{d\tau} \frac{\tau^{1-\beta}}{\beta} E_\alpha \left(\frac{-\alpha}{1-\alpha} (t-\tau)^\alpha \right) d\tau \right| \\ &\leq \frac{AB(\alpha)}{1-\alpha} \int_0^t \left| \frac{df(\tau)}{d\tau} \right| \frac{\tau^{1-\beta}}{\beta} E_\alpha \left(\frac{-\alpha}{1-\alpha} (t-\tau)^\alpha \right) d\tau \end{aligned} $$

$$< \frac{AB(\alpha)}{1-\alpha} \int_0^t \sup_{\tau\in[0,t]} \left| \frac{df(\tau)}{d\tau} \right| \frac{\tau^{1-\beta}}{\beta} E_\alpha \left(\frac{-\alpha}{1-\alpha} (t-\tau)^\alpha \right) d\tau$$

$$< \frac{AB(\alpha)}{1-\alpha} \left\| \frac{df}{dt} \right\|_\infty \int_0^t \frac{\tau^{1-\beta}}{\beta} E_\alpha \left(\frac{-\alpha}{1-\alpha} (t-\tau)^\alpha \right) d\tau$$

$$< \frac{AB(\alpha)}{\beta(1-\alpha)} \left\| \frac{df}{dt} \right\|_\infty \sum_{j=0}^{\infty} \frac{\left(\frac{-\alpha}{1-\alpha}\right)^j}{\Gamma(\alpha j+1)} \int_0^t \tau^{1-\beta}(t-\tau)^{j\alpha} d\tau.$$

We take $th = \tau$. Then, we reach

$$\left| {}_0^{FFMC}D_t^{\alpha,\beta} f(t) \right| < \frac{AB(\alpha)}{\beta(1-\alpha)} \left\| \frac{df}{dt} \right\|_\infty \sum_{j=0}^{\infty} \frac{\left(\frac{-\alpha}{1-\alpha}\right)^j}{\Gamma(\alpha j+1)} \int_0^1 h^{1-\beta}(1-h)^{j\alpha} dh t^{-\beta+j\alpha}$$

$$< \frac{AB(\alpha)}{\beta(1-\alpha)} \left\| \frac{df}{dt} \right\|_\infty \frac{1}{t^\beta} \sum_{j=0}^{\infty} \frac{\left(\frac{-\alpha}{1-\alpha}\right)^j}{\Gamma(\alpha j+1)} B(2-\beta, 1+j\alpha) t^{j\alpha}$$

$$< \frac{AB(\alpha)}{\beta(1-\alpha)} \left\| \frac{df}{dt} \right\|_\infty \frac{1}{t^\beta} \sum_{j=0}^{\infty} \frac{\left(\frac{-\alpha}{1-\alpha}\right)^j}{\Gamma(\alpha j+1)} \frac{\Gamma(2-\beta)\Gamma(1+j\alpha)}{\Gamma(2-\beta+1+j\alpha)} t^{j\alpha}$$

$$< \frac{AB(\alpha)}{\beta(1-\alpha)} \left\| \frac{df}{dt} \right\|_\infty \frac{\Gamma(2-\beta)}{t^\beta} \sum_{j=0}^{\infty} \frac{\left(\frac{-\alpha t^\alpha}{1-\alpha}\right)^j}{\Gamma(\alpha j+3-\beta)}$$

$$< \frac{AB(\alpha)}{\beta(1-\alpha)} \left\| \frac{df}{dt} \right\|_\infty \frac{\Gamma(2-\beta)}{t^\beta} E_{\alpha,3-\beta} \left(\frac{-\alpha}{1-\alpha} t^\alpha \right).$$

We investigate the following problem:

$${}_0^{FFEC}D_t^{\alpha,\beta} f(t) = \frac{M(\alpha)}{1-\alpha} \int_0^t \frac{df(\tau)}{d\tau^\beta} \exp\left(\frac{-\alpha}{1-\alpha}(t-\tau) \right) d\tau. \tag{3.5}$$

Assuming that f is differentiable. Then, we get

$${}_0^{FFEC}D_t^{\alpha,\beta} f(t) = \frac{M(\alpha)}{1-\alpha} \int_0^t \frac{df(\tau)}{d\tau} \frac{\tau^{1-\beta}}{\beta} \exp\left(\frac{-\alpha}{1-\alpha}(t-\tau) \right) d\tau. \tag{3.6}$$

Then, we obtain

$$\left| {}_0^{FFEC}D_t^{\alpha,\beta} f(t) \right| = \frac{M(\alpha)}{1-\alpha} \left| \int_0^t \frac{df(\tau)}{d\tau} \frac{\tau^{1-\beta}}{\beta} \exp\left(\frac{-\alpha}{1-\alpha}(t-\tau) \right) d\tau \right|$$

$$\leq \frac{M(\alpha)}{1-\alpha} \int_0^t \left| \frac{df(\tau)}{d\tau} \right| \frac{\tau^{1-\beta}}{\beta} \exp\left(\frac{-\alpha}{1-\alpha}(t-\tau) \right) d\tau$$

$$< \frac{M(\alpha)}{1-\alpha} \int_0^t \sup_{\tau\in[0,1]} \left| \frac{df(\tau)}{d\tau} \right| \frac{\tau^{1-\beta}}{\beta} \exp\left(\frac{-\alpha}{1-\alpha}(t-\tau) \right) d\tau$$

$$< \frac{M(\alpha)}{\beta(1-\alpha)} \left\| \frac{df}{dt} \right\|_\infty \int_0^t \tau^{1-\beta} \exp\left(\frac{-\alpha}{1-\alpha}(t-\tau) \right) d\tau$$

$$< \frac{M(\alpha)}{\beta(1-\alpha)}\left\|\frac{df}{dt}\right\|_{\infty}\int_0^t \tau^{1-\beta}\sum_{j=0}^{\infty}\frac{\left(\frac{-\alpha}{1-\alpha}(t-\tau)\right)^j}{j!}d\tau$$

$$< \frac{M(\alpha)}{\beta(1-\alpha)}\left\|\frac{df}{dt}\right\|_{\infty}\sum_{j=0}^{\infty}\frac{\left(\frac{-\alpha}{1-\alpha}\right)^j}{j!}\int_0^t (t-\tau)^j\tau^{1-\beta}d\tau$$

$$< \frac{M(\alpha)}{\beta(1-\alpha)}\left\|\frac{df}{dt}\right\|_{\infty}\sum_{j=0}^{\infty}\frac{\left(\frac{-\alpha}{1-\alpha}\right)^j}{j!}\int_0^1 (t-th)^j(th)^{1-\beta}t\,dh$$

$$< \frac{M(\alpha)t^{2-\beta}}{\beta(1-\alpha)}\left\|\frac{df}{dt}\right\|_{\infty}\sum_{j=0}^{\infty}\frac{\left(\frac{-\alpha}{1-\alpha}\right)^j}{j!}t^j\int_0^1 (1-h)^jh^{1-\beta}dh$$

$$< \frac{M(\alpha)t^{2-\beta}}{\beta(1-\alpha)}\left\|\frac{df}{dt}\right\|_{\infty}\sum_{j=0}^{\infty}\frac{\left(\frac{-\alpha}{1-\alpha}\right)^j}{j!}t^jB(2-\beta,1+j)$$

$$< \frac{M(\alpha)t^{2-\beta}}{\beta(1-\alpha)}\left\|\frac{df}{dt}\right\|_{\infty}\sum_{j=0}^{\infty}\frac{\left(\frac{-\alpha}{1-\alpha}\right)^j}{j!}t^j\frac{\Gamma(2-\beta)\Gamma(1+j)}{\Gamma(3-\beta+j)}$$

$$< \frac{M(\alpha)t^{2-\beta}\Gamma(2-\beta)}{\beta(1-\alpha)}\left\|\frac{df}{dt}\right\|_{\infty}\sum_{j=0}^{\infty}\frac{\left(\frac{-\alpha}{1-\alpha}t\right)^j}{\Gamma(3-\beta+j)}$$

$$< \frac{M(\alpha)t^{2-\beta}\Gamma(2-\beta)}{\beta(1-\alpha)}\left\|\frac{df}{dt}\right\|_{\infty}E_{1,3-\beta}\left(\frac{-\alpha}{1-\alpha}t\right).$$

Theorem 3.1 *If f is decreasing, increasing, and constant, we want to prove that the following fractal-fractional derivative is negative, positive, and zero, respectively.*

$${}_0^{FFPC}D_t^{\alpha,\beta}f(t) < 0. \tag{3.7}$$

$${}_0^{FFPC}D_t^{\alpha,\beta}f(t) > 0. \tag{3.8}$$

$${}_0^{FFPC}D_t^{\alpha,\beta}f(t) = 0. \tag{3.9}$$

Proof: We have

$${}_0^{FFPC}D_t^{\alpha,\beta}f(t) = \frac{1}{\Gamma(1-\alpha)}\int_0^t \frac{df(\tau)}{d\tau^\beta}(t-\tau)^{-\alpha}d\tau. \tag{3.10}$$

Assume that f is differentiable, then we have

$${}_0^{FFPC}D_t^{\alpha,\beta}f(t) = \frac{1}{\Gamma(1-\alpha)}\int_0^t \frac{df(\tau)}{d\tau}\frac{\tau^{1-\beta}}{\beta}(t-\tau)^{-\alpha}d\tau. \tag{3.11}$$

Since $t > \tau$, we have $(t-\tau)^{-\alpha} > 0$. Additionally, $\tau > 0$. This implies that

$$\tau^{\beta-1}(t-\tau)^{-\alpha} > 0.$$

On the other hand, f is decreasing. Therefore, we have $\frac{df(\tau)}{d\tau} < 0$. This implies that

$$\frac{df(\tau)}{d\tau}\tau^{\beta-1}(t-\tau)^{-\alpha} < 0 \quad \text{for all } \tau.$$

Therefore, we reach

$$\int_0^t \frac{df(\tau)}{d\tau}\frac{\tau^{1-\beta}}{\beta}(t-\tau)^{-\alpha}d\tau < 0.$$

We conclude that if f is decreasing, then the fractal-fractional derivative is negative. If f is increasing and constant, the proof is similar to this proof. Therefore, we omitted to prove for increasing and constant case. □

We consider the following problem:

$${}_0^{FFPC}D_t^{\alpha,\beta}f(t)g(t) = \frac{1}{\Gamma(1-\alpha)}\int_0^t \frac{df(\tau)g(\tau)}{d\tau^{\beta}}(t-\tau)^{-\alpha}d\tau. \tag{3.12}$$

Assuming that fg is differentiable. Then, we get

$${}_0^{FFPC}D_t^{\alpha,\beta}f(t)g(t) = \frac{1}{\Gamma(1-\alpha)}\int_0^t \frac{df(\tau)g(\tau)}{d\tau}\frac{\tau^{1-\beta}}{\beta}(t-\tau)^{-\alpha}d\tau. \tag{3.13}$$

If f and g are continuous and bounded, then we obtain

$${}_0^{FFPC}D_t^{\alpha,\beta}f(t)g(t) = \frac{1}{\Gamma(1-\alpha)}\int_0^t (f'(\tau)g(\tau)+f(\tau)g'(\tau))\frac{\tau^{1-\beta}}{\beta}(t-\tau)^{-\alpha}d\tau. \tag{3.14}$$

Theorem 3.2 *Assume that f and g are differentiable and bounded. Then, we obtain*

$$|{}_0^{FFPC}D_t^{\alpha,\beta}f(t)g(t)| < \|g\|_{\infty}{}_0^{FFPC}D_t^{\alpha,\beta}f(t) + \|f\|_{\infty}{}_0^{FFPC}D_t^{\alpha,\beta}g(t).$$

Proof: We obtain

$$\begin{aligned}|{}_0^{FFPC}D_t^{\alpha,\beta}f(t)g(t)| &= \frac{1}{\Gamma(1-\alpha)}\left|\int_0^t (f'(\tau)g(\tau)+f(\tau)g'(\tau))\frac{\tau^{1-\beta}}{\beta}(t-\tau)^{-\alpha}d\tau\right| \\ &< \frac{1}{\Gamma(1-\alpha)}\int_0^t \left(f'(\tau)\sup_{\tau\in[0,t]}|g(\tau)| + \sup_{\tau\in[0,t]}|f(\tau)|g'(\tau)\right) \\ &\quad \frac{\tau^{1-\beta}}{\beta}(t-\tau)^{-\alpha}d\tau\end{aligned}$$

$$
\begin{aligned}
&< \frac{1}{\Gamma(1-\alpha)} \int_0^t (f'(\tau) \sup_{\tau \in [0,t]} |g(\tau)| f'(\tau)) \frac{\tau^{1-\beta}}{\beta} (t-\tau)^{-\alpha} d\tau \\
&+ \frac{1}{\Gamma(1-\alpha)} \int_0^t (\sup_{\tau \in [0,t]} |f(\tau)| g'(\tau)) \frac{\tau^{1-\beta}}{\beta} (t-\tau)^{-\alpha} d\tau \\
&< \|g\|_{\infty} {}_0^{FFPC}D_t^{\alpha,\beta} f(t) + \|f\|_{\infty} {}_0^{FFPC}D_t^{\alpha,\beta} g(t).
\end{aligned}
$$

□

This result can also be proven for fractal-fractional derivatives with exponential decay and generalized Mittag-Leffler kernels.

Theorem 3.3 *If f and g are differentiable and satisfy the following condition*

$$\left\| \frac{df}{dt} - \frac{dg}{dt} \right\|_{\infty} < K \|f - g\|_{\infty}, \tag{3.15}$$

then we have

$$\|{}_0^{FFPC}D_t^{\alpha,\beta} f(t) - {}_0^{FFPC} D_t^{\alpha,\beta} g(t)\|_{\infty} < K \|f - g\|_{\infty}. \tag{3.16}$$

Proof: We have

$$
\begin{aligned}
\left\| \begin{matrix} {}_0^{FFPC}D_t^{\alpha,\beta} f(t) - \\ {}_0^{FFPC}D_t^{\alpha,\beta} g(t) \end{matrix} \right\|_{\infty} &= \left\| \frac{1}{\Gamma(1-\alpha)} \int_0^t \frac{df(\tau)}{d\tau} \frac{\tau^{1-\beta}}{\beta} (t-\tau)^{-\alpha} \right. \\
&\quad \left. - \frac{1}{\Gamma(1-\alpha)} \int_0^t \frac{dg(\tau)}{d\tau} \frac{\tau^{1-\beta}}{\beta} (t-\tau)^{-\alpha} \right\|_{\infty} \\
&= \frac{1}{\beta(\Gamma(1-\alpha))} \left\| \int_0^t \tau^{1-\beta} (t-\tau)^{-\alpha} \left[\frac{df(\tau)}{d\tau} - \frac{dg(\tau)}{d\tau} \right] d\tau \right\|_{\infty} \\
&\le \frac{1}{\beta(\Gamma(1-\alpha))} \int_0^t \tau^{1-\beta} (t-\tau)^{-\alpha} \left\| \frac{df(\tau)}{d\tau} - \frac{dg(\tau)}{d\tau} \right\|_{\infty} d\tau \\
&< \frac{1}{\beta(\Gamma(1-\alpha))} \left\| \frac{df}{dt} - \frac{dg}{dt} \right\|_{\infty} \int_0^t \tau^{1-\beta} (t-\tau)^{-\alpha} d\tau \\
&< \frac{1}{\beta(\Gamma(1-\alpha))} \left\| \frac{df}{dt} - \frac{dg}{dt} \right\|_{\infty} \int_0^1 (th)^{1-\beta} (t-th)^{-\alpha} t \, dh \\
&< \frac{t_0^{1-\alpha-\beta}}{\beta(\Gamma(1-\alpha))} \left\| \frac{df}{dt} - \frac{dg}{dt} \right\|_{\infty} B(2-\beta, 1-\alpha) \\
&< K \left\| \frac{df}{dt} - \frac{dg}{dt} \right\|_{\infty} \\
&< K \|f - g\|_{\infty}.
\end{aligned}
$$

□

Theorem 3.4 *Let f be analytic around 0, then we have*

$$ {}_{0}^{FFMC}D_t^{\alpha,\beta}f(t) = \frac{AB(\alpha)}{\beta(1-\alpha)}\sum_{j=0}^{\infty} a_j j t^{j+\beta-1}\Gamma(j+\beta-1)E_{\alpha,j+\beta}\left[\frac{-\alpha t^{\alpha}}{1-\alpha}\right]. \tag{3.17} $$

Proof: We have

$$ \begin{aligned} {}_{0}^{FFMC}D_t^{\alpha,\beta}f(t) &= \frac{AB(\alpha)}{1-\alpha}\int_0^t \frac{\tau^{\beta-1}}{\beta}\frac{d\left(\sum_{j=}^{\infty} a_j\tau^j\right)}{d\tau}E_{\alpha}\left(\frac{-\alpha(t-\tau)^{\alpha}}{1-\alpha}\right)d\tau \\ &= \frac{AB(\alpha)}{\beta(1-\alpha)}\int_0^t \tau^{\beta-1}\sum_{j=0}^{\infty} ja_j\tau^{j-1}E_{\alpha}\left(\frac{-\alpha(t-\tau)^{\alpha}}{1-\alpha}\right)d\tau \\ &= \frac{AB(\alpha)}{\beta(1-\alpha)}\int_0^t \sum_{j=0}^{\infty} ja_j\tau^{\beta+j-2}E_{\alpha}\left(\frac{-\alpha(t-\tau)^{\alpha}}{1-\alpha}\right)d\tau \\ &= \frac{AB(\alpha)}{\beta(1-\alpha)}\sum_{j=0}^{\infty} ja_j\int_0^t \tau^{\beta+j-2}E_{\alpha}\left(\frac{-\alpha(t-\tau)^{\alpha}}{1-\alpha}\right)d\tau \\ &= \frac{AB(\alpha)}{\beta(1-\alpha)}\sum_{j=0}^{\infty} ja_j\int_0^t \tau^{\beta+j-2}\sum_{k=0}^{\infty}\left(\frac{-\alpha}{1-\alpha}\right)^k\frac{(t-\tau)^{\alpha k}}{\Gamma(\alpha k+1)}d\tau \\ &= \frac{AB(\alpha)}{\beta(1-\alpha)}\sum_{j=0}^{\infty} ja_j\sum_{k=0}^{\infty}\left(\frac{-\alpha}{1-\alpha}\right)^k\frac{1}{\Gamma(\alpha k+1)}\int_0^t \tau^{\beta+j-2}(t-\tau)^{\alpha k}d\tau. \end{aligned} $$

We let $\tau = ht$. Then we obtain

$$ \begin{aligned} {}_{0}^{FFMC}D_t^{\alpha,\beta}f(t) &= \frac{AB(\alpha)}{\beta(1-\alpha)}\sum_{j=0}^{\infty} ja_j\sum_{k=0}^{\infty}\left(\frac{-\alpha}{1-\alpha}\right)^k\frac{1}{\Gamma(\alpha k+1)} \\ &\quad\times\int_0^t (ht)^{\beta+j-2}(t-ht)^{\alpha k}t\,dh \\ &= \frac{AB(\alpha)}{\beta(1-\alpha)}\sum_{j=0}^{\infty} ja_j\sum_{k=0}^{\infty}\left(\frac{-\alpha}{1-\alpha}\right)^k\frac{1}{\Gamma(\alpha k+1)}t^{j+\beta+k\alpha-1} \\ &\quad\times\int_0^t h^{\beta+j-2}(1-h)^{\alpha k}dh \\ &= \frac{AB(\alpha)}{\beta(1-\alpha)}\sum_{j=0}^{\infty} ja_j\sum_{k=0}^{\infty}\left(\frac{-\alpha}{1-\alpha}\right)^k\frac{1}{\Gamma(\alpha k+1)}t^{j+\beta+k\alpha-1} \\ &\quad\times B(j+\beta-1,\alpha k+1) \\ &= \frac{AB(\alpha)}{\beta(1-\alpha)}\sum_{j=0}^{\infty} ja_j\sum_{k=0}^{\infty}\left(\frac{-\alpha}{1-\alpha}\right)^k\frac{1}{\Gamma(\alpha k+1)}t^{j+\beta+k\alpha-1} \\ &\quad\times\frac{\Gamma(j+\beta-1)\Gamma(\alpha k+1)}{\Gamma(j+\beta+\alpha k)} \end{aligned} $$

$$= \frac{AB(\alpha)}{\beta(1-\alpha)} \sum_{j=0}^{\infty} j a_j t^{j+\beta-1} \Gamma(j+\beta-1) E_{\alpha,j+\beta}\left[\frac{-\alpha}{a-\alpha} t^{\alpha}\right].$$

This completes the proof. □

Theorem 3.5 *Let f be analytic around* 0, *then we get*

$${}_{0}^{FFPC}D_{t}^{\alpha,\beta} f(t) = \frac{t^{1-\alpha-\beta}}{\beta\Gamma(1-\alpha)} \sum_{j=0}^{\infty} a_j j t^j B(j-\beta+1, 1-\alpha). \tag{3.18}$$

Proof: We have

$${}_{0}^{FFPC}D_{t}^{\alpha,\beta} f(t) = \frac{1}{\Gamma(1-\alpha)} \int_0^t \frac{df(\tau)}{d\tau^{\beta}} (t-\tau)^{-\alpha} d\tau. \tag{3.19}$$

If f is differentiable, then we will get

$$\begin{aligned}
{}_{0}^{FFPC}D_{t}^{\alpha,\beta} f(t) &= \frac{1}{\beta\Gamma(1-\alpha)} \int_0^t \frac{df(\tau)}{d\tau} \tau^{1-\beta} (t-\tau)^{-\alpha} d\tau \\
&= \frac{1}{\beta\Gamma(1-\alpha)} \int_0^t \frac{d\left(\sum_{j=0}^{\infty} a_j \tau^j\right)}{d\tau} \tau^{1-\beta} (t-\tau)^{-\alpha} d\tau \\
&= \frac{1}{\beta\Gamma(1-\alpha)} \int_0^t \left(\sum_{j=0}^{\infty} a_j j \tau^{j-1}\right) \tau^{1-\beta} (t-\tau)^{-\alpha} d\tau \\
&= \frac{1}{\beta\Gamma(1-\alpha)} \sum_{j=0}^{\infty} a_j j \int_0^t \tau^{j-\beta} (t-\tau)^{-\alpha} d\tau.
\end{aligned}$$

Let choose $\tau = th$. Then, we will obtain

$$\begin{aligned}
{}_{0}^{FFPC}D_{t}^{\alpha,\beta} f(t) &= \frac{1}{\beta\Gamma(1-\alpha)} \sum_{j=0}^{\infty} a_j j \int_0^1 (th)^{j-\beta} (t-th)^{-\alpha} t \, dh \\
&= \frac{t^{1-\alpha-\beta}}{\beta\Gamma(1-\alpha)} \sum_{j=0}^{\infty} a_j j t^j \int_0^1 (h)^{j-\beta} (1-h)^{-\alpha} dh \\
&= \frac{t^{1-\alpha-\beta}}{\beta\Gamma(1-\alpha)} \sum_{j=0}^{\infty} a_j j t^j B(j-\beta+1, 1-\alpha) \\
&= \frac{t^{1-\alpha-\beta}}{\beta\Gamma(1-\alpha)} \sum_{j=0}^{\infty} a_j j t^j \frac{\Gamma(j-\beta+1)\Gamma(1-\alpha)}{\Gamma(2+j-\beta-\alpha)} \\
&= \frac{t^{1-\alpha-\beta}}{\beta} \sum_{j=0}^{\infty} a_j j t^j \frac{\Gamma(j-\beta+1)}{\Gamma(2+j-\beta-\alpha)}.
\end{aligned}$$

□

Theorem 3.6 *Let f be analytic around 0, then we get*

$$ {}_0^{FFEC}D_t^{\alpha,\beta}f(t) = \frac{M(\alpha)}{\beta(1-\alpha)}\sum_{j=0}^{\infty} a_j j t^{j-\beta+1}\Gamma(j-\beta+1)E_{1,2+j-\beta}\left(\frac{-\alpha}{1-\alpha}t\right). \tag{3.20} $$

Proof: We have

$$ {}_0^{FFEC}D_t^{\alpha,\beta}f(t) = \frac{M(\alpha)}{1-\alpha}\int_0^t \frac{df(\tau)}{d\tau^\beta}\exp\left(\frac{-\alpha}{1-\alpha}(t-\tau)\right)d\tau. \tag{3.21} $$

If f is differentiable, then we will get

$$ \begin{aligned} {}_0^{FFEC}D_t^{\alpha,\beta}f(t) &= \frac{M(\alpha)}{\beta(1-\alpha)}\int_0^t \frac{df(\tau)}{d\tau}\tau^{1-\beta}\exp\left(\frac{-\alpha}{1-\alpha}(t-\tau)\right)d\tau \\ &= \frac{M(\alpha)}{\beta(1-\alpha)}\int_0^t \frac{d\left(\sum_{j=0}^{\infty} a_j\tau^j\right)}{d\tau}\tau^{1-\beta}\exp\left(\frac{-\alpha}{1-\alpha}(t-\tau)\right)d\tau \\ &= \frac{M(\alpha)}{\beta(1-\alpha)}\int_0^t \left(\sum_{j=0}^{\infty} a_j j\tau^{j-1}\right)\tau^{1-\beta}\exp\left(\frac{-\alpha}{1-\alpha}(t-\tau)\right)d\tau \\ &= \frac{M(\alpha)}{\beta(1-\alpha)}\sum_{j=0}^{\infty} a_j j\int_0^t \tau^{j-\beta}\exp\left(\frac{-\alpha}{1-\alpha}(t-\tau)\right)d\tau \\ &= \frac{M(\alpha)}{\beta(1-\alpha)}\sum_{j=0}^{\infty} a_j j\int_0^t \tau^{j-\beta}\sum_{k=0}^{\infty}\frac{\left(\frac{-\alpha}{1-\alpha}\right)^k}{k!}(t-\tau)^k d\tau \\ &= \frac{M(\alpha)}{\beta(1-\alpha)}\sum_{j=0}^{\infty} a_j j\sum_{k=0}^{\infty}\frac{\left(\frac{-\alpha}{1-\alpha}\right)^k}{k!}\int_0^t \tau^{j-\beta}(t-\tau)^k d\tau. \end{aligned} $$

We let $\tau = th$. Then, we obtain

$$ \begin{aligned} {}_0^{FFEC}D_t^{\alpha,\beta}f(t) &= \frac{M(\alpha)}{\beta(1-\alpha)}\sum_{j=0}^{\infty} a_j j\sum_{k=0}^{\infty}\frac{\left(\frac{-\alpha}{1-\alpha}\right)^k}{k!}\int_0^1 (th)^{j-\beta}(t-th)^k t\,dh \\ &= \frac{M(\alpha)}{\beta(1-\alpha)}\sum_{j=0}^{\infty} a_j j t^{j-\beta+1}\sum_{k=0}^{\infty}\frac{\left(\frac{-\alpha t}{1-\alpha}\right)^k}{k!}\int_0^1 h^{j-\beta}(1-h)^k dh \\ &= \frac{M(\alpha)}{\beta(1-\alpha)}\sum_{j=0}^{\infty} a_j j t^{j-\beta+1}\sum_{k=0}^{\infty}\frac{\left(\frac{-\alpha t}{1-\alpha}\right)^k}{k!}B(j-\beta+1,k+1) \end{aligned} $$

$$= \frac{M(\alpha)}{\beta(1-\alpha)} \sum_{j=0}^{\infty} a_j j t^{j-\beta+1} \sum_{k=0}^{\infty} \frac{\left(\frac{-\alpha t}{1-\alpha}\right)^k}{k!} \frac{\Gamma(j-\beta+1)\Gamma(k+1)}{\Gamma(2+j-\beta+k)}$$

$$= \frac{M(\alpha)}{\beta(1-\alpha)} \sum_{j=0}^{\infty} a_j j t^{j-\beta+1} \Gamma(j-\beta+1) E_{1,2+j-\beta}\left(\frac{-\alpha}{1-\alpha} t\right).$$

□

3.3 Discretization

We consider the following fractal-fractional derivative with the power law kernel in the Caputo sense as:

$$ {}_0^{FFPC}D_t^{\alpha,\beta} f(t) = \frac{1}{\Gamma(1-\alpha)} \int_0^t \frac{df(\tau)}{d\tau^\beta} (t-\tau)^{-\alpha} d\tau. \tag{3.22}$$

We assume f to be differentiable, then we get

$$ {}_0^{FFPC}D_t^{\alpha,\beta} f(t) = \frac{1}{\beta\Gamma(1-\alpha)} \int_0^t \frac{df(\tau)}{d\tau} \tau^{1-\beta} (t-\tau)^{-\alpha} d\tau. \tag{3.23}$$

We put $t_n = n\Delta t$, then at t_{n+1}, we have

$$ {}_0^{FFPC}D_t^{\alpha,\beta} f(t_{n+1}) = \frac{1}{\beta\Gamma(1-\alpha)} \int_0^{t_{n+1}} \frac{df(\tau)}{d\tau} \tau^{1-\beta} (t_{n+1}-\tau)^{-\alpha} d\tau$$

$$= \frac{1}{\beta\Gamma(1-\alpha)} \sum_{j=0}^{n} \int_{t_j}^{t_{j+1}} \frac{f^{j+1}-f^j}{\Delta t} t_j^{1-\beta} (t_{n+1}-\tau)^{-\alpha} d\tau$$

$$= \frac{1}{\beta\Gamma(1-\alpha)} \sum_{j=0}^{n} \frac{f^{j+1}-f^j}{\Delta t} t_j^{1-\beta} \int_{t_j}^{t_{j+1}} (t_{n+1}-\tau)^{-\alpha} d\tau$$

$$= \frac{(\Delta t)^{-\alpha}}{\beta\Gamma(2-\alpha)} \sum_{j=0}^{n} t_j^{1-\beta} (f^{j+1}-f^j)[(n-j+1)^{1-\alpha} - (n-j)^{1-\alpha}].$$

We consider the following fractal-fractional derivative with the exponential decay kernel in the Caputo sense as:

$$ {}_0^{FFEC}D_t^{\alpha,\beta} f(t) = \frac{M(\alpha)}{1-\alpha} \int_0^t \frac{df(\tau)}{d\tau^\beta} \exp\left(\frac{-\alpha}{1-\alpha}(t-\tau)\right) d\tau. \tag{3.24}$$

We assume f to be differentiable, then we get

$$ {}_0^{FFEC}D_t^{\alpha,\beta} f(t) = \frac{M(\alpha)}{\beta(1-\alpha)} \int_0^t \frac{df(\tau)}{d\tau} \tau^{1-\beta} \exp\left(\frac{-\alpha}{1-\alpha}(t-\tau)\right) d\tau. \tag{3.25}$$

We put $t_n = n\Delta t$, then at t_{n+1}, we have

$$\begin{aligned}
{}_{0}^{FFEC}D_t^{\alpha,\beta}f(t_{n+1}) &= \frac{M(\alpha)}{\beta(1-\alpha)}\int_0^{t_{n+1}}\frac{df(\tau)}{d\tau}\tau^{1-\beta}\exp\left(\frac{-\alpha}{1-\alpha}(t_{n+1}-\tau)\right)d\tau\\
&= \frac{M(\alpha)}{\beta(1-\alpha)}\sum_{j=0}^{n}\int_{t_j}^{t_{j+1}}\frac{f^{j+1}-f^j}{\Delta t}t_j^{1-\beta}\exp\left(\frac{-\alpha}{1-\alpha}(t_{n+1}-\tau)\right)d\tau\\
&= \frac{M(\alpha)}{\beta(1-\alpha)}\sum_{j=0}^{n}\frac{f^{j+1}-f^j}{\Delta t}t_j^{1-\beta}\int_{t_j}^{t_{j+1}}\exp\left(\frac{-\alpha}{1-\alpha}(t_{n+1}-\tau)\right)d\tau\\
&= \frac{-M(\alpha)}{\beta}\sum_{j=0}^{n}\frac{f^{j+1}-f^j}{\Delta t}t_j^{1-\beta}\left[\exp\left(\frac{-\alpha}{1-\alpha}(n+1-j)\Delta t\right)\right.\\
&\quad\left.-\exp\left(\frac{-\alpha}{1-\alpha}(n-j)\Delta t\right)\right].
\end{aligned}$$

We consider the following fractal-fractional derivative with the generalized Mittag-Leffler kernel in the Caputo sense as:

$$ {}_{0}^{FFMC}D_t^{\alpha,\beta}f(t) = \frac{AB(\alpha)}{1-\alpha}\int_0^t\frac{df(\tau)}{d\tau^\beta}E_\alpha\left(\frac{-\alpha}{1-\alpha}(t-\tau)^\alpha\right)d\tau. \tag{3.26}$$

We assume f to be differentiable, then we get

$$ {}_{0}^{FFMC}D_t^{\alpha,\beta}f(t) = \frac{AB(\alpha)}{\beta(1-\alpha)}\int_0^t\frac{df(\tau)}{d\tau}\tau^{1-\beta}E_\alpha\left(\frac{-\alpha}{1-\alpha}(t-\tau)^\alpha\right)d\tau. \tag{3.27}$$

We put $t_n = n\Delta t$, then at t_{n+1}, we have

$$\begin{aligned}
{}_{0}^{FFMC}D_t^{\alpha,\beta}f(t_{n+1}) &= \frac{AB(\alpha)}{\beta(1-\alpha)}\int_0^{t_{n+1}}\frac{df(\tau)}{d\tau}\tau^{1-\beta}E_\alpha\left(\frac{-\alpha}{1-\alpha}(t_{n+1}-\tau)^\alpha\right)d\tau\\
&= \frac{AB(\alpha)}{\beta(1-\alpha)}\sum_{j=0}^{n}\int_{t_j}^{t_{j+1}}\frac{f^{j+1}-f^j}{\Delta t}t_j^{1-\beta}\\
&\quad\times\sum_{k=0}^{\infty}\frac{\left(\frac{-\alpha}{1-\alpha}\right)^k}{\Gamma(\alpha k+1)}(t_{n+1}-\tau)^{\alpha k}d\tau\\
&= \frac{AB(\alpha)}{\beta(1-\alpha)}\sum_{j=0}^{n}\frac{f^{j+1}-f^j}{\Delta t}t_j^{1-\beta}\\
&\quad\times\sum_{k=0}^{\infty}\frac{\left(\frac{-\alpha}{1-\alpha}\right)^k}{\Gamma(\alpha k+1)}\int_{t_j}^{t_{j+1}}(t_{n+1}-\tau)^{\alpha k}d\tau\\
&= \frac{AB(\alpha)}{\beta(1-\alpha)}\sum_{j=0}^{n}\frac{f^{j+1}-f^j}{\Delta t}t_j^{1-\beta}\sum_{k=0}^{\infty}\frac{\left(\frac{-\alpha}{1-\alpha}\right)^k}{\Gamma(\alpha k+2)}\\
&\quad\times[((n+1-j)\Delta t)^{\alpha k+1}-((n-j)\Delta t)^{\alpha k+1}]
\end{aligned}$$

$$= \frac{AB(\alpha)}{\beta(1-\alpha)} \sum_{j=0}^{n} \frac{f^{j+1} - f^{j}}{\Delta t} t_j^{1-\beta}$$

$$\left[(n+1-j)\Delta t E_{\alpha,2}\left(\frac{-\alpha}{1-\alpha}(n+1-j)\Delta t\right)\right.$$

$$\left.-(n-j)\Delta t E_{\alpha,2}\left(\frac{-\alpha}{1-\alpha}(n-j)\Delta t\right)\right]$$

$$= \frac{AB(\alpha)}{\beta(1-\alpha)} \sum_{j=0}^{n} (f^{j+1} - f^{j}) t_j^{1-\beta}$$

$$\left[(n+1-j) E_{\alpha,2}\left(\frac{-\alpha}{1-\alpha}(n+1-j)\Delta t\right)\right.$$

$$\left.-(n-j) E_{\alpha,2}\left(\frac{-\alpha}{1-\alpha}(n-j)\Delta t\right)\right].$$

3.4 Experiments

We consider the following problem with the power law kernel as:

$$ {}_{0}^{FFPC}D_t^{\alpha,\beta} u(x,t) = f(x,t,u(x,t)), \tag{3.28}$$

where $u(x,0) = g(x)$, $x_m - x_{m-1} = \Delta x$, $t_{n+1} - t_n = \Delta t$, $t_n = n\Delta t$, $x_m = m\Delta x$. The above equation can be approximated as:

$$\sum_{j=0}^{n} (u_m^{j+1} - u_m^{j}) \frac{(j\Delta t)^{1-\beta}(\Delta t)^{-\alpha}}{\beta\Gamma(2-\alpha)} [(n-j+1)^{1-\alpha} - (n-j)^{1-\alpha}] = f(x_m, t_{n+1}, u_m^{n+1}).$$

We consider the following problem with the exponential decay kernel as:

$$ {}_{0}^{FFEC}D_t^{\alpha,\beta} u(x,t) = f(x,t,u(x,t)). \tag{3.29}$$

The above equation can be approximated as:

$$\frac{-M(\alpha)}{\beta} \sum_{j=0}^{n} \frac{u_m^{j+1} - u_m^{j}}{\Delta t} (j\Delta t)^{1-\beta}$$

$$\left[\exp\left(\frac{-\alpha}{1-\alpha}(n+1-j)\Delta t\right) - \exp\left(\frac{-\alpha}{1-\alpha}(n-j)\Delta t\right)\right]$$

$$= f(x_m, t_{n+1}, u_m^{n+1}).$$

We consider the following problem with the generalized Mittag-Leffler kernel as:

$$ {}_{0}^{FFMC}D_t^{\alpha,\beta} u(x,t) = f(x,t,u(x,t)). \tag{3.30}$$

The above equation can be approximated as:

$$\frac{AB(\alpha)}{\beta(1-\alpha)}\sum_{j=0}^{n}(u_m^{j+1}-u_m^j)(j\Delta t)^{1-\beta}\left[(n+1-j)E_{\alpha,2}\left(\frac{-\alpha}{1-\alpha}(n+1-j)\Delta t\right)\right.$$
$$\left.-(n-j)E_{\alpha,2}\left(\frac{-\alpha}{1-\alpha}(n-j)\Delta t\right)\right]$$
$$=f(x_m,t_{n+1},u_m^{n+1}).$$

3.5 Stability Analysis

We discretize the following problems and investigate the stability analysis of them with the power law, exponential decay, and generalized Mittag-Leffler kernels. First, we consider the following fractal-fractional differential equation with the power law kernel.

$$_0^{FFPC}D_t^{\alpha,\beta}u(x,t)=\frac{\partial^2 u(x,t)}{\partial x^2}+x\frac{\partial u(x,t)}{\partial x}+\lambda u(x,t). \tag{3.31}$$

At (t_{n+1},x_m), we have

$$\sum_{j=0}^{n}(u_m^{j+1}-u_m^j)\frac{(j\Delta t)^{1-\beta}(\Delta t)^{-\alpha}}{\beta\Gamma(2-\alpha)}[(n-j+1)^{1-\alpha}-(n-j)^{1-\alpha}]$$
$$=\frac{u_{m+1}^{n+1}-2u_m^{n+1}+u_{m-1}^{n+1}}{(\Delta x)^2}+x_m\frac{u_{m+1}^{n+1}-u_{m-1}^{n+1}}{2\Delta x}+\lambda u_m^{n+1}.$$

We put $u_m^n=\delta_n\exp(ik_m x)$. Replacing in the above equation, we get

$$\sum_{j=0}^{n}(\delta_{j+1}\exp(ik_m x)-\delta_j\exp(ik_m x))\frac{(j\Delta t)^{1-\beta}(\Delta t)^{-\alpha}}{\beta\Gamma(2-\alpha)}[(n-j+1)^{1-\alpha}-(n-j)^{1-\alpha}]$$
$$=\frac{\delta_{n+1}\exp(ik_m(x+\Delta x))-2\delta_{n+1}\exp(ik_m x)+\delta_{j+1}\exp(ik_m(x-\Delta x))}{(\Delta x)^2}$$
$$+x_m\frac{\delta_{n+1}\exp(ik_m(x+\Delta x))-\delta_{n+1}\exp(ik_m(x-\Delta x))}{2\Delta x}+\lambda\delta_{n+1}\exp(ik_m x).$$

For simplicity, we take

$$\delta_{n,j}^{\alpha,\beta}=\frac{(j\Delta t)^{1-\beta}(\Delta t)^{-\alpha}}{\beta\Gamma(2-\alpha)}[(n-j+1)^{1-\alpha}-(n-j)^{1-\alpha}],\qquad a_1=\frac{1}{(\Delta x)^2},\quad a_2=\frac{x_m}{2\Delta x}.$$

Then, we get

$$\sum_{j=0}^{n}(\delta_{j+1}-\delta_j)\delta_{n,j}^{\alpha,\beta}=a_1(\delta_{n+1}\exp(ik_m\Delta x)-2\delta_{n+1}+\delta_{j+1}\exp(-ik_m\Delta x))$$
$$+a_2(\delta_{n+1}\exp(ik_m\Delta x)-\delta_{n+1}\exp(-ik_m\Delta x))+\lambda\delta_{n+1}.$$

Thus, we obtain

$$(\delta_{n+1}-\delta_n)\delta_{n,n}^{\alpha,\beta}+\sum_{j=0}^{n-1}(\delta_{j+1}-\delta_j)\delta_{nj}^{\alpha,\beta}$$
$$=\delta_{n+1}\left(-4a_1\sin^2\left(\frac{k_m\Delta x}{2}\right)+2a_2 i\sin(k_m\Delta x)+\lambda\right).$$

For $n=0$, we have

$$(\delta_1-\delta_0)\delta_{0,0}^{\alpha,\beta}=\delta_1\left(-4a_1\sin^2\left(\frac{k_m\Delta x}{2}\right)+2a_2 i\sin(k_m\Delta x)+\lambda\right),$$
$$\delta_1\left(\delta_{0,0}^{\alpha,\beta}+4a_1\sin^2\left(\frac{k_m\Delta x}{2}\right)-2a_2 i\sin(k_m\Delta x)-\lambda\right)=\delta_0\delta_{0,0}^{\alpha,\beta},$$

where $\left|\frac{\delta_1}{\delta_0}\right|<1$ implies

$$\frac{|\delta_{0,0}^{\alpha,\beta}|}{\sqrt{\left(\delta_{0,0}^{\alpha,\beta}+4a_1\sin^2\left(\frac{k_m\Delta x}{2}\right)-\lambda\right)^2+4a_2^2(\sin(k_m\Delta x))^2}}<1.$$

This is true for $\forall m$. Therefore, we obtain

$$\frac{|\delta_{0,0}^{\alpha,\beta}|}{\sqrt{(\delta_{0,0}^{\alpha,\beta}+4a_1-\lambda)^2+4a_2^2}}<1.$$

We assume that $\left|\frac{\delta_n}{\delta_0}\right|<1$. We need to show that $\left|\frac{\delta_{n+1}}{\delta_0}\right|<1$. We have

$$\delta_{n+1}\left(-4a_1\sin^2\left(\frac{k_m\Delta x}{2}\right)+2a_2 i\sin(k_m\Delta x)+\lambda\right)=\sum_{j=0}^{n}(\delta_{j+1}-\delta_j)\delta_{nj}^{\alpha,\beta}.$$

Then, we obtain

$$|\delta_{n+1}|\le\frac{\sum_{j=0}^{n}|\delta_{j+1}-\delta_j||\delta_{nj}^{\alpha,\beta}|}{\sqrt{\left(\lambda-4a_1\sin^2\left(\frac{k_m\Delta x}{2}\right)\right)^2+4a_2^2(\sin(k_m\Delta x))^2}}$$
$$<2\delta_0\frac{\sum_{j=0}^{n}|\delta_{nj}^{\alpha,\beta}|}{\sqrt{\left(\lambda-4a_1\sin^2\left(\frac{k_m\Delta x}{2}\right)\right)^2+4a_2^2(\sin(k_m\Delta x))^2}},$$

where $\left|\frac{\delta_{n+1}}{\delta_0}\right|<1$ implies

$$2\frac{\sum_{j=0}^{n}|\delta_{nj}^{\alpha,\beta}|}{\sqrt{\left(\lambda-4a_1\sin^2\left(\frac{k_m\Delta x}{2}\right)\right)^2+4a_2^2(\sin(k_m\Delta x))^2}}<1.$$

This is true for $\forall m$. Therefore, we obtain

$$2\frac{\sum_{j=0}^{n}\left|\delta_{n,j}^{\alpha,\beta}\right|}{\sqrt{(\lambda-4a_1)^2+4a_2^2}}<1.$$

The method is stable if

$$\min\left(\frac{\left|\delta_{0,0}^{\alpha,\beta}\right|}{\sqrt{\left(\delta_{0,0}^{\alpha,\beta}+4a_1-\lambda\right)^2+4a_2^2}},2\frac{\sum_{j=0}^{n}\left|\delta_{0,0}^{\alpha,\beta}\right|}{\sqrt{(\lambda-4a_1)^2+4a_2^2}}\right)<1.$$

We consider the following fractal-fractional differential equation with the exponential decay kernel.

$${}_{0}^{FFEC}D_t^{\alpha,\beta}u(x,t)=\frac{\partial^2 u(x,t)}{\partial x^2}+x\frac{\partial u(x,t)}{\partial x}+\lambda u(x,t). \tag{3.32}$$

At (t_{n+1},x_m), we have

$$\begin{aligned}&\frac{-M(\alpha)}{\beta}\sum_{j=0}^{n}\frac{u_m^{j+1}-u_m^j}{\Delta t}(j\Delta t)^{1-\beta}\\&\left[\exp\left(\frac{-\alpha}{1-\alpha}(n+1-j)\Delta t\right)-\exp\left(\frac{-\alpha}{1-\alpha}(n-j)\Delta t\right)\right]\\&=\frac{u_{m+1}^{n+1}-2u_m^{n+1}+u_{m-1}^{n+1}}{(\Delta x)^2}+x_m\frac{u_{m+1}^{n+1}-u_{m-1}^{n+1}}{2\Delta x}+\lambda u_m^{n+1}.\end{aligned}$$

We put $u_m^n=\delta_n\exp(ik_mx)$. Replacing in the above equation, we get

$$\begin{aligned}&\frac{-M(\alpha)}{\beta}\sum_{j=0}^{n}\frac{\delta_{n+1}\exp(ik_mx)-\delta_n\exp(ik_mx)}{\Delta t}(j\Delta t)^{1-\beta}\\&\times\left[\exp\left(\frac{-\alpha}{1-\alpha}(n+1-j)\Delta t\right)-\exp\left(\frac{-\alpha}{1-\alpha}(n-j)\Delta t\right)\right]\\&=\frac{\delta_{n+1}\exp(ik_m(x+\Delta x))-2\delta_{n+1}\exp(ik_mx)+\delta_{j+1}\exp(ik_m(x-\Delta x))}{(\Delta x)^2}\\&+x_m\frac{\delta_{n+1}\exp(ik_m(x+\Delta x))-\delta_{n+1}\exp(ik_m(x-\Delta x))}{2\Delta x}+\lambda\delta_{n+1}\exp(ik_mx).\end{aligned}$$

For simplicity, we take

$$\begin{aligned}A_{n,j}^{\alpha,\beta}&=\left[\exp\left(\frac{-\alpha}{1-\alpha}(n+1-j)\Delta t\right)-\exp\left(\frac{-\alpha}{1-\alpha}(n-j)\Delta t\right)\right]\\&\frac{-M(\alpha)}{\beta}\frac{1}{\Delta t}(j\Delta t)^{1-\beta},\\b_1&=\frac{1}{(\Delta x)^2},\quad b_2=\frac{x_m}{2\Delta x}.\end{aligned}$$

Then, we get

$$\sum_{j=0}^{n}(\delta_{j+1}-\delta_j)A_{n,j}^{\alpha,\beta}=b_1(\delta_{n+1}\exp(ik_m\Delta x)-2\delta_{n+1}+\delta_{j+1}\exp(-ik_m\Delta x))$$
$$+b_2(\delta_{n+1}\exp(ik_m\Delta x)-\delta_{n+1}\exp(-ik_m\Delta x))+\lambda\delta_{n+1}.$$

Thus, we obtain

$$(\delta_{n+1}-\delta_n)A_{n,n}^{\alpha,\beta}+\sum_{j=0}^{n-1}(\delta_{j+1}-\delta_j)A_{n,j}^{\alpha,\beta}$$
$$=\delta_{n+1}\left(-4b_1\sin^2\left(\frac{k_m\Delta x}{2}\right)+2b_2 i\sin(k_m\Delta x)+\lambda\right).$$

For $n=0$, we have

$$(\delta_1-\delta_0)A_{0,0}^{\alpha,\beta}=\delta_1\left(-4b_1\sin^2\left(\frac{k_m\Delta x}{2}\right)+2b_2 i\sin(k_m\Delta x)+\lambda\right).$$
$$\delta_1\left(A_{0,0}^{\alpha,\beta}+4b_1\sin^2\left(\frac{k_m\Delta x}{2}\right)-2b_2 i\sin(k_m\Delta x)-\lambda\right)=\delta_0A_{0,0}^{\alpha,\beta},$$

where $\left|\frac{\delta_1}{\delta_0}\right|<1$ implies

$$\frac{\left|A_{0,0}^{\alpha,\beta}\right|}{\sqrt{\left(A_{0,0}^{\alpha,\beta}+4b_1\sin^2\left(\frac{k_m\Delta x}{2}\right)-\lambda\right)^2+4b_2^2(\sin(k_m\Delta x))^2}}<1.$$

This is true for $\forall m$. Therefore, we obtain

$$\frac{|A_{0,0}^{\alpha,\beta}|}{\sqrt{(\delta_{0,0}^{\alpha,\beta}+4b_1-\lambda)^2+4b_2^2}}<1.$$

We assume that $\left|\frac{\delta_n}{\delta_0}\right|<1$. We need to show that $\left|\frac{\delta_{n+1}}{\delta_0}\right|<1$. We have

$$\delta_{n+1}\left(-4b_1\sin^2\left(\frac{k_m\Delta x}{2}\right)+2b_2 i\sin(k_m\Delta x)+\lambda\right)=\sum_{j=0}^{n}(\delta_{j+1}-\delta_j)A_{n,j}^{\alpha,\beta}.$$

Then, we obtain

$$|\delta_{n+1}|\le\frac{\sum_{j=0}^{n}|\delta_{j+1}-\delta_j|\left|A_{n,j}^{\alpha,\beta}\right|}{\sqrt{\left(\lambda-4b_1\sin^2\left(\frac{k_m\Delta x}{2}\right)\right)^2+4b_2^2(\sin(k_m\Delta x))^2}}$$
$$<2\delta_0\frac{\sum_{j=0}^{n}|A_{n,j}^{\alpha,\beta}|}{\sqrt{\left(\lambda-4b_1\sin^2\left(\frac{k_m\Delta x}{2}\right)\right)^2+4b_2^2(\sin(k_m\Delta x))^2}},$$

where $\left|\frac{\delta_{n+1}}{\delta_0}\right| < 1$ implies

$$2\frac{\sum_{j=0}^{n}|A_{n,j}^{\alpha,\beta}|}{\sqrt{\left(\lambda - 4b_1\sin^2\left(\frac{k_m\Delta x}{2}\right)\right)^2 + 4b_2^2(\sin(k_m\Delta x))^2}} < 1.$$

This is true for $\forall m$. Therefore, we obtain

$$2\frac{\sum_{j=0}^{n}|A_{n,j}^{\alpha,\beta}|}{\sqrt{(\lambda - 4b_1)^2 + 4b_2^2}} < 1.$$

The method is stable if

$$\min\left(\frac{\left|A_{0,0}^{\alpha,\beta}\right|}{\sqrt{(A_{0,0}^{\alpha,\beta} + 4b_1 - \lambda)^2 + 4b_2^2}}, 2\frac{\sum_{j=0}^{n}\left|A_{n,j}^{\alpha,\beta}\right|}{\sqrt{(\lambda - 4b_1)^2 + 4b_2^2}}\right) < 1.$$

We consider the following fractal–fractional differential equation with the generalized Mittag-Leffler kernel.

$$ {}_{0}^{FFMC}D_t^{\alpha,\beta}u(x,t) = \frac{\partial^2 u(x,t)}{\partial x^2} + x\frac{\partial u(x,t)}{\partial x} + \lambda u(x,t). \tag{3.33}$$

At (t_{n+1}, x_m), we have

$$\begin{aligned}
&\frac{AB(\alpha)}{\beta(1-\alpha)}\sum_{j=0}^{n}(u_m^{j+1} - u_m^j)(j\Delta t)^{1-\beta}\left[(n+1-j)E_{\alpha,2}\left(\frac{-\alpha}{1-\alpha}(n+1-j)\Delta t\right)\right.\\
&\quad\left.-(n-j)E_{\alpha,2}\left(\frac{-\alpha}{1-\alpha}(n-j)\Delta t\right)\right]\\
&= \frac{u_{m+1}^{n+1} - 2u_m^{n+1} + u_{m-1}^{n+1}}{(\Delta x)^2} + x_m\frac{u_{m+1}^{n+1} - u_{m-1}^{n+1}}{2\Delta x} + \lambda u_m^{n+1}.
\end{aligned}$$

We put $u_m^n = \delta_n \exp(ik_m x)$. Replacing in the above equation, we get

$$\begin{aligned}
&\frac{AB(\alpha)}{\beta(1-\alpha)}\sum_{j=0}^{n}(\delta_{n+1}\exp(ik_m x) - \delta_n\exp(ik_m x))(j\Delta t)^{1-\beta}\\
&\quad\left[(n+1-j)E_{\alpha,2}\left(\frac{-\alpha}{1-\alpha}(n+1-j)\Delta t\right) -(n-j)E_{\alpha,2}\left(\frac{-\alpha}{1-\alpha}(n-j)\Delta t\right)\right]\\
&= \frac{\delta_{n+1}\exp(ik_m(x+\Delta x)) - 2\delta_{n+1}\exp(ik_m x) + \delta_{j+1}\exp(ik_m(x-\Delta x))}{(\Delta x)^2}\\
&\quad + x_m\frac{\delta_{n+1}\exp(ik_m(x+\Delta x)) - \delta_{n+1}\exp(ik_m(x-\Delta x))}{2\Delta x} + \lambda\delta_{n+1}\exp(ik_m x).
\end{aligned}$$

For simplicity, we take

$$K_{n,j}^{\alpha,\beta} = \frac{AB(\alpha)}{\beta(1-\alpha)}(j\Delta t)^{1-\beta}\left[(n+1-j)E_{\alpha,2}\left(\frac{-\alpha}{1-\alpha}(n+1-j)\Delta t\right)\right.$$
$$\left.-(n-j)E_{\alpha,2}\left(\frac{-\alpha}{1-\alpha}(n-j)\Delta t\right)\right]$$
$$c_1 = \frac{1}{(\Delta x)^2}, \quad c_2 = \frac{x_m}{2\Delta x}.$$

Then, we get

$$\sum_{j=0}^{n}(\delta_{j+1}-\delta_j)K_{n,j}^{\alpha,\beta} = c_1(\delta_{n+1}\exp(ik_m\Delta x) - 2\delta_{n+1} + \delta_{j+1}\exp(-ik_m\Delta x))$$
$$+c_2(\delta_{n+1}\exp(ik_m\Delta x) - \delta_{n+1}\exp(-ik_m\Delta x)) + \lambda\delta_{n+1}.$$

Thus, we obtain

$$(\delta_{n+1}-\delta_n)K_{n,n}^{\alpha,\beta} + \sum_{j=0}^{n-1}(\delta_{j+1}-\delta_j)K_{n,j}^{\alpha,\beta}$$
$$= \delta_{n+1}\left(-4c_1\sin^2\left(\frac{k_m\Delta x}{2}\right) + 2c_2 i\sin(k_m\Delta x) + \lambda\right).$$

For $n = 0$, we have

$$(\delta_1-\delta_0)K_{0,0}^{\alpha,\beta} = \delta_1\left(-4c_1\sin^2\left(\frac{k_m\Delta x}{2}\right) + 2c_2 i\sin(k_m\Delta x) + \lambda\right),$$
$$\delta_1\left(K_{0,0}^{\alpha,\beta} + 4c_1\sin^2\left(\frac{k_m\Delta x}{2}\right) - 2c_2 i\sin(k_m\Delta x) - \lambda\right) = \delta_0 K_{0,0}^{\alpha,\beta},$$

where $\left|\frac{\delta_1}{\delta_0}\right| < 1$ implies

$$\frac{|K_{0,0}^{\alpha,\beta}|}{\sqrt{\left(K_{0,0}^{\alpha,\beta} + 4c_1\sin^2\left(\frac{k_m\Delta x}{2}\right) - \lambda\right)^2 + 4c_2^2(\sin(k_m\Delta x))^2}} < 1.$$

This is true for $\forall m$. Therefore, we obtain

$$\frac{|K_{0,0}^{\alpha,\beta}|}{\sqrt{(\delta_{0,0}^{\alpha,\beta} + 4c_1 - \lambda)^2 + 4c_2^2}} < 1.$$

We assume that $\left|\frac{\delta_n}{\delta_0}\right| < 1$. We need to show that $\left|\frac{\delta_{n+1}}{\delta_0}\right| < 1$. We have

$$\delta_{n+1}\left(-4c_1\sin^2\left(\frac{k_m\Delta x}{2}\right) + 2c_2 i\sin(k_m\Delta x) + \lambda\right) = \sum_{j=0}^{n}(\delta_{j+1}-\delta_j)K_{n,j}^{\alpha,\beta}.$$

Then, we obtain

$$|\delta_{n+1}| \leq \frac{\sum_{j=0}^{n} |\delta_{j+1} - \delta_j| \left|K_{n,j}^{\alpha,\beta}\right|}{\sqrt{\left(\lambda - 4c_1 \sin^2\left(\frac{k_m \Delta x}{2}\right)\right)^2 + 4c_2^2(\sin(k_m \Delta x))^2}}$$

$$< 2\delta_0 \frac{\sum_{j=0}^{n} \left|K_{n,j}^{\alpha,\beta}\right|}{\sqrt{\left(\lambda - 4c_1 \sin^2\left(\frac{k_m \Delta x}{2}\right)\right)^2 + 4c_2^2(\sin(k_m \Delta x))^2}},$$

where $\left|\frac{\delta_{n+1}}{\delta_0}\right| < 1$ implies

$$2\frac{\sum_{j=0}^{n} |K_{n,j}^{\alpha,\beta}|}{\sqrt{\left(\lambda - 4c_1 \sin^2\left(\frac{k_m \Delta x}{2}\right)\right)^2 + 4c_2^2(\sin(k_m \Delta x))^2}} < 1.$$

This is true for $\forall m$. Therefore, we obtain

$$2\frac{\sum_{j=0}^{n} |K_{n,j}^{\alpha,\beta}|}{\sqrt{(\lambda - 4c_1)^2 + 4c_2^2}} < 1.$$

The method is stable if

$$\min\left(\frac{|K_{0,0}^{\alpha,\beta}|}{\sqrt{(K_{0,0}^{\alpha,\beta} + 4c_1 - \lambda)^2 + 4c_2^2}}, 2\frac{\sum_{j=0}^{n} |K_{n,j}^{\alpha,\beta}|}{\sqrt{(\lambda - 4c_1)^2 + 4c_2^2}}\right) < 1.$$

3.6 Conclusion

Differential operators have been recognized as powerful mathematical tools to modeling real world problems facing humankind every day. Mathematicians while trying to depict these physical problems rely on ordinary or partial differential equation mostly. Very recently, novel differential operators were suggested. The differential operators were defined fractal derivative of convolution of continuous with kernels including the power law, exponential decay, and the generalized Mittag-Leffler function. This work considered these differentials to be convolution of fractal derivative of differentiable functions and the three mainly used kernels. This gave birth to new class of partial differential and ordinary differential equations that were studied in this work analytically and numerically.

Bibliography

1 Owolabi, K.M. and Atangana, A. (2018). Chaotic behaviour in system of noninteger-order ordinary differential equations. *Chaos, Solitons Fractals* **115**: 362–370.

2 Owolabi, K.M. (2018). Analysis and numerical simulation of multicomponent system with Atangana–Baleanu fractional derivative. *Chaos, Solitons Fractals* **115**: 127–134.

3 Podlubny, I. (1999). *Fractional Differential Equations*. New York: Academic Press.

4 Karaagac, B. (2018). Analysis of the cable equation with non-local and non-singular kernel fractional derivative. *Eur. Phys. J. Plus* **133**: 54.

5 Kilbas, A.A., Srivastava, H.M., and Trujillo, J.J. (2006). *Theory and Applications of Fractional Differential Equations*. San Diego, CA: Elsevier.

6 Goufo, E.F.D. (2019). Strange attractor existence for non-local operators applied to four-dimensional chaotic systems with two equilibrium points. *Chaos* **29** (2): 023117.

7 Goufo, E.F.D., Mbehou, M., and Pene, M.M.K. (2018). A peculiar application of Atangana–Baleanu fractional derivative in neuroscience: chaotic burst dynamics. *Chaos, Solitons Fractals* **115**: 170–176.

8 Matsumoto, T. (1984). A chaotic attractor from Chua's circuit. *IEEE Trans. Circuits Syst.* **31** (12): 1055–1058.

9 Chua, L., Komuro, M., and Matsumoto, T. (1986). The double scroll family. *IEEE Trans. Circ. Syst.* **33** (11): 1072–1118.

10 Lü, J. and Chen, G. (2006). Generating multiscroll chaotic attractors: theories, methods and applications. *Int. J. Bifurcation Chaos* **16** (04): 775–858.

11 Akgül, A. (2018). A novel method for a fractional derivative with non-local and non-singular kernel. *Chaos, Solitons Fractals* **114**: 478–482.

12 Akgül, E.K. (2019). Solutions of the linear and nonlinear differential equations within the generalized fractional derivatives. *Chaos* **29**: 023108.

13 Atangana, A. and Baleanu, D. (2016). New fractional derivatives with nonlocal and non-singular kernel, theory and application to heat transfer model. *Therm. Sci.* **20**: 763–769.

14 Atangana, A. (2017). Fractal-fractional differentiation and integration: connecting fractal calculus and fractional calculus to predict complex system. *Chaos, Solitons Fractals* **102**: 396–406.

4

New Estimations for Exponentially Convexity via Conformable Fractional Operators

Alper Ekinci[1] *and Sever S. Dragomir*[2]

[1] *Department of International Trade, Bandirma Vocational High School, Bandirma Onyedi Eylul University, Balikesir, Turkey*
[2] *Theory of Inequality, School of Engineering and Science, Victoria University, Melbourne, VIC, Australia*

4.1 Introduction

We will give the following well-known definitions in the fractional calculus.

Definition 4.1 **([1])** Let $f \in L_1[a,b]$. The Riemann–Liouville integrals $J^{\alpha}_{a+}f$ and $J^{\alpha}_{b-}f$ of order $\alpha > 0$ are defined by

$$J^{\alpha}_{a+}f(t) = \frac{1}{\Gamma(\alpha)} \int_a^t (t-x)^{\alpha-1} f(x)\, dx, \quad t > a$$

and

$$J^{\alpha}_{b-}f(t) = \frac{1}{\Gamma(\alpha)} \int_t^b (x-t)^{\alpha-1} f(x)\, dx, \quad t < b,$$

Respectively, where $\Gamma(\alpha) = \int_0^{\infty} e^{-t} t^{\alpha-1}\, dt$.

Here, $J^{0}_{a+}f(t) = J^{0}_{b-}f(t) = f(t)$. In the case of $\alpha = 1$, the fractional integral reduces to classical integral.

In [2], Abdeljawad gave the following definitions of Right–Left conformable fractional integrals:

Definition 4.2 **([2])** Let $\alpha \in (n, n+1]$, $n = 0, 1, 2, \ldots$ and set $\beta = \alpha - n$. Then, the left conformable fractional integral of any order $\alpha > 0$ is defined by

$$(I^{\alpha}_{a}f)(t) = \frac{1}{n!} \int_a^t (t-x)^n (x-a)^{\beta-1} f(x)\, dx.$$

Fractional Order Analysis: Theory, Methods and Applications, First Edition.
Edited by Hemen Dutta, Ahmet Ocak Akdemir, and Abdon Atangana.

Definition 4.3 Analogously, the right conformable fractional integral of any order $\alpha > 0$ is defined by

$$({}^{b}I_{\alpha}f)(t) = \frac{1}{n!}\int_{t}^{b}(x-t)^{n}(b-x)^{\beta-1}f(x)\,dx.$$

Notice that if $\alpha = n+1$, then $\beta = \alpha - n = n+1-n = 1$; hence, $(I^{a}_{n+1}f)(t) = (J^{n+1}_{a+}f)(t)$ and $({}^{b}I_{n+1}f)(t) = (J^{n+1}_{b-}f)(t)$.

In [3], Khalil et al. gave the following definition.

Definition 4.4 ([3]) (Conformable fractional integral) Let $\alpha \in (0,1]$, $0 \le \kappa_1 < \kappa_2$. A function $h : [\kappa_1, \kappa_2] \to \mathbb{R}$ is α -fractional integrable on $[\kappa_1, \kappa_2]$ if the integral

$$\int_{\kappa_1}^{\kappa_2} h(x)\,d_{\alpha}x = \int_{\kappa_1}^{\kappa_2} h(x)x^{\alpha-1}\,dx$$

exists and is finite. All α-fractional integrable functions on $[\kappa_1, \kappa_2]$ are indicated by $L_{\alpha}([\kappa_1, \kappa_2])$.

Remark 4.1 ([3])

$$I^{\kappa_1}_{\alpha}(h_1)(s) = I^{\kappa_1}_{1}(s^{\alpha-1}h_1) = \int_{\kappa_1}^{s}\frac{h_1(x)}{x^{1-\alpha}}\,dx,$$

where the integral is the usual Riemann improper integral, and $\alpha \in (0,1]$.

In [4], Jarad et al. have defined a new fractional integral operator. Also, they gave some properties and relations between some other fractional integral operators, as Riemann–Liouville fractional integral, Hadamard fractional integrals, generalized fractional integral operators, etc., with this operator.

Definition 4.5 ([4]) Let $\beta \in \mathbb{C}$, $Re(\beta) > 0$, then the left- and right-sided fractional conformable integral operators have defined, respectively, as follows:

$$ {}^{\beta}_{a}\mathfrak{J}^{\alpha}f(x) = \frac{1}{\Gamma(\beta)}\int_{a}^{x}\left(\frac{(x-a)^{\alpha}-(t-a)^{\alpha}}{\alpha}\right)^{\beta-1}\frac{f(t)}{(t-a)^{1-\alpha}}\,dt;$$

$$ {}^{\beta}\mathfrak{J}^{\alpha}_{b}f(x) = \frac{1}{\Gamma(\beta)}\int_{x}^{b}\left(\frac{(b-x)^{\alpha}-(b-t)^{\alpha}}{\alpha}\right)^{\beta-1}\frac{f(t)}{(b-t)^{1-\alpha}}\,dt.$$

The results presented here, being general, can be reduced to yield many relatively simple inequalities and identities for functions associated with certain fractional integral operators. For example, the case $\alpha = 1$ in the obtained results is found to yield the same results involving Riemann–Liouville fractional integrals, given before, in literature. Furthermore, getting more knowledge, see Ref. [4].

Katugampola (see [5] and [6]) considered the following iterative process in 2011:

$$\int_a^x t_1^\rho \, dt_1 \int_a^{t_1} t_2^\rho \, dt_2 \dots \int_a^{t_{n-1}} t_n^\rho f(t_n) \, dt_n$$
$$= \frac{(\rho+1)^{1-n}}{(n-1)!} \int_a^x (t^{\rho+1} - \tau^{\rho+1})^{n-1} \tau^\rho f(\tau) \, d\tau,$$

for $n \in \mathbb{N}$. This generates Katugampola's concept of fractional integral, defined in [5] and also in [6].

Definition 4.6 **[5]** Let $f \in [a, b]$, the left-sided Katugampola fractional integral ${}^\rho I^\alpha_{a+} f$ of order $\alpha \in \mathbb{C}$, $\mathrm{Re}(\alpha) > 0$ is defined by

$${}^\rho I^\alpha_{a+} f(x) = \frac{\rho^{1-\alpha}}{\Gamma(\alpha)} \int_a^x \frac{t^{\rho-1}}{(x^\rho - t^\rho)^{1-\alpha}} f(t) \, dt, \quad x > a,$$

the right-sided Katugampola fractional integral ${}^\rho I^\alpha_{b-} f$ of order $\alpha \in \mathbb{C}$, $\mathrm{Re}(\alpha) > 0$ is defined by

$${}^\rho I^\alpha_{b-} f(x) = \frac{\rho^{1-\alpha}}{\Gamma(\alpha)} \int_x^b \frac{t^{\rho-1}}{(t^\rho - x^\rho)^{1-\alpha}} f(t) \, dt, \quad x < b.$$

Katugampola's operators are generalizations of A. Erdélyi and H. Kober operators introduced in 1940 (see [7] and [8]), as well. Other similar approaches on moving iterative integrals and derivatives into fractional framework in connection with theoretic and practical applications are in the mathematical literature of the last decade.

Remark 4.2 If $\rho = 1$, then the Katugampola fractional integrals become Riemann–Liouville fractional integrals.

In [9], Raina introduced a class of functions defined formally by

$$\mathcal{F}^\sigma_{\rho,\lambda}(x) = \mathcal{F}^{\sigma(0),\sigma(1),\dots}_{\rho,\lambda}(x) = \sum_{k=0}^{\infty} \frac{\sigma(k)}{\Gamma(\rho k + \lambda)} x^k \quad (\rho, \lambda > 0; |x| < \mathbb{R}), \tag{4.1}$$

where the coefficients $\sigma(k)$ $(k \in \mathbb{N} = \mathbb{N} \cup \{0\})$ is a bounded sequence of positive real numbers and $\mathbb{R}$ is the set of real numbers. With the help of (4.1), Raina [9] defined the following left-sided and right-sided fractional integral operators, respectively, as follows:

$$(\mathcal{J}^\sigma_{\rho,\lambda,a+;w}\varphi)(x) = \int_a^x (x-t)^{\lambda-1} \mathcal{F}^\sigma_{\rho,\lambda}[w(x-t)^\rho]\varphi(t) \, dt \quad (x > a > 0), \tag{4.2}$$

$$(\mathcal{J}^\sigma_{\rho,\lambda,b-;w}\varphi)(x) = \int_x^b (t-x)^{\lambda-1} \mathcal{F}^\sigma_{\rho,\lambda}[w(t-x)^\rho]\varphi(t) \, dt \quad (0 < x < b), \tag{4.3}$$

where $\lambda, \rho > 0$, $w \in \mathbb{R}$, and $\varphi(t)$ is such that the integral on the right side exits.

It is easy to verify that $\mathcal{J}^{\sigma}_{\rho,\lambda,a+;w}\varphi(x)$ and $\mathcal{J}^{\sigma}_{\rho,\lambda,b-;w}\varphi(x)$ are bounded integral operators on $L(a,b)$, if

$$\mathfrak{M} := \mathcal{F}^{\sigma}_{\rho,\lambda+1}[w(b-a)^{\rho}] < \infty. \tag{4.4}$$

In fact, for $\varphi \in L(a,b)$, we have

$$||\mathcal{J}^{\sigma}_{\rho,\lambda,a+;w}\varphi(x)||_1 \leq \mathfrak{M}(b-a)^{\lambda}||\varphi||_1 \tag{4.5}$$

and

$$||\mathcal{J}^{\sigma}_{\rho,\lambda,b-;w}\varphi(x)||_1 \leq \mathfrak{M}(b-a)^{\lambda}||\varphi||_1, \tag{4.6}$$

where

$$||\varphi||_p := \left(\int_a^b |\varphi(t)|^p \, dt\right)^{\frac{1}{p}}.$$

Here, many useful fractional integral operators can be obtained by specializing the coefficient $\sigma(k)$. For instance, the classical Riemann–Liouville fractional integrals J^{α}_{a+} and J^{α}_{b-} of order α follow easily by setting $\lambda = \alpha$, $\sigma(0) = 1$, and $w = 0$ in (4.2) and (4.3).

The following definition is well-known in the literature as convex function.

Definition 4.7 A function $f : I \subset \mathbb{R} \to \mathbb{R}$ is said to be convex function, if

$$f(tx + (1-t)y) \leq tf(x) + (1-t)f(y), \quad \forall x, y \in I, \ t \in [0,1].$$

Recall the definition of exponentially convex function, which is mainly due to Antczak [7] and Dragomir and Gomm [10].

Definition 4.8 (See [7, 10]) Let $f : I \subset \mathbb{R} \to \mathbb{R}$ is exponentially convex function, if f is positive, $\forall a, b \in I$ and $t \in [0,1]$, we have

$$e^{f((1-t)a+tb)} \leq [(1-t)e^{f(a)} + te^{f(b)}], \quad a, b \in I, \quad t \in [0,1].$$

Several new results related to exponentially convex functions can be found in [7, 10–13].

In [14], Fejér established a so-called Hermite–Hadamard–Fejér inequality related to the integral mean of a convex function f that is a weighted generalization of the well-known Hermite–Hadamard inequality, which is recalled in the following theorem.

Theorem 4.1 *Let $f : [a,b] \to \mathbb{R}$ $(a < b)$ be a convex function and $f \in L_1(a,b)$. Also let $g : [a,b] \to \mathbb{R}$ be non-negative, integrable, and symmetric to $(a+b)/2$.*

Then,

$$f\left(\frac{a+b}{2}\right)\int_a^b g(x)\,dx \le \frac{1}{b-a}\int_a^b f(x)g(x)\,dx \le \frac{f(a)+f(b)}{2}\int_a^b g(x)\,dx.$$

An important fractional integral operator has been defined by Khalil et al. in [3], as the following:

Definition 4.9 (Conformable fractional integral) Let $\alpha \in (0,1]$, $0 \le \kappa_1 < \kappa_2$. A function $h : [\kappa_1, \kappa_2] \to \mathbb{R}$ is α-fractional integrable on $[\kappa_1, \kappa_2]$ if the integral

$$\int_{\kappa_1}^{\kappa_2} h(x)\,d_\alpha x = \int_{\kappa_1}^{\kappa_2} h(x)x^{\alpha-1}\,dx$$

exists and is finite. All α-fractional integrable functions on $[\kappa_1, \kappa_2]$ are indicated by $L_\alpha([\kappa_1, \kappa_2])$.

Recently, several researchers have obtained inequalities by using this definition, see Refs. [2, 3, 15–18].

Remark 4.3 (See [3])

$$I_\alpha^{\kappa_1}(h_1)\,(s) = I_1^{\kappa_1}(s^{\alpha-1}h_1) = \int_{\kappa_1}^{s} \frac{h_1(x)}{x^{1-\alpha}}\,dx,$$

where the integral is the usual Riemann improper integral, and $\alpha \in (0,1]$.

The main aims of this chapter are to give a new variant of Hermite–Hadamard–Fejér inequality for exponentially convex functions, to obtain an integral identity and prove some Hermite–Hadamard type inequalities via α-fractional integral operators.

4.2 Main Results

The exponentially convex version of Hermite–Hadamard–Fejér inequality can be represented in conformable fractional integrals forms as follows:

Theorem 4.2 *Suppose that $a, b \in I$, $b > a$, $f : I = [a,b] \to (0,\infty)$ is an exponentially convex function and symmetric with respect to $\frac{a+b}{2}$ and $w : [a,b] \to \mathbb{R}$ is a non-negative integrable function. Then the following inequality*

$$e^{f\left(\frac{a+b}{2}\right)}\int_a^b e^{g(x)}\,d_\alpha x \le \int_a^b e^{f(x)}e^{g(x)}\,d_\alpha x \le \frac{e^{f(a)}+e^{f(b)}}{2}\int_a^b e^{g(x)}\,d_\alpha x \tag{4.7}$$

holds for any $\alpha \in (0,1]$.

Proof: Since $f : I \to \mathbb{R}$ is an exponentially convex function and is symmetric with respect to $\frac{a+b}{2}$, then for any $x, y \in I$ and $t = \frac{1}{2}$, we have

$$e^{f\left(\frac{x+y}{2}\right)} \leq \frac{e^{f(x)} + e^{f(y)}}{2}. \tag{4.8}$$

That is, with $x = ta + (1-t)b$ and $y = (1-t)a + tb$, the inequality (4.8) becomes

$$\begin{aligned} e^{f\left(\frac{a+b}{2}\right)} &\leq \frac{e^{f(ta+(1-t)b)} + e^{f((1-t)a+tb)}}{2} \\ &= e^{f((1-t)a+tb)} \quad (f \text{ is symmetric}). \end{aligned}$$

By using the change of variable, we get

$$\begin{aligned} & e^{f\left(\frac{a+b}{2}\right)} \int_a^b e^{g(x)}\, d_\alpha x \\ &= (b-a) e^{f\left(\frac{a+b}{2}\right)} \int_0^1 e^{g((1-t)a+tb)} ((1-t)a+tb)^{\alpha-1}\, dt \\ &\leq (b-a) \int_0^1 e^{f((1-t)a+tb)} e^{g((1-t)a+tb)} ((1-t)a+tb)^{\alpha-1}\, dt \\ &= \int_a^b e^{f(x)} e^{g(x)}\, d_\alpha x, \end{aligned}$$

namely,

$$e^{f\left(\frac{a+b}{2}\right)} \int_a^b e^{g(x)}\, d_\alpha x \leq \int_a^b e^{f(x)} e^{g(x)}\, d_\alpha x,$$

which completes the proof of first inequality.

To prove the second inequality in (4.7), by using f exponentially convexity, so we can write

$$e^{f((1-t)a+tb)} \leq (1-t)e^{f(a)} + te^{f(b)},$$

and similarly

$$e^{f(ta+(1-t)b)} \leq te^{f(a)} + (1-t)e^{f(b)}.$$

Now, by choosing $x = (1-t)a + tb$, we have

$$\begin{aligned} & \int_a^b e^{f(x)} e^{g(x)} x^{\alpha-1}\, dx \\ &= \int_0^1 e^{f((1-t)a+tb)} e^{g((1-t)a+tb)} ((1-t)a+tb)^{\alpha-1}\, dt \\ &\leq (b-a) \cdot \int_0^1 [(1-t)e^{f(a)} + te^{f(b)}] e^{g((1-t)a+tb)} ((1-t)a+tb)^{\alpha-1}\, dt. \end{aligned} \tag{4.9}$$

By a similar argument, it is easy to see

$$\int_a^b e^{f(x)} e^{g(x)} x^{\alpha-1}\, dx$$

$$= \int_0^1 e^{f((1-t)a+tb)} e^{g(((1-t)a+tb)}((1-t)a+tb)^{\alpha-1}\, dt$$

$$\leq \int_0^1 e^{f(ta+(1-t)b)} e^{g(((1-t)a+tb)}((1-t)a+tb)^{\alpha-1}\, dt. \tag{4.10}$$

If we add (4.9) and (4.10) , we obtain

$$2\int_a^b e^{f(x)} e^{g(x)}\, d_\alpha x \leq (b-a)[e^{f(a)} + e^{f(b)}] \int_0^1 e^{g(a(1-t)+tb)} (a(1-t)+tb)^{\alpha-1}\, dt$$

$$= [e^{f(a)} + e^{f(b)}] \int_a^b e^{g(x)} x^{\alpha-1}\, dx.$$

So, we can write

$$\int_a^b e^{f(x)} e^{g(x)}\,{}_d\alpha x \leq \frac{e^{f(a)} + e^{f(b)}}{2} \int_a^b e^{g(x)}\, d_\alpha x,$$

which is the required result. □

Corollary 4.1 *If we set* $e^{g(x)} = 1$ *in (4.7), then we get the following inequality:*

$$e^{f\left(\frac{a+b}{2}\right)} \leq \frac{a}{b^\alpha - a^\alpha} \int_a^b e^{f(x)}\, d_\alpha x \leq \frac{e^{f(a)} + e^{f(b)}}{2}.$$

In order to obtain Hermite–Hadamard type inequalities, we establish the following identity.

Lemma 4.1 *Let* $a, b \in I$ *with* $b > a$ *and* $f : I = [a, b] \to (0, \infty)$ *be an* α*-fractional differentiable exponentially convex function on* (a, b)*, then one has the following equality:*

$$\frac{e^{f(a)} + e^{f(b)}}{2} - \frac{\alpha}{b^\alpha - a^\alpha} \int_a^b e^{f(x)}\, d_\alpha x$$

$$= \frac{b-a}{2(b^\alpha - a^\alpha)} \left[\left\{ \int_0^1 (a(1-t)+tb)^{2\alpha-1} - a^\alpha (a(1-t)+tb)^{\alpha-1} \right\} \right.$$

$$\times D_\alpha(e^f)(a(1-t)+tb) f^{(\alpha)}(a(1-t)+tb) t^{1-\alpha}\,{}_d\alpha t$$

$$+ \int_0^1 \{(a(1-t)+tb)^{2\alpha-1} - b^\alpha (a(1-t)+tb)^{\alpha-1}\}$$

$$\left. \times\, D_\alpha(e^f)(a(1-t)+tb) f^{(\alpha)}(a(1-t)+tb) t^{1-\alpha}\,{}_d\alpha t \right]. \tag{4.11}$$

Proof: Integrating by parts, we have

$$I$$

$$= \left[\left\{ \int_0^1 (a(1-t)+tb)^{2\alpha-1} - a^\alpha (a(1-t)+tb)^{\alpha-1} \right\} \right.$$

$$\times D_\alpha(e^f)(a(1-t)+tb)f^{(\alpha)}(a(1-t)+tb)t^{1-\alpha}d_\alpha t$$
$$+\int_0^1 \{(a(1-t)+tb)^{2\alpha-1} - b^\alpha(a(1-t)+tb)^{\alpha-1}\}$$
$$\left.\times D_\alpha(e^f)(a(1-t)+tb)f^{(\alpha)}(a(1-t)+tb)t^{1-\alpha}d_\alpha t\right]$$
$$=\int_0^1 \{(a(1-t)+tb)^\alpha - a^\alpha(a(1-t)+tb)^{\alpha-1}\}e^{f(a(1-t)+tb)}f'(a(1-t)+tb)\,dt$$
$$+\int_0^1 \{(a(1-t)+tb)^\alpha - b^\alpha(a(1-t)+tb)^{\alpha-1}\}e^{f(a(1-t)+tb)}f'(a(1-t)+tb)\,dt$$

$$I$$
$$=[(a(1-t)+tb)^\alpha - a^\alpha]\frac{e^{f(a(1-t)+tb)}}{b-a}|_0^1 - \int_0^1 \alpha(a(1-t)+tb)^{\alpha-1}e^{f(a(1-t)+tb)}\,dt$$
$$+[(a(1-t)+tb)^\alpha - b^\alpha]\frac{e^{f(a(1-t)+tb)}}{b-a}|_0^1 - \int_0^1 \alpha(a(1-t)+tb)^{\alpha-1}e^{f(a(1-t)+tb)}\,dt$$
$$=\frac{1}{b-a}\left[(b^\alpha - a^\alpha)e^{f(b)} - \alpha\int_a^b e^{f(x)}\,d_\alpha x\right]$$
$$+\frac{1}{b-a}\left[(b^\alpha - a^\alpha)e^{f(a)} - \alpha\int_a^b e^{f(x)}\,d_\alpha x\right]$$
$$=\frac{b^\alpha - a^\alpha}{b-a}[e^{f(a)}+e^{f(a)}] - \frac{2\alpha}{b-a}\int_a^b e^{f(x)}\,d_\alpha x,$$

where we have used the change of variable $x = a(1-t)+tb$ and then multiplied both sides by $\frac{b-a}{2(b^\alpha-a^\alpha)}$ to get the desired result. □

Remark 4.4 If we set $\alpha = 1$ in (4.11), then we obtain the new result:

$$\frac{e^{f(a)}+e^{f(b)}}{2} - \frac{1}{b-a}\int_0^1 e^{f(x)}\,dx = \frac{1}{2}\int_0^1 (2t-1)e^{f(a(1-t)+tb)}f'(a(1-t)+tb)\,dt.$$

Now, we are ready to prove new Hermite–Hadamard type inequalities, which are embodied the following theorems.

Theorem 4.3 *Let $a, b \in I$ with $b > a$ and $f : I = [a,b] \to (0,\infty)$ be an α-fractional differentiable exponentially convex function on (a,b) for $\alpha(0,1]$ such that $D_\alpha(f) \in L_\alpha[a,b]$. If $|f'|$ is convex, then we have*

$$\left|\frac{e^{f(a)}+e^{f(b)}}{2} - \frac{\alpha}{b^\alpha - a^\alpha}\int_a^b e^{f(x)}\,d_\alpha x\right|$$
$$\le \frac{b-a}{6}[|e^{f(a)}f'(a)| + |e^{f(b)}f'(b)| + \Delta(a,b)],$$

where

$$\Delta(a,b) = |e^{f(a)}f'(b)| + |e^{f(b)}f'(a)|.$$

Proof: From Lemma 4.1, using the property of the modulus and exponentially convexity of $|f'|$, we have

$$\left|\frac{e^{f(a)}+e^{f(b)}}{2} - \frac{\alpha}{b^\alpha - a^\alpha}\int_a^b e^{f(x)}\, d_\alpha x\right|$$

$$= \left|\frac{b-a}{2(b^\alpha - a^\alpha)}\left[\left\{\int_0^1 (a(1-t)+tb)^{2\alpha-1} - a^\alpha(a(1-t)+tb)^{\alpha-1}\right\}\right.\right.$$

$$\times D_\alpha(e^f)(a(1-t)+tb)f^{(\alpha)}(a(1-t)+tb)t^{1-\alpha}\, d_\alpha t$$

$$+ \int_0^1 \{(a(1-t)+tb)^{2\alpha-1} - b^\alpha(a(1-t)+tb)^{\alpha-1}\}$$

$$\left.\left.\times\ D_\alpha(e^f)(a(1-t)+tb)f^{(\alpha)}(a(1-t)+tb)t^{1-\alpha}d_\alpha t\right]\right|$$

$$\leq \frac{b-a}{2(b^\alpha - a^\alpha)}\left[\int_0^1 [(a(1-t)+tb)^\alpha - a^\alpha]|e^{f(a(1-t)+tb)}f'(a(1-t)+tb)|\, dt\right.$$

$$\left.+ \int_0^1 [b^\alpha - (a(1-t)+tb)^\alpha]|e^{f(a(1-t)+tb)}f'(a(1-t)+tb)|\, dt\right]$$

$$\leq \frac{b-a}{2(b^\alpha - a^\alpha)}\left[\int_0^1 [(a(1-t)+tb)^\alpha - a^\alpha]\{(1-t)|e^{f(a)}| + te^{f(b)}\}\right.$$

$$\times \{(1-t)|f'(a)| + tf'(b)\}\, dt$$

$$+ \int_0^1 [b^\alpha - (a(1-t)+tb)^\alpha]\{(1-t)|e^{f(a)}| + te^{f(b)}\}\{(1-t)|f'(a)| + tf'(b)\}\, dt]$$

$$= \frac{b-a}{2(b^\alpha - a^\alpha)}\left[\int_0^1 [(a(1-t)+tb)^\alpha - a^\alpha]\right.$$

$$\{(1-t)^2|e^{f(a)}f'(a)| + t^2|e^{f(b)}f'(b)| + t(1-t)[|e^{f(a)}f'(b)| + |e^{f(b)}f'(a)|]\}\, dt$$

$$+ \int_0^1 [b^\alpha - (a(1-t)+tb)^\alpha]$$

$$\left.\{(1-t)^2|e^{f(a)}f'(a)| + t|e^{f(b)}f'(b)| + t(1-t)[|e^{f(a)}f'(b)| + |e^{f(b)}f'(a)|]\, dt\}\right]$$

$$= \frac{b-a}{2(b^\alpha - a^\alpha)}[(b^\alpha - a^\alpha)\{(1-t)^2|e^{f(a)}f'(a)| + t^2|e^{f(b)}f'(b)| + t(1-t)\Delta(a,b)\}\, dt]$$

$$= \frac{b-a}{6}[|e^{f(a)}f'(a)| + |e^{f(b)}f'(b)| + \Delta(a,b)].$$

This completes the proof. □

Theorem 4.4 *Let $a, b \in I$ with $b > a$ and $f : I = [a, b] \to (0, \infty)$ be an α-fractional differentiable exponentially convex function on (a, b) for $\alpha \in (0, 1]$ such that $D_\alpha(f) \in L_\alpha[a, b]$. If $|f'|^q$ is convex for $q > 1$ and $p^{-1} + q^{-1} = 1$, then we prove*

$$\left|\frac{e^{f(a)} + e^{f(b)}}{2} - \frac{\alpha}{b^\alpha - a^\alpha}\int_a^b e^{f(x)}\, d_\alpha x\right|$$
$$\leq \frac{b-a}{2(b^\alpha - a^\alpha)}\left[(\zeta_1(\alpha, p))^{\frac{1}{p}}\right.$$
$$\left. + (\zeta_2(\alpha, p))^{\frac{1}{p}}\right]\left\{\frac{|e^{f(a)}f'(a)|^q + |e^{f(b)}f'(b)|^q + \Delta_q(a, b)}{3}\right\}^{\frac{1}{q}}, \tag{4.12}$$

where

$$\Delta_q(a, b) = |e^{f(a)}f'(b)|^q + |e^{f(b)}f'(a)|^q.$$

Proof: From Lemma 4.1, using the property of modulus and exponentially convexity of $|f'|$, we have

$$\left|\frac{e^{f(a)} + e^{f(b)}}{2} - \frac{\alpha}{b^\alpha - a^\alpha}\int_a^b e^{f(x)}\, d_\alpha x\right| \leq \frac{b-a}{2(b^\alpha - a^\alpha)}$$
$$\left[\int_0^1 \{(a(1-t) + tb)^\alpha - a^\alpha\}|e^{f(a(1-t)+tb)}f'(a(1-t) + tb)|\, dt\right.$$
$$\left. + \int_0^1 \{(a(1-t) + tb)^\alpha - b^\alpha\}|e^{f(a(1-t)+tb)}f'(a(1-t) + tb)|\, dt\right].$$

Now by using the Hölder inequality, we can write

$$\int_0^1 [(a(1-t) + tb)^\alpha - a^\alpha]|e^{f(a(1-t)+tb)}f'(a(1-t) + tb)|\, dt$$
$$\leq \left(\int_0^1 [(a(1-t) + tb)^\alpha - a^\alpha]^p\, dt\right)^{\frac{1}{p}}\left(\int_0^1 |e^{f(a(1-t)+tb)}f'(a(1-t) + tb)|\right)^{\frac{1}{q}}$$
$$\leq \left(\int_0^1 [(a(1-t) + tb)^\alpha - a^\alpha]^p\, dt\right)^{\frac{1}{p}}$$
$$\left(\int_0^1 [(1-t)|e^{f(a)}|^q + t|e^{f(b)}|^q][(1-t)|f'(a)|^q + t|f'(b)|^q]\, dt\right)^{\frac{1}{q}}$$

$$\int_0^1 [(a(1-t) + tb)^\alpha - a^\alpha]|e^{f(a(1-t)+tb)}f'(a(1-t) + tb)|\, dt$$
$$\leq \left(\int_0^1 [(a(1-t) + tb)^\alpha - a^\alpha]^p\, dt\right)^{\frac{1}{p}}$$

$$
\left(\int_0^1 [(1-t)^2|e^{f(a)}f'(a)|^q + t^2|e^{f(b)}f'(b)|^q + t(1-t)\right.
$$

$$
\left.\times\{|e^{f(a)}f'(b)|^q + |e^{f(b)}f'(a)|^q\}]\,dt\right)^{\frac{1}{q}}
$$

$$
= \left(\int_0^1 [(a(1-t)+tb)^\alpha - a^\alpha]^p\,dt\right)^{\frac{1}{p}}
$$

$$
\left(\int_0^1 [(1-t)^2|e^{f(a)}f'(a)|^q + t^2|e^{f(b)}f'(b)|^q + t(1-t)\Delta_q(a,b)]\,dt\right)^{\frac{1}{q}}
$$

$$
= (\zeta_1(\alpha,p))^{\frac{1}{p}}\left(\frac{|e^{f(a)}f'(a)|^q + |e^{f(b)}f'(b)|^q + \Delta_q(a,b)}{3}\right)^{\frac{1}{q}}. \tag{4.13}
$$

Similarly, we have

$$
\int_0^1 [b^\alpha - (a(1-t)+tb)^\alpha]|e^{f(a(1-t)+tb)}f'(a(1-t)+tb)|\,dt
$$

$$
\leq \left(\int_0^1 [b^\alpha - (a(1-t)+tb)^\alpha]^p\,dt\right)^{\frac{1}{p}}\left(\int_0^1 |e^{f(a(1-t)+tb)}f'(a(1-t)+tb)|\,dt\right)^{\frac{1}{q}}
$$

$$
\leq (\zeta_2(\alpha,p))^{\frac{1}{p}}\left(\frac{|e^{f(a)}f'(a)|^q + |e^{f(b)}f'(b)|^q + \Delta_q(a,b)}{3}\right)^{\frac{1}{q}}. \tag{4.14}
$$

By using the inequalities (4.13) and (4.14), we have the result in (4.12). □

Corollary 4.2 *If we set $\alpha = 1$ in Theorem 4.1, then we have the new inequality:*

$$
\left|\frac{e^{f(a)} + e^{f(b)}}{2} - \frac{1}{b-a}\int_a^b e^{f(x)}\,dx\right|
$$

$$
\leq \frac{b-a}{(p+1)^{\frac{1}{p}}}\left(\frac{|e^{f(a)}f'(a)|^q + |e^{f(b)}f'(b)|^q + \Delta_q(a,b)}{3}\right)^{\frac{1}{q}}.
$$

Theorem 4.5 *Let $a, b \in I$ with $b > a$ and $f : I = [a,b] \to (0,\infty)$ be an α-fractional differentiable exponentially convex function on (a,b) for $\alpha \in (0,1]$ such that $D_\alpha(f) \in L_\alpha[a,b]$. If $|f'|^q$ is convex for $q > 1$ and $p^{-1} + q^{-1} = 1$, then we have*

$$
\left|\frac{e^{f(a)} + e^{f(b)}}{2} - \frac{\alpha}{b^\alpha - a^\alpha}\int_a^b e^{f(x)}\,d_\alpha x\right|
$$

$$
\leq \frac{b-a}{2(b^\alpha - a^\alpha)}\left[(\eta_1(\alpha))^{1-\frac{1}{q}}\left\{\eta_2(a,b)|e^{f(a)}f'(a)|^q + \eta_3(a,b)|e^{f(b)}f'(b)|^q\right.\right.
$$

$$
\left.+\eta_4(a,b)\Delta(a,b)\right\}^{\frac{1}{q}} + (\theta_1(\alpha))^{1-\frac{1}{q}}
$$

$$\times \left\{\theta_2(a,b)|e^{f(a)}f'(a)|^q + \theta_3(a,b)|e^{f(b)}f'(b)|^q + \theta_4(a,b)\Delta_q(a,b)\right\}^{\frac{1}{q}}\Bigg], \quad (4.15)$$

where

$$\eta_2(\alpha) = \frac{1}{3(b-a)^3(\alpha+1)(\alpha+2)(\alpha+3)} \times [3\{2(b^{\alpha+3} - a^{\alpha+3}) - 2(b-a)(\alpha+3)a^{\alpha+2} - (b-a)^2 a^{\alpha+1}(\alpha+2)(\alpha+3)\} - a^\alpha(b-a)^3(\alpha+1)(\alpha+2)(\alpha+3)],$$

$$\eta_3(\alpha) = \frac{1}{3(b-a)^3(\alpha+1)(\alpha+2)(\alpha+3)} \times [3\{2(b^{\alpha+3} - a^{\alpha+3}) - 2(b-a)(\alpha+3)b^{\alpha+2} - (b-a)^2 b^{\alpha+1}(\alpha+2)(\alpha+3)\} - a^\alpha(b-a)^3(\alpha+1)(\alpha+2)(\alpha+3)],$$

$$\eta_4(\alpha) = \frac{1}{6(b-a)^3(\alpha+1)(\alpha+2)(\alpha+3)} \times [6\{(b-a)(\alpha+3)(a^{\alpha+2} + b^{\alpha+2}) + 2(a^{\alpha+3} - b^{\alpha+3})\} - a^\alpha(b-a)^3(\alpha+1)(\alpha+2)(\alpha+3)],$$

$$\theta_2(\alpha) = \frac{1}{3(b-a)^3(\alpha+1)(\alpha+2)(\alpha+3)} \times [-3\{2(b^{\alpha+3} - a^{\alpha+3}) - 2(b-a)(\alpha+3)a^{\alpha+2} - (b-a)^2 a^{\alpha+1}(\alpha+2)(\alpha+3)\} + b^\alpha(b-a)^3(\alpha+1)(\alpha+2)(\alpha+3)],$$

$$\theta_3(\alpha) = \frac{1}{3(b-a)^3(\alpha+1)(\alpha+2)(\alpha+3)} \times [-3\{2(b^{\alpha+3} - a^{\alpha+3}) - 2(b-a)(\alpha+3)b^{\alpha+2} - (b-a)^2 b^{\alpha+1}(\alpha+2)(\alpha+3)\} + b^\alpha(b-a)^3(\alpha+1)(\alpha+2)(\alpha+3)],$$

$$\theta_4(\alpha) = \frac{1}{-6(b-a)^3(\alpha+1)(\alpha+2)(\alpha+3)} \times [6\{(b-a)(\alpha+3)(a^{\alpha+2} + b^{\alpha+2}) + 2(a^{\alpha+3} - b^{\alpha+3})\} + b^\alpha(b-a)^3(\alpha+1)(\alpha+2)(\alpha+3)],$$

and $\Delta_q(a,b)$ *is given in* (4.12).

Proof: From Lemma 4.1, using the property of modulus and exponentially convexity of $|f'|$, we can write

$$\left|\frac{e^{f(a)} + e^{f(b)}}{2} - \frac{\alpha}{b^\alpha - a^\alpha}\int_a^b e^{f(x)}\, d_\alpha x\right| \leq \frac{b-a}{2(b^\alpha - a^\alpha)}$$

$$\left[\int_0^1 [(a(1-t)+tb)^\alpha - a^\alpha]|e^{f(a(1-t)+tb)}f'(a(1-t)+tb)|\, dt + \int_0^1 \{b^\alpha - (a(1-t)+tb)^\alpha\}|e^{f(a(1-t)+tb)}f'(a(1-t)+tb)|\, dt\right]. \quad (4.16)$$

Now by applying the power-mean inequality to the resulting inequality, we get

$$\begin{aligned}
&\int_0^1 [(a(1-t)+tb)^\alpha - a^\alpha]|e^{f(a(1-t)+tb)}f'(a(1-t)+tb)|\, dt \\
&\leq \left(\int_0^1 [(a(1-t)+tb)^\alpha - a^\alpha]\, dt\right)^{1-\frac{1}{q}} \\
&\times\left(\int_0^1 [(a(1-t)+tb)^\alpha - a^\alpha]|e^{f(a(1-t)+tb)}f'(a(1-t)+tb)|^q\right)^{\frac{1}{q}},
\end{aligned} \tag{4.17}$$

and similarly, we have

$$\begin{aligned}
&\int_0^1 [b^\alpha - (a(1-t)+tb)^\alpha]|e^{f(a(1-t)+tb)}f'(a(1-t)+tb)|\, dt \\
&\leq \left(\int_0^1 [b^\alpha - (a(1-t)+tb)^\alpha]\, dt\right)^{1-\frac{1}{q}} \\
&\times\left(\int_0^1 [b^\alpha - (a(1-t)+tb)^\alpha]|e^{f(a(1-t)+tb)}f'(a(1-t)+tb)|^q\right)^{\frac{1}{q}}.
\end{aligned} \tag{4.18}$$

Now by the exponentially convexity of $|f'|^q$ from above, we have

$$\begin{aligned}
&\int_0^1 [(a(1-t)+tb)^\alpha - a^\alpha]|e^{f(a(1-t)+tb)}f'(a(1-t)+tb)|\, dt \\
&\leq \int_0^1 [(a(1-t)+tb)^\alpha - a^\alpha]\{(1-t)|e^{f(a)}|^q + t|e^{f(b)}|^q\} \\
&\quad \times\{(1-t)|f'(a)|^q + t|f'(b)|^q\}\, dt \\
&= \int_0^1 [(a(1-t)+tb)^\alpha - a^\alpha]\{(1-t)^2|e^{f(a)}f'(a)|^q + t^2|e^{f(b)}f'(b)|^q \\
&\quad + t(1-t)[|e^{f(a)}f'(b)|^q + |e^{f(b)}f'(a)|^q]\, dt\} \\
&= \int_0^1 [(a(1-t)+tb)^\alpha - a^\alpha]\{(1-t)^2|e^{f(a)}f'(a)|^q + t^2|e^{f(b)}f'(b)|^q \\
&\quad + t(1-t)\Delta_q(a,b)\}\, dt \\
&= \eta_2(a,b)|e^{f(a)}f'(a)|^q + \eta_3(a,b)|e^{f(b)}f'(b)|^q + \eta_4(a,b)\Delta_q(a,b)
\end{aligned}$$

and

$$\begin{aligned}
&\int_0^1 [b^\alpha - (a(1-t)+tb)^\alpha]|e^{f(a(1-t)+tb)}f'(a(1-t)+tb)|\, dt \\
&\leq \theta_2(a,b)|e^{f(a)}f'(a)|^q + \theta_3(a,b)|e^{f(b)}f'(b)|^q + \theta_4(a,b)\Delta_q(a,b),
\end{aligned} \tag{4.19}$$

where we have the following

$$\begin{aligned}
\eta_1(\alpha) &= \int_0^1 [(a(1-t)+tb)^\alpha - a^\alpha]\, dt \\
&= \frac{b^{\alpha+1} - a^{\alpha+1}}{(\alpha+1)(b-a)} - a^\alpha,
\end{aligned}$$

$$\begin{aligned}\theta_1(\alpha) &= \int_0^1 [b^\alpha - (a(1-t)+tb)^\alpha]\,dt \\ &= b^\alpha - \frac{b^{\alpha+1} - a^{\alpha+1}}{(\alpha+1)(b-a)}.\end{aligned}$$

Hence, we have the result in (4.15). □

Corollary 4.3 *If we set $q = 1$, then under the assumption of Theorem 4.5, we have a new result*

$$\begin{aligned}&\left|\frac{e^{f(a)} + e^{f(b)}}{2} - \frac{\alpha}{b^\alpha - a^\alpha}\int_a^b e^{f(x)}\, d_\alpha x\right| \\ &\leq \frac{b-a}{2(b^\alpha - a^\alpha)} \times [(\eta_2(a,b) + \theta_2(a,b))|e^{f(a)}f'(a)| + (\eta_3(a,b) + \theta_3(a,b))|e^{f(b)}f'(b)| \\ &+ (\eta_4(a,b) + \theta_4(a,b))\Delta(a,b)].\end{aligned}$$

Lemma 4.2 *Let $I \subset \mathbb{R}$ be an open interval, $\varsigma, \zeta \in I$ with $\varsigma < \zeta$ and $\varphi : [\varsigma, \zeta] \to \mathbb{R}$ be a differentiable function such that $(e^\varphi)'$ is integrable and $0 < \alpha \leq 1$ on (ς, ζ) with $\varsigma < \zeta$. If $|(e^\varphi)'|$ is convex on $[\varsigma, \zeta]$, then the following identity holds*

$$\begin{aligned}&\left[\left(\frac{(\zeta-\varsigma)^\alpha + (\zeta - x)^\alpha - (x-\varsigma)^\alpha}{(\zeta-\varsigma)^\alpha}\right)\frac{e^\varphi(\zeta)}{2} + \left(\frac{(\zeta-\varsigma)^\alpha - (\zeta-x)^\alpha + (x-\varsigma)^\alpha}{(\zeta-\varsigma)^\alpha}\right)\frac{e^\varphi(\varsigma)}{2}\right. \\ &\left. - \frac{\Gamma(\alpha+1)}{2(\zeta-\varsigma)^\alpha}[J^\alpha_{\varsigma^+} e^{\varphi(\zeta)} + J^\alpha_{\zeta^-} e^{\varphi(\varsigma)}]\right] = \frac{1}{2}\sum_{i=1}^{4} I_i,\end{aligned}$$

where:

$$\begin{aligned}I_1 &= \frac{(x-\varsigma)^{\alpha+1}}{(\zeta-\varsigma)^\alpha}\int_0^1 (\kappa^\alpha - 1)e^{\varphi(\kappa x + (1-\kappa)\varsigma)}\varphi'(\kappa x + (1-\kappa)\varsigma)\,d\kappa, \\ I_2 &= \frac{(\zeta - x)^{\alpha+1}}{(\zeta-\varsigma)^\alpha}\int_0^1 \left\{\left(\frac{\varsigma-\zeta}{x-\zeta} - \kappa\right)^\alpha - \left(\frac{\varsigma - x}{x-\zeta}\right)^\alpha\right\} e^{\varphi(\kappa x + (1-\kappa)\zeta)} \\ &\quad\times \varphi'(\kappa x + (1-\kappa)\zeta)\,d\kappa, \\ I_3 &= \frac{(\zeta - x)^{\alpha+1}}{(\zeta-\varsigma)^\alpha}\int_0^1 (\kappa^\alpha - 1)e^{\varphi(\kappa x + (1-\kappa)\zeta)}\varphi'(\kappa x + (1-\kappa)\zeta)\,d\kappa, \\ I_4 &= \frac{(x-\varsigma)^{\alpha+1}}{(\zeta-\varsigma)^\alpha}\int_0^1 \left\{\left(\frac{\zeta - x}{x-\varsigma}\right)^\alpha - \left(\frac{\zeta-\varsigma}{x-\varsigma} - \kappa\right)^\alpha\right\} e^{\varphi(\kappa x + (1-\kappa)\varsigma)} \\ &\quad\times \varphi'(\kappa x + (1-\kappa)\varsigma)\,d\kappa.\end{aligned}$$

Theorem 4.6 *Let $\varphi : I = [\varsigma, \zeta] \to \mathbb{R}$ be a differentiable function such that $(e^\varphi)'$ $\in (L[\varsigma, \zeta])$ and $0 < \alpha \leq 1$ on (ς, ζ) and if $|(e^\varphi)'|$ is convex on $[\varsigma, \zeta]$, then the following* *fractional integral inequality holds*

$$\left|\left(\frac{(\zeta-\varsigma)^{\alpha}+(\zeta-x)^{\alpha}-(x-\varsigma)^{\alpha}}{(\zeta-\varsigma)^{\alpha}}\right)\frac{e^{\varphi}(\zeta)}{2}+\left(\frac{(\zeta-\varsigma)^{\alpha}-(\zeta-x)^{\alpha}+(x-\varsigma)^{\alpha}}{(\zeta-\varsigma)^{\alpha}}\right)\frac{e^{\varphi}(\varsigma)}{2}\right.$$
$$\left.-\frac{\Gamma(\alpha+1)}{2(\zeta-\varsigma)^{\alpha}}[J^{\alpha}_{\varsigma+}e^{\varphi(\zeta)}+J^{\alpha}_{\zeta-}e^{\varphi(\varsigma)}]\right|$$
$$\leq\frac{(x-\varsigma)^{\alpha+1}}{(\zeta-\varsigma)^{\alpha+1}}[(\theta_1+\delta_1)|e^{\varphi(x)}\varphi'(x)|+(\theta_2+\delta_2)|e^{\varphi(\varsigma)}\varphi'(\varsigma)|+(\theta_3+\delta_3)\Delta(x,\varsigma)]$$
$$+\frac{(\zeta-x)^{\alpha+1}}{(\zeta-\varsigma)^{\alpha}}[(\theta_1+\rho_1)|e^{\varphi(x)}\varphi'(x)|+(\theta_2+\rho_2)|e^{\varphi(\zeta)}\varphi'(\zeta)|+(\theta_3+\rho_3)\Delta(x,\zeta)],$$

where

$$\theta_1=\int_0^1|1-\kappa^{\alpha}|\kappa^2\,d\kappa=\frac{\alpha}{3(\alpha+3)},$$
$$\theta_2=\int_0^1|1-\kappa^{\alpha}|(1-\kappa)^2\,d\kappa=\frac{\alpha(\alpha^2+6\alpha+5)}{3(\alpha+1)(\alpha+2)(\alpha+3)},$$
$$\theta_3=\int_0^1|1-\kappa^{\alpha}|\kappa(1-\kappa)\,d\kappa=\frac{\alpha(\alpha+5)}{6(\alpha+2)(\alpha+3)},$$
$$=\frac{1}{6}\left(\frac{\zeta-x}{x-\varsigma}\right)^{\alpha}+\frac{(\zeta-x)^{\alpha+2}+(\zeta-\varsigma)^{\alpha+2}}{(\alpha+1)(\alpha+2)(x-\varsigma)^{\alpha+2}}+\frac{2[(\zeta-x)^{\alpha+3}-(\zeta-\varsigma)^{\alpha+3}]}{(\alpha+1)(\alpha+2)(\alpha+3)(x-\varsigma)^{\alpha+3}},$$

$$\delta_1=\int_0^1\left|\left(\frac{\zeta-x}{x-\varsigma}\right)^{\alpha}-\left(\frac{\zeta-\varsigma}{x-\varsigma}-\kappa\right)^{\alpha}\right|\kappa^2\,d\kappa$$
$$=\frac{2[(\zeta-x)^{\alpha+3}-(\varsigma-\zeta)^{\alpha+3}]}{(\alpha+1)(\alpha+2)(\alpha+3)(x-\varsigma)^{\alpha+3}}-\frac{2(\zeta-x)^{\alpha+2}}{(\alpha+1)(\alpha+2)(x-\varsigma)^{\alpha+2}}$$
$$-\frac{(\zeta-x)^{\alpha+1}}{(\alpha+1)(x-\varsigma)^{\alpha+1}}+\frac{(\zeta-x)^{\alpha}}{3(x-\varsigma)^{\alpha}},$$
$$\delta_2=\int_0^1\left|\left(\frac{\zeta-x}{x-\varsigma}\right)^{\alpha}-\left(\frac{\zeta-\varsigma}{x-\varsigma}-\kappa\right)^{\alpha}\right|(1-\kappa)^2\,d\kappa$$
$$=\frac{1}{3}\left(\frac{\zeta-x}{x-\varsigma}\right)^{\alpha}-\frac{(\zeta-\varsigma)^{\alpha+1}}{(x-\varsigma)^{\alpha+1}(\alpha+1)}-\frac{2(\zeta-\varsigma)^{\alpha+2}}{(\alpha+1)(\alpha+2)(x-\varsigma)^{\alpha+2}}$$
$$+\frac{2[(\zeta-x)^{\alpha+3}-(\zeta-\varsigma)^{\alpha+3}]}{(\alpha+1)(\alpha+2)(\alpha+3)(x-\varsigma)^{\alpha+3}},$$
$$\delta_3=\int_0^1\left|\left(\frac{\zeta-x}{x-\varsigma}\right)^{\alpha}-\left(\frac{\zeta-\varsigma}{x-\varsigma}-\kappa\right)^{\alpha}\right|\kappa(1-\kappa)\,d\kappa,$$

$$\rho_1 = \int_0^1 \left| \left(\frac{\varsigma - \zeta}{x - \zeta} - \kappa \right)^\alpha - \left(\frac{\varsigma - x}{x - \zeta} \right)^\alpha \right| \kappa^2 \, d\kappa$$
$$= \frac{2[(\varsigma - \zeta)^{\alpha+3} - (\varsigma - x)^{\alpha+3}]}{(\alpha+1)(\alpha+2)(\alpha+3)(x-\zeta)^{\alpha+3}} - \frac{2(\varsigma - x)^{\alpha+2}}{(\alpha+1)(\alpha+2)(x-\zeta)^{\alpha+2}}$$
$$- \frac{(\varsigma - x)^{\alpha+1}}{(\alpha+1)(x-\zeta)^{\alpha+1}} - \frac{(\varsigma - x)^{\alpha}}{3(x-\zeta)^{\alpha}},$$
$$\rho_2 = \int_0^1 \left| \left(\frac{\varsigma - \zeta}{x - \zeta} - \kappa \right)^\alpha - \left(\frac{\varsigma - x}{x - \zeta} \right)^\alpha \right| (1-\kappa)^2 \, d\kappa$$
$$= \frac{(\varsigma - \zeta)^{\alpha+1}}{(x-\zeta)^{\alpha+1}(\alpha+1)} - \frac{2(\varsigma - \zeta)^{\alpha+2}}{(x-\zeta)^{\alpha+2}(\alpha+1)(\alpha+2)}$$
$$- \frac{2[(\varsigma - x)^{\alpha+3} - (\varsigma - \zeta)^{\alpha+3}]}{(x-\zeta)^{\alpha+3}(\alpha+1)(\alpha+2)(\alpha+3)} - \frac{1}{3}\left(\frac{\varsigma - x}{x - \zeta} \right)^\alpha,$$
$$\rho_3 = \int_0^1 \left| \left(\frac{\varsigma - \zeta}{x - \zeta} - \kappa \right)^\alpha - \left(\frac{\varsigma - x}{x - \zeta} \right)^\alpha \right| \kappa(1-\kappa) \, d\kappa$$
$$= \frac{(\varsigma - x)^{\alpha+2} + (\varsigma - \zeta)^{\alpha+2}}{(\alpha+1)(\alpha+2)(x-\zeta)^{\alpha+2}} - \frac{2[(\varsigma - x)^{\alpha+3} - (\varsigma - \zeta)^{\alpha+3}]}{(x-\zeta)^{\alpha+3}(\alpha+1)(\alpha+2)(\alpha+3)}$$
$$- - \frac{1}{6}\left(\frac{\varsigma - x}{x - \zeta} \right)^\alpha,$$

and

$$\Delta(x, \varsigma) = |e^{\varphi(\varsigma)}\varphi'(x)| + |e^{\varphi(x)}\varphi'(\varsigma)|,$$
$$\Delta(x, \zeta) = |e^{\varphi(\zeta)}\varphi'(x)| + |e^{\varphi(x)}\varphi'(\zeta)|.$$

Theorem 4.7 *Let $\varphi : I = [\varsigma, \zeta] \to \mathbb{R}$ be a differentiable function on the interior I° of I with $\varsigma < \zeta$. If $(e^\varphi)' \in (L[\varsigma, \zeta])$ and $0 < \alpha \leq 1$ on (ς, ζ) and if $|(e^\varphi)'|^q$ is convex on $[\varsigma, \zeta]$ for $q \geq 1$, then the following fractional integral inequality holds*

$$\left| \left(\frac{(\zeta - \varsigma)^\alpha + (\zeta - x)^\alpha - (x - \varsigma)^\alpha}{(\zeta - \varsigma)^\alpha} \right) \frac{e^\varphi(\zeta)}{2} + \left(\frac{(\zeta - \varsigma)^\alpha - (\zeta - x)^\alpha + (x - \varsigma)^\alpha}{(\zeta - \varsigma)^\alpha} \right) \frac{e^\varphi(\varsigma)}{2} \right.$$
$$\left. - \frac{\Gamma(\alpha+1)}{2(\zeta - \varsigma)^\alpha} [J^\alpha_{\varsigma+} e^{\varphi(\zeta)} + J^\alpha_{\zeta-} e^{\varphi(\varsigma)}] \right|$$
$$\leq \frac{(x - \varsigma)^{\alpha+1}}{(\zeta - \varsigma)^{\alpha+1}} (\phi_1)^{1-\frac{1}{q}} [\theta_1 |e^{\varphi(x)}\varphi'(x)| + \theta_2 |e^{\varphi(\varsigma)}\varphi'(\varsigma)| + \Delta_1(x, \varsigma)]$$
$$+ \frac{(\zeta - x)^{\alpha+1}}{(\zeta - \varsigma)^{\alpha+1}} (\phi_2)^{1-\frac{1}{q}} [\theta_1 |e^{\varphi(x)}\varphi'(x)| + \theta_2 |e^{\varphi(\zeta)}\varphi'(\zeta)| + \Delta_2(x, \zeta)]$$

$$+\frac{(x-\varsigma)^{\alpha+1}}{(\zeta-\varsigma)^{\alpha+1}}(\phi_3)^{1-\frac{1}{q}}[\delta_1|e^{\varphi(x)}\varphi'(x)|+\delta_2|e^{\varphi(\varsigma)}\varphi'(\varsigma)|+\Delta_1(x,\varsigma)]$$
$$+\frac{(\zeta-x)^{\alpha+1}}{(\zeta-\varsigma)^{\alpha+1}}(\phi_4)^{1-\frac{1}{q}}[\rho_1|e^{\varphi(x)}\varphi'(x)|+\rho_2|e^{\varphi(\zeta)}\varphi'(\zeta)|+\Delta_2(x,\zeta)],$$

where θ_i, δ_i, and ρ_i, for $i = 1, 2, 3$ are given in Theorem 4.6 and

$$\phi_1=\int_0^1|\kappa^\alpha-1|\,d\kappa=\frac{\alpha}{\alpha+1},$$
$$\phi_2=\int_0^1|1-\kappa^\alpha|\,d\kappa=\frac{\alpha}{\alpha+1},$$
$$\phi_3=\int_0^1\left|\left(\frac{\zeta-x}{x-\varsigma}-\kappa\right)^\alpha-\left(\frac{\zeta-x}{x-\varsigma}\right)^\alpha\right|d\kappa$$
$$=\frac{(\zeta-x)^{\alpha+1}-(\zeta-\varsigma)^{\alpha+1}-(\zeta-x)^\alpha(\alpha+1)(x-\varsigma)}{(\alpha+1)(x-\varsigma)^{\alpha+1}},$$
$$\phi_4=\int_0^1\left|\left(\frac{\varsigma-\zeta}{x-\varsigma}-\kappa\right)^\alpha-\left(\frac{\varsigma-x}{x-\zeta}\right)^\alpha\right|d\kappa$$
$$=\frac{(x-\zeta)^{\alpha+1}-(\varsigma-x)^{\alpha+1}-(\varsigma-x)^\alpha(\alpha+1)(x-\zeta)}{(\alpha+1)(x-\zeta)^{\alpha+1}},$$

$$\Delta_1(x,\varsigma)=[|e^{\varphi(x)}\varphi'(\varsigma)|^q+|e^{\varphi(\varsigma)}\varphi'(x)|^q],$$
$$\Delta_2(x,\zeta)=[|e^{\varphi(x)}\varphi'(\zeta)|^q+|e^{\varphi(\zeta)}\varphi'(x)|^q].$$

Theorem 4.8 *Let $\varphi : I = [\varsigma, \zeta] \to \mathbb{R}$ be a differentiable function on the interior I° of I with $\varsigma < \zeta$. If $(e^\varphi)' \in (L[\varsigma, \zeta])$ and $0 < \alpha \le 1$ on (ς, ζ) and if $|(e^\varphi)'|^q$ is concave on $[\varsigma, \zeta]$ for some fixed $p > 1$ with $q = \frac{p}{p-1}$, then the following fractional integral inequality holds*

$$\Bigg|\left(\frac{(\zeta-\varsigma)^\alpha+(\zeta-x)^\alpha-(x-\varsigma)^\alpha}{(\zeta-\varsigma)^\alpha}\right)\frac{e^{\varphi}(\zeta)}{2}+\left(\frac{(\zeta-\varsigma)^\alpha-(\zeta-x)^\alpha+(x-\varsigma)^\alpha}{(\zeta-\varsigma)^\alpha}\right)\frac{e^{\varphi}(\varsigma)}{2}$$
$$-\frac{\Gamma(\alpha+1)}{2(\zeta-\varsigma)^\alpha}[J^\alpha_{\varsigma+}e^{\varphi(\zeta)}+J^\alpha_{\zeta-}e^{\varphi(\varsigma)}]\Bigg|$$
$$\le\frac{(x-\varsigma)^{\alpha+1}}{(\zeta-\varsigma)^{\alpha+1}}\left[\left|\phi_1e^{\phi\left(\frac{Q_1}{\phi_1}\right)}\phi'\left(\frac{Q_1}{\phi_1}\right)\right|+\left|\phi_3e^{\phi\left(\frac{Q_3}{\phi_3}\right)}\phi'\left(\frac{Q_3}{\phi_3}\right)\right|\right]$$
$$+\frac{(\zeta-x)^{\alpha+1}}{(\zeta-\varsigma)^{\alpha+1}}\left[\left|\phi_2e^{\phi\left(\frac{Q_2}{\phi_2}\right)}\phi'\left(\frac{Q_2}{\phi_2}\right)\right|+\left|\phi_4e^{\phi\left(\frac{Q_4}{\phi_4}\right)}\phi'\left(\frac{Q_4}{\phi_4}\right)\right|\right],$$

where

$$Q_1=\frac{\alpha^2(x+\varsigma)+\alpha(x+3\varsigma)}{2(\alpha+1)(\alpha+2)},$$

$$Q_2 = \frac{-\alpha^2(x+\varsigma) - \alpha(x+3\varsigma)}{2(\alpha+1)(\alpha+2)},$$

$$Q_3 = \frac{(x+\varsigma)(\zeta-x)^\alpha}{2(x-\varsigma)^\alpha} + \frac{x(\zeta-\varsigma)^{\alpha+1} - \varsigma(\zeta-\varsigma)^{\alpha+1}}{(x-\varsigma)^{\alpha+1}(\alpha+1)} - \frac{(\zeta-x)^{\alpha+2} - (\zeta-\varsigma)^{\alpha+2}}{(x-\varsigma)^{\alpha+1}(\alpha+1)(\alpha+2)},$$

$$Q_4 = \frac{\zeta(\varsigma-\zeta)^{\alpha+1} - x(\varsigma-x)^{\alpha+1}}{(\alpha+1)(x-\zeta)^{\alpha+1}} - \frac{(\varsigma-x)^{\alpha+2} - (\varsigma-\zeta)^{\alpha+2}}{(\alpha+2)(x-\zeta)^{\alpha+1}} - \frac{(\varsigma-x)^\alpha(x+\zeta)}{2(x-\zeta)^\alpha}.$$

Bibliography

1 Kilbas, A.A., Sirivastava, H.M., and Trujillo, J.J. (2006). *Theory and Applications of Fractional Differential Equations, North-Holland Mathematics Studies*, vol. **204**. Amsterdam: Elsevier Sci. B.V.

2 Abdeljawad, T. (2015). On conformable fractional calculus. *J. Comput. Appl. Math.* **279**: 57–66.

3 Khalil, R., Al Horani, M., Yousef, A., and Sababheh, M. (2014). A new definition of fractional derivative. *J. Comput. Appl. Math.* **264**: 65–70.

4 Jarad, F., Uğurlu, E., Abdeljawad,T., and Baleanu, D. (2017). On a new class of fractional operators. *Adv. Differ. Equ.* **2017**(1): 247. https://doi.org/10.1186/s13662-017-1306-z.

5 Katugampola, U.N. (2011). New approach to a generalized fractional integrals. *Appl. Math. Comput.* **218** (4): 860–865.

6 Katugampola, U.N. (2014). New approach to a generalized fractional derivatives. *Bull. Math.Anal. Appl.* **6** (4): 1–15.

7 Antczak, T. (2001). On (p,r)-invex sets and functions. *J. Math. Anal. Appl.* **263**: 355–379.

8 Kiryakova, V. (1993). *Generalized Fractional Calculus and Applications*. Harlow, England: Longman Scientific and Technical.

9 Raina, R.K. (2005). On generalized Wright's hypergeometric functions and fractional calculus operators. *East Asian Math. J.* **21** (2): 191–203.

10 Dragomir, S.S. and Gomm, I. (2015). Some Hermite–Hadamard type inequalities for functions whose exponentials are convex. *Stud. Univ. Babes-Bolyai Math.* **60** (4): 527–534.

11 Alirezaei, G. and Mathar, R. (2018). On exponentially concave functions and their impact in information theory. *J. Inf. Theory Appl.* **9**(5): 265–274.

12 Pal, S. and Wong, T.K.L. (2018). On exponentially concave functions and a new information geometry. *Ann. Probab.*1605.05819.

13 Pecaric, J. and Jaksetic, J. (2013). On exponential convexity, EulerRadau expansions and stolarsky means. *Rad Hrvat. Mat. Znan.* **17** (515): 81–94.

14 Fejer, L. (1906). Uberdie fourierreihen. *Math. Naturwise. Anz Ungar. Akad., Wiss* **24**: 369–390(in Hungarian).

15 Set, E., Akdemir, A.O., and Mumcu, I. (2017). Ostrowski type inequalities involving special functions via conformable fractional integrals. *J. Adv. Math. Stud.* **10** (3): 386–395.

16 Set, E., Mumcu, I., and Özdemir, M.E. (2017). On the more general Hermite-Hadamard type inequalities for convex functions via conformable fractional integrals. *Topol. Algebra Appl.* **5** (1): 67–73.

17 Akdemir, A.O., Ekinci, A., and Set, E. (2017). Conformable fractional integrals and related new integral inequalities. *J. Nonlinear Convex Anal.* **18** (4): 661–674.

18 Akdemir, A.O., Set,E., and Mumcu, I. (2018). Hadamard's inequality and its extensions for conformable fractional integrals of any order $\alpha > 0$. *Creat. Math.Inf.* **27**(2): 197–206.

5

Lyapunov-type Inequalities for Local Fractional Proportional Derivatives

Thabet Abdeljawad[1,2,3]

[1] *Department of Mathematics and General Sciences, Prince Sultan University, Riyadh, Saudi Arabia*
[2] *Department of Medical Research, China Medical University, Taichung, Taiwan*
[3] *Department of Computer Science and Information Engineering, Asia University, Taichung, Taiwan*

5.1 Introduction

Fractional calculus has proved to be an active field of research [1–4]. Many researchers use integration and differentiation with respect to arbitrary order to generalize many dynamic equations and integral inequalities [5–18]. The concepts of local and nonlocal differentiation and integration are topics that started to be discussed in the community of fractional calculus. Both concepts allow us to integrate or differentiate with respect to arbitrary order. The existence of the delay in the structure of nonlocal fractional derivatives appear in the form of convolution type. Such structure makes nonlocal fractional derivatives more suitable to describe or model complex systems. Both local and nonlocal fractional derivatives are important and have many applications. Moreover, the theory of local fractional differentiation and integration is essential to develop the nonlocal one. In fact, local fractional integrals and derivatives can always be iterated to generate nonlocal ones. The authors in [19] iterated a type of conformable derivatives [20, 21] (as type of local fractional) to generate generalized nonlocal type of fractional derivatives. Following the same line, the authors in [22] recently used a modified type of conformable derivatives as a kind of local fractional derivatives to generate nonlocal fractional proportional derivatives and integrals whereas their discrete versions within nabla have been announced in [23].

Integral inequalities are effective tools in studying and developing the theory of dynamic equations which contribute in many real world applications. Many types of integral inequalities have been announced and investigated in literature such as Gronwal, Lyapunov, Hadamard, and Minkowski. One of the most important inequalities is the so-called Lyapunov inequality, which will be under investigation

Fractional Order Analysis: Theory, Methods and Applications, First Edition.
Edited by Hemen Dutta, Ahmet Ocak Akdemir, and Abdon Atangana.

in the current work. Such inequality was announced when the author in [24] proved that if the boundary value problem (BVP)

$$\psi''(t) + q(t)\psi(t) = 0, \quad t \in (a, b), \ \psi(a) = \psi(b) = 0$$

has a nontrivial solution, where q is a real-valued continuous function, then

$$\int_a^b |q(s)|ds > \frac{4}{b-a}. \tag{5.1}$$

After then, many researchers started to generalize and modify Lyapunov-type inequalities by studying different types of BVPs [25–32]. An attractive way of the generalization of Lyapunov inequalities has appeared in the frame of fractional calculus [33–46], discrete fractional calculus [47–49] and q-fractional calculus [50].

Motivated by what we mentioned above, in this work, our main goal is to consider BVPs involving local fractional proportional derivatives [20, 21] which are considered as modification of the conformable derivatives [51–53]. We shall prove and relate a certain Lyapunov-type inequality in the frame of sequential local fractional proportional BVPs to the Lyapunov inequality, which has been proved recently in [42] for the nonlocal proportional ones. Then, we present an open problem for a more general sequential BVP within local fractional proportional operators. To make the relation between the local and local-type inequalities clear to the readers, we review some results about local and nonlocal fractional Lyapunov inequalities. Finally, we shall motivate the researchers by extending local fractional proportional derivatives and integrals to the higher order and presenting a corresponding Lyapunov open problem.

This chapter is organized as follows. In Section 5.2, we present notations and basic concepts related to local and nonlocal fractional proportional operators. In Section 5.3, we review some recent results about Lyapunov-type inequalities for BVPs in the frame of some local and nonlocal fractional operators, where the classical Riemann–Liouville and Caputo fractional derivatives are considered nonlocal fractional derivatives that are generated by the local usual derivatives and integrals. In Section 5.4, we present a Lyapunov inequality for sequential local fractional proportional operators as a special case of that in the nonlocal case and present an open problem for a more general sequential BVP that cannot be considered as special case of the nonlocal fractional proportional one. In Section 5.5, we extend the local fractional proportional calculus to higher order and present a Lyapunov-type inequality open problem. Finally, in Section 5.6, we list our conclusions.

5.2 The Local Fractional Proportional Derivatives and Their Generated Nonlocal Fractional Proportional Integrals and Derivatives

In [20], Anderson et al. defined the modified conformable derivative (called proportional derivative as well) by

Definition 5.1 (Modified conformable derivatives) Let $\rho \in [0,1]$, and let the functions $\kappa_0, \kappa_1 : [0,1] \times \mathbb{R} \to [0,\infty)$ be continuous such that for all $t \in \mathbb{R}$, we have

$$\lim_{\rho\to 0^+} \kappa_1(\rho,t) = 1,\ \lim_{\rho\to 0^+} \kappa_0(\rho,t) = 0,\ \lim_{\rho\to 1^-} \kappa_1(\rho,t) = 0,\ \lim_{\rho\to 1^-} \kappa_0(\rho,t) = 1,$$

and $\kappa_1(\rho,t) \neq 0, \rho \in [0,1), \kappa_0(\rho,t) \neq 0, \rho \in (0,1]$. Then, the modified conformable differential operator of order ρ is defined by

$$D^\rho \psi(t) = \kappa_1(\rho,t)\psi(t) + \kappa_0(\rho,t)\psi'(t).$$

Of special interest, the authors in [22] considered the case $\kappa_1(\rho,t) = 1-\rho$ and $\kappa_0(\rho,t) = \rho$. That is

$$D^\rho \psi(t) = (1-\rho)\psi(t) + \rho\psi'(t). \tag{5.2}$$

We shall call this derivative a local fractional proportional derivative with constant reference functions. Then, they defined the proportional integral associated to D^ρ by

$${}_aI^{1,\rho}\psi(t) = {}_aI^\rho\psi(t) = \frac{1}{\rho}\int_a^t e^{\frac{\rho-1}{\rho}(t-s)}\psi(s)ds,$$

so that the following action becomes valid.

Proposition 5.1 [22] *For f defined on* $[a,\infty)$ *and differentiable on* (a,∞) *and* $\rho \in (0,1)$, *we have*

$${}_aI^{1,\rho}D^\rho\psi(t) = {}_aI^\rho D^\rho\psi(t) = \psi(t) - e^{\frac{\rho-1}{\rho}(t-a)}\psi(a).$$

The above proportional integrals were iterated in [22] to generate the following generalized type fractional integral:

Definition 5.2 [22] For $\rho > 0$ and $\alpha \in \mathbb{C}, Re(\alpha) > 0$, we define the (left) nonlocal fractional proportional integral of ψ by

$$({}_aI^{\alpha,\rho}\psi)(t) = \frac{1}{\rho^\alpha\Gamma(\alpha)}\int_a^t e^{\frac{\rho-1}{\rho}(t-\tau)}(t-\tau)^{\alpha-1}\psi(\tau)d\tau. \tag{5.3}$$

For $\rho = 1$, the nonlocal fractional proportional integral coincides with the Riemann-Liouville integral. That is

$$({}_aI^{\alpha,1}\psi)(t) = (\ {}_a\mathbf{I}^\alpha\psi)(t) = \frac{1}{\Gamma(\alpha)}\int_a^t (t-s)^{\alpha-1}\psi(s)ds. \tag{5.4}$$

For the definitions of Riemann–Liouville (${}_aD^\alpha$) and Caputo fractional derivative (${}_a^CD^\alpha$) and their properties, we refer the reader to Refs. [1–4].

Definition 5.3 [22] For $\rho > 0$ and $\alpha \in \mathbb{C}, Re(\alpha) > 0$, we define the (left) nonlocal fractional proportional derivative of ψ by

$$({}_aD^{\alpha,\rho}\psi)(t) = D_a^{n,\rho}I^{n-\alpha,\rho}f(t) = \frac{D_t^{n,\rho}}{\rho^{n-\alpha}\Gamma(n-\alpha)}\int_a^t e^{\frac{\rho-1}{\rho}(t-\tau)}(t-\tau)^{n-\alpha-1}\psi(\tau)d\tau, \tag{5.5}$$

where $n = [Re(\alpha)] + 1$ and $D^{n,\rho}$ means the iteration of the operator D^ρ n-times.

Clearly, if we let $\rho = 1$ in Definition 5.3, then we obtain the left Riemann–Liouville fractional derivative. Also, $\lim_{\alpha\to 0}(D^{\alpha,\rho}f)(t) = f(t)$ and $\lim_{\alpha\to 1}(D^{\alpha,\rho}f)(t) = (D^\rho f)(t)$.

The proportional integral operators ${}_aI^{1,\rho}$ clearly do not have the semigroup property in ρ. However, for the generated fractional proportional integrals, we can state:

Theorem 5.1 [22] **(The semigroup property for the fractional proportional integrals)** *If $\rho > 0, Re(\alpha) > 0$, and $Re(\beta) > 0$, then, for f continuous and defined for $t \geq a$, we have*

$$I^{\alpha,\rho}({}_aI^{\beta,\rho}f)(t) = I^{\beta,\rho}({}_aI^{\alpha,\rho}f)(t) = ({}_aI^{\alpha+\beta,\rho}f)(t). \tag{5.6}$$

For the local fractional conformable derivatives, we refer the reader to [51, 52] and for the fundamental conformable Sturm–Liouville problem, we refer to [53].

Definition 5.4 [22] For $\rho \in [0, 1]$ and $\alpha \in \mathbb{C}$ with $Re(\alpha) > 0$, we define the left generalized Caputo fractional (proportional) derivative starting at a by

$$\begin{aligned}({}_a^CD^{\alpha,\rho}\psi)(t) &= {}_aI^{n-\alpha,\rho}(D^{n,\rho}\psi)(t)\\ &= \frac{1}{\rho^{n-\alpha}\Gamma(n-\alpha)}\int_a^t e^{\frac{\rho-1}{\rho}(t-s)}(t-s)^{n-\alpha-1}(D^{n,\rho}\psi)(s)ds,\end{aligned} \tag{5.7}$$

where $n = [Re(\alpha)] + 1$.

Theorem 5.2 [22] *For $\rho > 0$ and $n = [Re(\alpha] + 1$, we have*

$$ {}_aI^{\alpha,\rho}({}_a^CD^{\alpha,\rho}f)(t) = f(t) - \sum_{k=0}^{n-1} \frac{(D^{k,\rho}f)(a)}{\rho^k k!}(t-a)^k e^{\frac{\rho-1}{\rho}(t-a)}. \tag{5.8} $$

Proposition 5.2 *For any $\alpha \in \mathbb{C}$ with $Re(\alpha) > 0$ and $\rho > 0$, $n = [Re(\alpha)] + 1$, we have*

$$ ({}_a^CD^{\alpha,\rho}f)(t) = ({}_aD^{\alpha,\rho}f)(t) - \sum_{k=0}^{n-1} \frac{\rho^{\alpha-k}}{\Gamma(k+1-\alpha)}(t-a)^{k-\alpha} e^{\frac{\rho-1}{\rho}(t-a)}(D^{k,\rho}f)(a). \tag{5.9} $$

For more about local and local fractional proportional derivatives and integrals and the their discrete versions, we refer to [22, 23].

5.3 Lyapunov-Type Inequalities for Some Nonlocal and Local Fractional Operators

In this section, we review some Lyapunov-type inequalities for some local and nonlocal fractional derivatives.

The author in [33–35] proved the following fractional versions of Lyapunov-type inequalities.

Theorem 5.3 [33] *If the following Riemann–Liouville fractional BVP:*

$$ {}_aD^{\alpha}\psi(t) + q(t)\psi(t) = 0, \quad t \in (a,b),\ \psi(a) = \psi(b) = 0 \tag{5.10} $$

has a nontrivial solution with $q(t)$ is continuous real-valued function, then

$$ \int_a^b |q(s)|ds > \Gamma(\alpha)\left(\frac{4}{b-a}\right)^{\alpha-1}. \tag{5.11} $$

As an application of Theorem 5.12, it was proved that for $\alpha \in (1,2]$, the Mittag-Leffler function $E_\alpha(w)$ has no real zeros for $|w| \leq \Gamma(\alpha)4^{\alpha-1}$. The discrete version of Theorem 5.12 in the sense of delta was proved in [47]. A generalization of Theorem 5.12 has been proved in [54].

Theorem 5.4 [35] *If the following Caputo fractional BVP:*

$$ {}_a^CD^{\alpha}\psi(t) + q(t)\psi(t) = 0, \quad t \in (a,b),\ \psi(a) = \psi(b) = 0 \tag{5.12} $$

has a nontrivial solution with $q(t)$ is continuous real-valued function, then

$$\int_a^b |q(s)|ds > \frac{\Gamma(\alpha)\alpha^\alpha}{[(\alpha-1)(b-a)]^{\alpha-1}}. \tag{5.13}$$

As an application of Theorem 5.4, it was proved that for $\alpha \in (1,2]$, the Mittag-Leffler function $E_{\alpha,2}(w)$ has no real zeros for $w \in [-\Gamma(\alpha)\frac{\alpha^\alpha}{(\alpha-1)^{\alpha-1}})$. A generalization of Theorem 5.4 has been obtained in [55].

Consider the sequential Riemann–Liouville fractional BVP

$$\begin{aligned} &{}_aD_a^{\alpha_1} D^{\alpha_2}\psi(t) + q(t)\psi(t) = 0, \quad t \in (a,b),\ 0 < \alpha_1, \alpha_2 \leq 1, \\ &1 < \mu = \alpha_1 + \alpha_2 \leq 2,\ \psi(a) = \psi(b) = 0. \end{aligned} \tag{5.14}$$

Making use of the fact that ((5.14)) can be reduced to

$${}_aD^\mu\psi(t) + q(t)\psi(t) = 0, \quad t \in (a,b),\ \mu = \alpha_1 + \alpha_2,\ \psi(a) = \psi(b) = 0, \tag{5.15}$$

then the following result follows easily by applying Theorem 5.3.

Theorem 5.5 [34] *If the sequential Riemann–Liouville fractional BVP* (5.14) *has a nontrivial continuous real solution with $q(t)$ is continuous real-valued function, then*

$$\int_a^b |q(s)|ds > \Gamma(\alpha)\left(\frac{4}{b-a}\right)^{\gamma-1}, \quad \gamma = \alpha_1 + \alpha_2. \tag{5.16}$$

For that reason, the author in the same work [34] proved the following sequential Caputo Lyapunov-type inequality.

Theorem 5.6 [34] *If the BVP*

$$\begin{aligned} &{}_a^CD^{\alpha_1}\,{}_a^CD^{\alpha_2}\psi(t) + q(t)\psi(t) = 0,\ t \in (a,b), 0 < \alpha_1, \alpha_2 \leq 1, \\ &1 < \mu = \alpha_1 + \alpha_2 \leq 2,\ \psi(a) = \psi(b) = 0 \end{aligned} \tag{5.17}$$

has a nontrivial continuous solution with $q(t)$ continuous real function, then

$$\int_a^b |q(s)|ds > \frac{\Gamma(\alpha_1+\alpha_2)(\alpha_1+2\alpha_2-1)^{\alpha_1+2\alpha_2-1}}{(b-a)^{\alpha_1+\alpha_2-1}(\alpha_1+\alpha_2-1)^{\alpha_1+\alpha_2-1}\alpha_2^{\alpha_2}}. \tag{5.18}$$

Consider the local conformable fractional BVP:

$$\mathbb{T}_a^\rho\psi(t) + q(t)\psi(t) = 0,\ t \in (a,b),\ 1 < \rho \leq 2,\ \psi(a) = \psi(b) = 0. \tag{5.19}$$

The authors in [44] proved the following Lyapunov inequality

Theorem 5.7 [44] *If the BVP (5.19) has a nontrivial solution, where q is a real-valued continuous function on [a, b], then*

$$\int_a^b |q(t)|dt > \frac{\rho^\rho}{(\rho-1)^{\rho-1}(b-a)^{\rho-1}}, \quad 1<\rho\leq 2. \tag{5.20}$$

Consider the following sequential conformable BVP:

$$(\mathbb{T}_a^\rho \mathbb{T}_a^\rho \psi)(t) + q(t)\psi(t) = 0,\ a<t<b,\ \frac{1}{2}<\rho\leq 1,\ \psi(a)=\psi(b)=0. \tag{5.21}$$

In [44], the following inequality was proved as well.

Theorem 5.8 *If the BVP (5.21) has a nontrivial solution, where q is a real-valued continuous function on [a, b], then*

$$\int_a^b |q(t)|dt > \frac{3\rho-1}{(b-a)^{2\rho-1}}\left(\frac{3\rho-1}{2\rho-1}\right)^{\frac{2\rho-1}{\rho}}, \quad \frac{1}{2}<\rho\leq 1. \tag{5.22}$$

Consider the following fractional proportional BVP:

$$({}_aD^{\alpha,\rho}\psi)(t)+q(t)\psi(t)=0,\ 1<\alpha\leq 2, \rho\in[0,1],\ t\in(a,b),\ \psi(a)=\psi(b)=0. \tag{5.23}$$

Theorem 5.9 [42] *Assume the fractional proportional BVP (5.23) has a nontrivial solution $\psi(t)$ in the space*

$$C[a,b] = \{f : [a,b]\to\mathbb{R} : f \quad \text{is continuous}\},$$

and that the condition

$$\frac{(\alpha-1)\rho}{1-\rho}\geq b-a, \quad a\geq 0, \tag{5.24}$$

is satisfied, then we must have

$$\int_a^b |q(s)|ds > \frac{\rho^\alpha\Gamma(\alpha)4^{\alpha-1}}{(b-a)^{\alpha-1}}. \tag{5.25}$$

Consider the following fractional proportional BVP:

$$({}_a^CD^{\alpha,\rho}\psi)(t)+q(t)\psi(t)=0,\ 1<\alpha\leq 2, \rho\in[0,1],\ t\in(a,b),\ \psi(a)=\psi(b)=0. \tag{5.26}$$

Theorem 5.10 [42] *If the BVP (5.26) has a continuous nontrivial solution, where $\frac{\rho}{1-\rho} \geq b - a$ and $q(t)$ is continuous as well, then the following inequality holds*

$$\int_a^b |q(\tau)| d\tau > \frac{(\rho\alpha)^\alpha \Gamma(\alpha)}{(\alpha - 1)^{\alpha-1}(b - a)^{\alpha-1}}. \tag{5.27}$$

Consider the following weighted fractional proportional BVP:

$$\begin{aligned} &({}_aD^{\alpha,\rho}y)(t) + (t-a)^\beta q(t)y(t) = 0,\ 1 < \alpha \leq 2,\ \beta \geq 2-\alpha, \rho \in [0,1], \\ &t \in (a,b),\ y(a^+) = y(b) = 0, \end{aligned} \tag{5.28}$$

where $q(t) \in C[a,b]$ and

$$\begin{aligned} y \in \Omega = C^{\alpha,\rho}_{2-\alpha}[a,b] = \{f : (a,b] : (x-a)^{2-\alpha} f(x) \in C[a,b], \\ (x-a)^{2-\alpha}\, {}_aD^{\alpha,\rho} f(x) \in C[a,b]\}. \end{aligned}$$

The authors used the following Lyapunov result of the above weighted BVP (5.28) to estimate an upper bound for the free zero disk of the Kilbas-Saigo Mittag-Leffler functions of three parameters.

Theorem 5.11 [42] *Assume the fractional proportional BVP (5.28), with the condition*

$$\frac{\rho}{1-\rho}(\beta + \alpha - 1) \geq b - a,\ \beta \geq 2 - \alpha \tag{5.29}$$

is satisfied, has a nontrivial solution $\psi(t)$ in the space $\Omega_\beta \supseteq \Omega$, then we must have

$$\int_a^b |q(s)| ds > \frac{1}{h_\beta(s^*)}, \tag{5.30}$$

where $h_\beta(t) = \frac{(t-a)^{\beta+\alpha-1}(b-t)^{\alpha-1}}{\rho^\alpha \Gamma(\alpha)(b-a)^{\alpha-1}}$, $s^* = \frac{(\alpha-1)a+(\beta+\alpha-1)b}{\alpha}$, and

$$\begin{aligned} \Omega = C^{\alpha,\rho}_{2-\alpha}[a,b] = \{f : (a,b] : (x-a)^{2-\alpha} f(x) \in C[a,b], \\ (x-a)^{2-\alpha}\, {}_aD^{\alpha,\rho} f(x) \in C[a,b] \quad \}. \end{aligned}$$

The special case $\beta = 2 - \alpha$ yields the following result.

Theorem 5.12 *Assume the fractional proportional BVP (5.28) with $\beta = 2 - \alpha$ has a nontrivial solution $\psi(t)$ in the space Ω and the condition*

$$\frac{\rho}{1-\rho} \geq b - a,\ a \geq 0, \tag{5.31}$$

is satisfied, then we must have

$$\int_a^b |q(s)| ds > \frac{\alpha^\alpha \Gamma(\alpha) \rho^\alpha}{(\alpha - 1)^{\alpha-1}(b - a)}. \tag{5.32}$$

For the extension to higher-order and Lyapunov-type inequalities of fractional operators with exponential (*CF* or Caputo–Fabrizio) and Mittag-Leffler kernels (*AB* or Atangana–Baleanu) and the discrete versions, we refer the reader to [43–49]. Such types of operators have nonsingular convolution type kernels and in the confirmation in the limiting cases $\alpha \to 0$ or $\alpha \to 1$ the theory of delta Dirac functions is used. More precisely, such fractional derivatives and their corresponding integrals do not depend on an iterative process through the so-called Cauchy–Leibniz formula. There is debate regarding such operators, specially the *CF*-type, whether they are nonlocal or not. In fact, since the exponential kernel can be separated then under certain transformation the *CF*-dynamic equation can be transformed to an ordinary one [56–58]. However, it is not possible to separate the Mittag-Leffler kernel and apply similar transformations.

5.4 The Lyapunov Inequality for the Sequential Local Fractional Proportional Boundary Value Problem

In this section, we present outline of the proof steps of the Lyapunov inequality regarding a special case of Theorem 5.9. In fact, the special case $\alpha = 2$ will reduce to the below sequential local fractional proportional BVP.

$$(D^\rho D^\rho \psi)(t) + q(t)\psi(t) = 0, \ \ \rho \in (1/2, 1], \ \ t \in (a, b), \ \ \psi(a) = \psi(b) = 0. \tag{5.33}$$

Even though, the next results regarding (5.33) are the special cases of Theorem 5.9 with $\alpha = 2$, for the sake of showing detailed steps in proving Lyapunov-type inequalities for the sequential local fractional derivatives, we shall go through them.

Lemma 5.1 *A continuous function $\psi(t)$ is a solution of the BVP* (5.33) *if and only if it satisfies the integral equation*

$$\psi(t) = \int_a^b \mathcal{G}(t, s)q(s)\psi(s)ds, \tag{5.34}$$

where $\mathcal{G}$ is the Green function for (5.33) *defined by*

$$\mathcal{G}(t, s) = \frac{1}{\rho^2}\begin{cases} \left[\dfrac{(t-a)e^{\gamma(t-a)}e^{\gamma(b-s)}(b-s)}{(b-a)e^{\gamma(b-a)}} - (t-s)e^{\gamma(t-s)}\right], & a \le s \le t \le b, \\ \dfrac{(t-a)e^{\gamma(t-a)}e^{\gamma(b-s)}(b-s)}{(b-a)e^{\gamma(b-a)}}, & a \le t \le s \le b, \end{cases}$$

where $\gamma = \frac{\rho-1}{\rho}$.

Proof: Apply the local fractional proportional integral operator ${}_aI^\rho$ to both sides of (5.33) and make use of Proposition 5.1 to get

$$D^\rho\psi(t) = e^{\gamma(t-a)}(D^\rho\psi)(a) + {}_aI^\rho[-q(t)\psi(t)]. \tag{5.35}$$

If we apply the integral operator ${}_aI^\rho$ once again, then we have

$$\psi(t) - e^{\gamma(t-a)}\psi(a) = {}_aI^\rho[e^{\gamma(t-a)}(D^\rho\psi)(a)] + I^{2,\rho}[-q(t)\psi(t)]. \tag{5.36}$$

By noting that

$${}_aI^\rho[e^{\gamma(t-a)}(D^\rho\psi)(a)] = \frac{(D^\rho\psi)(a)}{\rho}e^{\gamma(t-a)}(t-a), \tag{5.37}$$

we conclude that

$$\psi(t) = \psi(a)e^{\gamma(t-a)} + \frac{(D^\rho\psi)(a)}{\rho}(t-a)e^{\gamma(t-a)} - \frac{1}{\rho^2}\int_a^t e^{\gamma(t-s)}(t-s)q(s)\psi(s)ds. \tag{5.38}$$

That is,

$$\psi(t) = d_1e^{\gamma(t-a)} + d_2(t-a)e^{\gamma(t-a)} - \frac{1}{\rho^2}\int_a^t e^{\gamma(t-s)}(t-s)q(s)\psi(s)ds. \tag{5.39}$$

The boundary condition $\psi(a) = 0$ implies that $d_1 = 0$ and after that the boundary condition $\psi(b) = 0$ will result in

$$d_2 = \frac{1}{\rho^2(b-a)e^{\gamma(b-a)}}\int_a^b e^{\gamma(b-s)}(b-sq(s)\psi(s)ds). \tag{5.40}$$

Therefore,

$$\psi(t) = \frac{(t-a)e^{\gamma(t-a)}}{(b-a)\rho^2e^{\gamma(b-a)}}\int_a^b e^{\gamma(b-s)}(b-s)q(s)\psi(s)ds$$
$$-\frac{1}{\rho^2}\int_a^t e^{\gamma(t-s)}(t-s)q(s)\psi(s)ds. \tag{5.41}$$

Finally, if we split the first integral in (5.41), we obtain the representation of the Green's function as stated in the theorem. □

Lemma 5.2 *Under the assumption that*

$$\frac{\rho}{1-\rho} \geq b-a, \quad a \geq 0, \tag{5.42}$$

the Green function $\mathcal{G}$ formulated above in Lemma 5.1 has the following properties:

i) $\mathcal{G}(t,s) \geq 0$, *for all* $a \leq t, s \leq b$.
ii) $\max_{t\in[a,b]} \mathcal{G}(t,s) = \mathcal{G}(s,s) = \frac{(s-a)(b-s)}{\rho^2(b-a)}$ *for* $s \in [a,b]$.

iii) $g(s) = \mathcal{G}(s,s)$ *has a unique maximum, given by*

$$\max_{s\in[a,b]} \mathcal{G}(s,s) = g\left(\frac{a+b}{2}\right) = \frac{(b-a)}{4\rho^2}.$$

Proof: The first part is clear and the proof of the second part is straight forward. The proof of the third part can be achieved by noting that $g'(t) = 0$ at $t = \frac{a+b}{2}$. In fact, the details of the proof are the special case $\alpha = 2$ of Lemma 2 in [42]. □

The proof of the following theorem is the special case $\alpha = 2$ of the proof of Theorem 5.9 as was shown in [42].

Theorem 5.13 *Assume the fractional proportional BVP (5.33) has a nontrivial solution $\psi(t)$ in the space*

$$C[a,b] = \{f : [a,b] \to \mathbb{R} : f \quad \text{is continuous}\},$$

and that the condition

$$\frac{\rho}{1-\rho} \geq b-a, \quad a \geq 0, \tag{5.43}$$

is satisfied, then we must have

$$\int_a^b |q(s)|ds > \frac{4\rho^2}{(b-a)}. \tag{5.44}$$

Now, we consider the following more general sequential local fractional proportional BVP.

$$(D^{\rho_1} D^{\rho_2}\psi)(t) + q(t)\psi(t) = 0,\ 0 < \rho_1, \rho_2 \leq 1, \quad 1 < \rho_1 + \rho_2 \leq 2, \tag{5.45}$$

$$t \in (a,b), \quad \psi(a) = \psi(b) = 0.$$

Open Problem 5.1: Noting that ${}_aI^{\rho}{}_aI^{\rho}\psi(t) = {}_aI^{2,\rho}$ and that

$${}_aI^{\rho_1}{}_aI^{\rho_2}\psi(t) \neq {}_aI^{\rho_1+\rho_2}\psi(t)$$

in general, it is of logic to ask whether it is possible or complicated to obtain a Lyapunov-type inequality for the problem (5.45)!

Also, such an open problem is still open for this more general sequential BVP in case of conformable derivatives!

Remark 5.1 Notice that when we apply the local fractional proportional integral operators ${}_aI^{\rho_1}$ and ${}_aI^{\rho_2}$, respectively, to Eq. (5.45), for $\rho_1 \neq \rho_2 \gamma_1 \neq \gamma_2$ we get

$$\begin{aligned}\psi(t) &= e^{\gamma_2(t-a)}\psi(a) + {}_aI^{\rho_2}[e^{\gamma_1(t-a)}({}_aD^{\rho_1}\psi)(a)] + {}_aI^{\rho_2}{}_aI^{\rho_1}[-q(t)\psi(t)] \\ &= e^{\gamma_2(t-a)}\psi(a) + \frac{(D^{\rho_1}\psi)(a)}{\rho_2(\gamma_1-\gamma_2)}[e^{\gamma_1(t-a)} - e^{\gamma_2(t-a)}] + {}_aI_a^{\rho_2}I^{\rho_1}[-q(t)\psi(t)] \\ &= c_1 + c_2[e^{\gamma_1(t-a)} - e^{\gamma_2(t-a)}] + {}_aI^{\rho_2}{}_aI^{\rho_1}[-q(t)\psi(t)],\end{aligned} \tag{5.46}$$

where $\gamma_1 = \frac{\rho_1-1}{\rho_1}$ and $\gamma_2 = \frac{\rho_2-1}{\rho_2}$. Also, it is evident that the condition $\psi(a) = 0$ implies that $c_1 = 0$. In Open Problem 5.1, it remains to use the boundary condition $\psi(b) = 0$ to find c_2, find the corresponding Green's function, study its properties and finally to formulate, if possible, a Lyapunov-type inequality.

5.5 A Higher-Order Extension of the Local Fractional Proportional Operators and an Associate Lyapunov Open Problem

In this section, we define the higher-order local fractional proportional derivatives and integrals, give a particular case example, and present a Lyapunov-type open problem.

Definition 5.5 Let $n < \rho \leq n+1, n = 0, 1, 2, \ldots$, and $\sigma = \rho - n$. Then, the local fractional proportional derivative and integral of order ρ are defined by

$${}_aD^{\rho}\psi(t) = {}_aD^{\sigma}D^{n}\psi(t) = (1-\sigma)D^{n}\psi(t) + \sigma D^{n+1}\psi(t) \tag{5.47}$$

and

$${}_aI^{\rho}\psi(t) = {}_a\mathbf{I}^{n}{}_aI^{\sigma}\psi(t) = \frac{1}{(n-1)!}\int_a^t (t-s)^{n-1}({}_aI^{\sigma}\psi)(s)ds. \tag{5.48}$$

Above, ${}_a\mathbf{I}^n$ is the nth order Riemann–Liouville fractional integral and D^n is the nth order differential operator.

Example 5.1 Let $1 < \rho \leq 2$. Then, $\sigma = \rho - 1, n = 1, \gamma_\sigma = \frac{\sigma-1}{\sigma} = \frac{\rho-1}{\rho-2}$, and hence

$$\begin{aligned}{}_aI^{\rho}\psi(t) &= \frac{1}{\rho-1}\int_a^t \left(\int_a^s e^{\gamma_\sigma(s-\tau)}\psi(\tau)d\tau\right) ds \\ &= \frac{1}{\rho-1}\int_a^t \psi(\tau)\left(\int_\tau^t e^{\gamma_\sigma(s-\tau)ds}\right) d\tau \\ &= \frac{1}{(\rho-1)\gamma_\sigma}\int_a^t \psi(\tau)[e^{\gamma_\sigma(t-\tau)} - 1]d\tau.\end{aligned} \tag{5.49}$$

Theorem 5.14 *For any $\rho > 0$ and integrable function ψ on $[a, t]$, we have*

$$({}_aI^\rho D^\rho \psi)(t) = \psi(t) - \sum_{k=0}^{n-1} \frac{(D^k\psi)(a)}{k!}(t-a)^k - (D^n\psi)(a)E_{1,n+1}(\gamma_\sigma, t-a). \tag{5.50}$$

Proof: Following Definition 5.5, Proposition 5.1 and using the fact that

$${}_a\mathbf{I}^n D^n \psi(t)\psi(t) = \psi(t) - \sum_{k=0}^{n-1} \frac{(D^k\psi)(a)}{k!}(t-a)^k,$$

we have

$$\begin{aligned}
({}_aI^\rho D^\rho \psi)(t) &= = {}_a\mathbf{I}^n {}_aI^\sigma D^\sigma D^n \psi(t) \\
&= {}_a\mathbf{I}^n [D^n\psi(t) - e^{\gamma_\sigma(t-a)}(D^n\psi)(a)] \\
&= \psi(t) - \sum_{k=0}^{n-1} \frac{(D^k\psi)(a)}{k!}(t-a)^k - (D^n\psi)(a){}_a\mathbf{I}^n[e^{\gamma_\sigma(t-a)}] \\
&= \psi(t) - \sum_{k=0}^{n-1} \frac{(D^k\psi)(a)}{k!}(t-a)^k - (D^n\psi)(a)E_{1,n+1}(\gamma_\sigma, t-a), \quad (5.51)
\end{aligned}$$

where $\sigma = n - \rho$ and $\gamma_\sigma = \frac{\sigma-1}{\sigma} = \frac{n-\rho-1}{n-\rho}$. The notion $e^{\gamma_\sigma(t-a)} = E_{1,1}(\gamma_\sigma, t-a)$ and the fact that ${}_a\mathbf{I}^\mu E_{\alpha,1}(\lambda, z) = E_{\alpha,\mu+1}(\lambda, z)$ have been used above. □

Open Problem 5.2: Is possible to prove a Lyapunov-type inequality for the higher local fractional proportional BVP:

$$(D^\rho\psi)(t) + q(t)\psi(t) = 0, \; 1 < \rho \le 2, \; t \in (a, b), \; \psi(a) = \psi(b) = 0. \tag{5.52}$$

Remark 5.2 Regarding the above open problem, we remind that it was possible to prove a Lyapunov-type inequality for the higher-order conformable derivatives and that the local fractional proportional derivatives are modifications of the conformable derivatives as a kind of local fractional derivatives. The conformable analogue of Open Problem 5.2 was done in Theorem 5.7 or the reference [44] included therein.

Remark 5.3 If we apply the integral operator ${}_aI^\rho, 1 < \rho \le 2$ on the problem (5.52) and make use of Theorem 5.14 and Example 5.1, then we have

$$\begin{aligned}
\psi(t) &= c_1 + c_2E_{1,2}(\gamma_\sigma, t-a) + {}_aI^\rho[-q(t)\psi(t)] \\
&= c_1 + c_2E_{1,2}(\gamma_\sigma, t-a) + \frac{1}{(\rho-1)\gamma_\sigma}\int_a^t [-q(\tau)\psi(\tau)][e^{\gamma_\sigma(t-\tau)} - 1]d\tau \\
&= c_1 + c_2E_{1,2}(\gamma_\sigma, t-a) + \frac{1}{(\rho-2)}\int_a^t [-q(\tau)\psi(\tau)][e^{\gamma_\sigma(t-\tau)} - 1]d\tau, \quad (5.53)
\end{aligned}$$

where $\sigma = \rho - 1$ and $\gamma_\sigma = \frac{\sigma-1}{\sigma} = \frac{\rho-2}{\rho-1}$. Since $E_{1,2}(\gamma_\sigma, 0) = 0$, then it is clear that the boundary condition $\psi(a) = 0$ will imply that $c_1 = 0$. In Open Problem 5.2, it remains to use the boundary condition $\psi(b) = 0$ to find c_2, find the corresponding Green's function, study its properties and finally to formulate, if possible, a Lyapunov-type inequality.

5.6 Conclusion

Our conclusion comments are summarized as follows:

1) Ordinary derivatives and integrals are local operators that used to generate nonlocal fractional operators such as Riemann–Liouville and Caputo derivatives. Conformable and proportional derivatives are types of local fractional operators, which can be used to generate more general nonlocal fractional operators. We reviewed some Lyapunov-type inequalities for some local fractional operators and their corresponding nonlocal ones to stress on the relation between them.
2) We have presented a Lyapunov inequality for a sequential local fractional proportional BVP and showed that it is a special case of the nonlocal one by letting $\alpha = 2$. On the other hand, we proposed a Lyapunov open problem for a more sequential problem that cannot be a special case of the nonlocal one. In fact, this open problem is also still open in the frame of conformable derivatives.
3) We have extended local fractional proportional operators to higher order and presented a Lyapunov open problem.
4) All the proven Lyapunov-type inequalities for either local or nonlocal fractional operators tend to the classical or ordinary Lyapunov inequality (5.1) as $\alpha \to 2$ in the higher-order case and as $\alpha \to 1$ in the sequential case.
5) It is important to generalize Lyapunov inequalities for both local and nonlocal fractional operators.

Acknowledgement

The author would like to thank Prince Sultan University for funding this work through research group Nonlinear Analysis Methods in Applied Mathematics (NAMAM) group number RG-DES-2017-01-17.

Bibliography

1 Hilfer, R. (2000). *Applications of Fractional Calculus in Physics*. Singapore: Word Scientific.

2 Kilbas, A., Srivastava, H.M., and Trujillo, J.J. (2006). *Theory and Application of Fractional Differential Equations, North Holland Mathematics Studies*, vol. **204**. Amsterdam, The Netherland: Elsevier.

3 Podlubny, I. (1999). *Fractional Differential Equations*. San Diego, CA: Academic Press.

4 Samko, S.G., Kilbas, A.A., and Marichev, O.I. (1993). *Fractional Integrals and Derivatives: Theory and Applications*. Yverdon: Gordon and Breach.

5 Rashid, S., Akdemir, A.O., Jarad, F. et al. (2019). Simpson's type integral inequalities for κ-fractional integrals and their applications. *AIMS Math.* **4** (4): 10871100.

6 Rashid, S., Noor, M.A., Noor, K.I., and Akdemir, A.O. (2019). Some new generalizations for exponentially s-convex functions and inequalities via fractional operators. *Fractal Fract.* **3** (2): 24. https://doi.org/10.3390/fractalfract3020024.

7 Set, E., Akdemir, A.O., and Mumcu, I. (2017). Ostrowski type inequalities involving special functions via conformable fractional integrals. *J. Adv. Math. Stud.* **10** (3): 386395.

8 Set, E., Akdemir, A.O., and Çelik, B. (2018). On generalization of Fejér type inequalities via fractional integral operator. *Filomat* **32** (16):5537–5547.

9 Akdemir, A.O., Set, E., and Mumcu, H.Code.Idot. (2018). Hadamard's inequality and its extensions for conformable fractional integrals of any order α ¿ 0, *Creat. Math. Infor.* **27** (2):197–206.

10 Yaldız, H. and Akdemir, A.O. (2018). Katugampola fractional integrals within the class of convex functions. *Turk. J. Sci.* **3** (1): 4050.

11 Set, E., Akdemir, A.O., and Alan, E.A. (2019). Hermite–Hadamard and Hermite–Hadamard–Fejer type inequalities involving fractional integral operators. *FILOMAT* **33** (8).

12 Akdemir, A.O., Set, E., and Ekinci, A. (2019). On new conformable fractional integral inequalities for product of different kinds of convexity. *TWMS J. Appl. Eng. Math.* **9** (1): 142150.

13 Rashid, S., Abdeljawad, T., Jarad, F., and Noor, M.A. (2019). Some estimates for generalized RiemannLiouville fractional integrals of exponentially convex functions and their applications. *Mathematics* **7**: 807. https://doi.org/10.3390/math7090807.

14 Rahman, G., Khan, A., Abdeljawad, T., and Nisar, K.S. (2019). The Minkowski inequalities via generalized proportional fractional integral operators. *Adv. Differ. Equ.* **2019** (1): 287.

15 Khan, H., Abdeljawad, T., Tunç, C. et al. (2019). Minkowski's inequality for the AB-fractional integral operator. *J. Inequal. Appl.* **2019**: 96.

16 Alzabut, J., Abdeljawad, T., Jarad, F., and Sudsutad, W. (2019). A Gronwall inequality via the generalized proportional fractional derivative with applications. *J. Inequal. Appl.* **2019**: 101.

17 Adjabi, Y., Jarad, F., and Abdeljawad, T. (2017). On generalized fractional operators and a Gronwall type inequality with applications. *Filomat* **31** (17): 54575473.

18 Ma, Q.-H., Ma, C., and Wang, J. (2017). A Lyapunov-type inequality for a fractional differential equation with Hadamard derivative. *J. Math. Inequal.* **11** (1): 135141.

19 Jarad, F., Ugurlu, E., Abdeljawad, T., and Baleanu, D. (2017). On a new class of fractional operators. *Adv. Differ. Equ.* **2017**: 247.

20 Anderson, D.R. and Ulness, D.J. (2015). Newly defined conformable derivatives. *Adv. Dyn. Sys. Appl.* **10** (2): 109137.

21 Anderson, D.R. (2017). Second-order self-adjoint differential equations using a proportional-derivative controller. *Commun. Appl. Nonlinear Anal.* **24**: 1748.

22 Jarad, F., Abdeljawad, T., and Alzabut, J. (2017). Generalized fractional derivatives generated by a class of local proportional derivatives. *Eur. Phys. J. Spec. Top.* **226** (1618): 34573471, online 25 July 2018.

23 Abdeljawad, T., Jarad, F., and Alzabut, J. (2017). Fractional proportional differences with memory. *Eur. Phys. J. Spec. Top.* **226** (1618): 33333354 (online 25 July 2018).

24 Lyapunov, A.M. (1907). Probleme général de la stabilité du mouvment. *Ann. Fac. Sci. Univ. Toulouse* **2**: 227247; Reprinted in: *Ann. Math. Stud.* No. 17, Princeton, 1947.

25 Pinasco, J.P. (2013). *Lyapunov-Type Inequalities, Springer Briefs in Mathematics*. New York: Springer.

26 Çakmak, D. (2010). Lyapunov-type integral inequalities for certain higher order differential equations. *Appl. Math. Comput.* **216**: 368373.

27 Clark, S. and Hinton, D.B. (2010). A Lyapunov-type inequality for linear Hamiltonian systems. *Math. Inequal. Appl.* **1**: 201209.

28 Parhi, N. and Panigrahi, S. (2002). A Lyapunov-type integral inequality for higher order differential equations. *Math. Slovaca* **52**: 3146.

29 Yang, X. (2003). On Lyapunov-type inequality for certain higher-order differential equations. *Appl. Math. Comput.* **134** (23): 307317.

30 Yang, X. and Lo, K. (2010). Lyapunov-type inequality for a class of even-order differential equations. *Appl. Math. Comput.* **215** (11): 38843890.

31 Tiryaki, A. (2010). Recent developments of Lyapunov type inequalities. *Adv. Dyn. Syst. Appl.* **5** (2): 231248.

32 Agarwal, R. and Ozbekler, A. (2016). Lyapunov type inequalities for second -order differential equations with mixed nonlinearities. *Analysis* **36** (4): 245252.

33 Ferreira, R.A.C. (2013). A Lyapunov-type inequality for a fractional initial value problem. *Fract. Calc. Appl. Anal.* **16** (4): 978984.

34 Ferreira, R.A.C. (2016). Lyapunov-type inequalities for some sequential fractional initial value problems. *Adv. Dyn. Syst. Appl.* **11** (1): 33–43.

35 Ferreira, R.A.C. (2014). On a Lyapunov-type inequality and the zeros of a certain Mittag-Leffler function. *J. Math. Anal. Appl.* **412** (2): 10581063.

36 Jleli, M. and Samet, B. (2015). Lyapunov-type inequalities for fractional boundary value problems. *Electron. J. Differ. Equ.* **2015** (88): 111.

37 Khaldi, R. and Guezane-Lakoud, A. (2017). Lyapunov inequality for a boundary value problem involving conformable derivative. *Progr. Fract. Differ. Appl.* **3** (4): 323329.

38 Chidouh, A. and Torres, D. (2017). A generalized Lyapunov's inequality for fractional boundary value problem. *J. Comput. Appl. Math.* **312**: 192197.

39 Rong, J. and Bai, Ch. (2015). Lyapunov type inequality for fractional differential equation with fractional boundary conditions. *Adv. Differ. Equ.* **2015**: 82.

40 Sitho, S., Ntouyas, S.K., Yukunthorn, W., and Tariboon, J. (2016). Lyapunov's type inequalities for hybrid fractional differential equations. *J. Inequal. Appl.* **2016**: 170.

41 Arifi, N.A., Altun, I., Jleli, M. et al. (2016). Lyapunov-type inequalities for fractional p=Laplacian equation. *J. Inequal. Appl.* **2016**: 189.

42 Abdeljawad, T., Jarad, F., Mallak, S., and Alzabut, J. (2019). Lyapunov type inequalities via fractional proportional derivatives and application on the free zero disc of KilbasSaigo generalized MittagLeffler functions. *Eur. Phys. J. Plus* **134**: 247. https://doi.org/10.1140/epjp/i2019-12772-1.

43 Abdeljawad, T., Agarwal, R.P., Alzabut, J. et al. (2018). Lyapunov-type inequalities for mixed non-linear forced differential equations within conformable derivatives. *J. Inequal. Appl.* **2018**: 143.

44 Abdeljawad, T., Alzabut, J., and Jarad, F. (2017). A generalized Lyapunov-type inequality in the frame of conformable derivatives. *Adv. Differ. Equ.* Article Number: 321 Published: OCT 11.

45 Abdeljawad, T. (2017). Fractional operators with exponential kernels and a Lyapunov type inequality. *Advances in Difference Equations* **6**: Article Number: 313.

46 Abdeljawad, T. (2017). A Lyapunov type inequality for fractional operators with nonsingular Mittag-Leffler kernel. *J. Inequal. Appl.* **2017**: 130. https://doi.org/10.1186/s13660-017-1400-5.

47 Ferreira, R.A.C. (2015). Some discrete fractional Lyapunov-type inequalities. *Fract. Differ. Calc.* **5** (1): 8792.

48 Abdeljawad, T. and Madjidi, F. (2017). Lyapunov type inequalities for fractional difference operators with discrete Mittag-Leffler kernels of order 2¡ α ¡ 5/2. *Eur. Phys. J. Spec. Top.* **226** (1618): 33553368.

49 Abdeljawad, T., Al-Mdallal, Q.M., and Hajji, M.A. (2017). Arbitrary order fractional difference operators with discrete exponential kernels and applications. *Discrete Dyn. Nat. Soc.* Article ID 4149320, 8 pages.

50 Jleli, M. and Samet, B. (2016). A Lyapunov-type inequality for a fractional q-difference boundary value problem. *J. Nonlinear. Sci. Appl.* **9**: 19651976.

51 Abdeljawad, T. (2013). On conformable fractional calculus. *J. Comput. Appl. Math.* **279**: 5766.

52 Khalil, R., Al Horani, M., Yousef, A., and Sababheh, M. (2014). A new definition of fractional derivative. *J. Comput. Appl. Math.* **264**: 6570.

53 Al-Refai, M. and Abdeljawad, T. (2017). Fundamental results of conformable SturmLiouville eigenvalue problems. *Complexity* **2017**: Article ID 3720471, 7 pages.

54 Jleli, M., Kirane, M., and Samet, B. (2017). HartmanWintner-type inequality for a fractional boundary value problem via a fractional derivative with respect to another function. *Discrete Dyn. Nat. Soc.* **2017**: Article ID 5123240, 8 pages. https://doi.org/10.1155/2017/5123240.

55 Jarad, F., Adjabi, Y., Abdeljawad, T. et al. Lyapunov type inequality in the frame of generalized Caputo derivatives. *Discrete Contin. Dyn. Syst., Ser. S.* https://doi.org/10.3934/dcdss.2020212.

56 Al-Refai, M. (2018). Reduction of order formula and fundamental set of solutions for linear fractional differential equations. *Appl. Math. Lett.* **82**: 813.

57 Al-Refai, M. and Pal, K. (2019). New aspects of CaputoFabrizio fractional derivative. *Progr. Fract. Differ. Appl.* **5** (2): 157166.

58 Al-Refai, M. and Jarrah, A.M. (2019). Fundamental results on weighted CaputoFabrizio fractional derivative. *Chaos, Solitons Fractals* **126**: 711.

6

Minkowski-Type Inequalities for Mixed Conformable Fractional Integrals

Erhan Set[1] *and Muhamet E. Özdemir*[2]

[1] *Department of Mathematics, Faculty of Science and Arts, Ordu University, Campus, Ordu, Turkey*
[2] *Department of Mathematics Education, Education Faculty, Uludağ University, Görükle Campus, Bursa, Turkey*

6.1 Introduction and Preliminaries

In the mathematical literature (see, e.g. [1]), the well-known Minkowski integral inequality have been stated as follows.

Theorem 6.1 *Let* $p \geq 1$, $0 < \int_a^b f^p(x)dx < \infty$ *and* $0 < \int_a^b g^p(x)dx < \infty$. *Then,*

$$\left(\int_a^b (f(x)+g(x))^p dx\right)^{\frac{1}{p}} \leq \left(\int_a^b f^p(x)dx\right)^{\frac{1}{p}} + \left(\int_a^b g^p(x)\right)^{\frac{1}{p}}. \tag{6.1}$$

The inequality (6.1) has attracted many researchers attention, and several of extensions, generalizations, variants of this inequality have been presented in the literature. In this context, L. Bougoffa obtained the reverse Minkowski inequality for classical Riemann integrals in 2000 as the following.

Theorem 6.2 [2] *Let f and g be positive functions satisfying*

$$0 < m \leq \frac{f(x)}{g(x)} \leq M, \;\; \forall x \in [a,b].$$

Then,

$$\left(\int_a^b f^p(x)dx\right)^{\frac{1}{p}} + \left(\int_a^b g^p(x)\right)^{\frac{1}{p}} \leq c\left(\int_a^b (f(x)+g(x))^p dx\right)^{\frac{1}{p}},$$

where $c = \frac{M(m+1)+(M+1)}{(m+1)(M+1)}$.

Fractional Order Analysis: Theory, Methods and Applications, First Edition.
Edited by Hemen Dutta, Ahmet Ocak Akdemir, and Abdon Atangana.

Fractional calculus, which deals with integration and differentiation according to arbitrary order, is one of the most interesting subjects in mathematics in the last 20 years. Many mathematicians like Liouville, Riemann, Erdelyi, Kober, Hadamard, and Weyl made major contributions to the theory of fractional calculus and have defined fractional integral operators that called with names of them. Riemann–Liouville fractional integral operators come at the beginning of this type of operator. Recently, many generalizations of Riemann–Liouville fractional integral operators have been obtained, one of which is mixed conformable fractional integral operator obtained using conformable fractional operators defined by T. Abdeljawad (see [3, 4]). On the other hand, one of the ways to be obtain generalizations of the classical integral inequalities existing in literature is also to use fractional integral operators. By using different types of fractional integrals, many generalizations, extension and variations of inequalities such as Hermite–Hadamard, Ostrowski, Grüss, Chebyshev, Simpson, and Minkowski have been obtained. The aim of this study is to obtain new generalizations of the Minkowski-type inequalities using mixed conformable fractional integrals.

Now, some fractional integral operators and Minkowski-type inequalities obtained with the help of these operators will be given in the following order.

Definition 6.1 Let $f \in L[a,b]$. The Riemann–Liouville integrals $J^{\alpha}_{a+}f$ and $J^{\alpha}_{b-}f$ of order $\alpha > 0$ with $a \geq 0$ are defined by

$$J^{\alpha}_{a+}f(x) = \frac{1}{\Gamma(\alpha)} \int_a^x (x-t)^{\alpha-1} f(t)dt, \; x > a$$

and

$$J^{\alpha}_{b-}f(x) = \frac{1}{\Gamma(\alpha)} \int_x^b (t-x)^{\alpha-1} f(t)dt, \; x < b$$

respectively, where $\Gamma(\alpha) = \int_0^{\infty} e^{-t}u^{\alpha-1}du$. Here is $J^{0}_{a+}f(x) = J^{0}_{b-}f(x) = f(x)$.

Theorem 6.3 [5] *Let $\alpha > 0$, $p \geq 1$ and let f, g be two positive functions on $[0, \infty)$, such that for all $t > 0$, $J^{\alpha}f^{p}(t) < \infty$, $J^{\alpha}g^{p}(t) < \infty$. If $0 < m \leq \frac{f(\tau)}{g(\tau)} \leq M$, $\tau \in [0,t]$, then we have*

$$\left[J^{\alpha}f^{p}(t)\right]^{\frac{1}{p}} + \left[J^{\alpha}g^{p}(t)\right]^{\frac{1}{p}} \leq \frac{1+M(m+2)}{(m+1)(M+1)} \left[J^{\alpha}(f+g)^{p}(t)\right]^{\frac{1}{p}}.$$

Theorem 6.4 [5] *Let $\alpha > 0$, $p \geq 1$ and let f, g be two positive functions on $[0, \infty)$, such that for all $t > 0$, $J^{\alpha}f^{p}(t) < \infty$, $J^{\alpha}g^{p}(t) < \infty$. If $0 < m \leq \frac{f(\tau)}{g(\tau)} \leq M$, $\tau \in [0,t]$, then we have*

$$\left[J^{\alpha}f^{p}(t)\right]^{\frac{2}{p}} + \left[J^{\alpha}g^{p}(t)\right]^{\frac{2}{p}} \geq \left(\frac{(M+1)(m+1)}{M} - 2\right) \left[J^{\alpha}f^{p}(t)\right]^{\frac{1}{p}} \left[J^{\alpha}g^{p}(t)\right]^{\frac{1}{p}}.$$

Definition 6.2 [6] The Hadamard fractional integral of order $\alpha > 0$ of function $f(x)$, for all $x > 1$ is defined as

$$_H D_{1,x}^{-\alpha} f(x) = \frac{1}{\Gamma(\alpha)} \int_1^x \ln\left(\frac{x}{t}\right)^{\alpha-1} f(t)\frac{dt}{t},$$

where $\Gamma(\alpha) = \int_0^\infty e^{-u}u^{\alpha-1}du$.

Theorem 6.5 *[7, 8] Let $\alpha > 0$, $p \geq 1$ and let f, g be two positive functions on $[0, \infty)$, such that for all $t > 0$, $_H D_{1,t}^{-\alpha} f^p(t) < \infty$, $_H D_{1,t}^{-\alpha} g^p(t) < \infty$. If $0 < m \leq \frac{f(\tau)}{g(\tau)} \leq M$, $\tau \in [0, t]$, then we have*

$$\left[{_H D_{1,t}^{-\alpha}} f^p(t)\right]^{\frac{1}{p}} + \left[{_H D_{1,t}^{-\alpha}} g^p(t)\right]^{\frac{1}{p}} \leq \frac{1+M(m+2)}{(m+1)(M+1)}\left[{_H D_{1,t}^{-\alpha}}(f+g)^p(t)\right]^{\frac{1}{p}}.$$

Theorem 6.6 *[7, 8] Let $\alpha > 0$, $p \geq 1$ and let f, g be two positive functions on $[0, \infty)$, such that for all $t > 0$, $_H D_{1,t}^{-\alpha} f^p(t) < \infty$, $_H D_{1,t}^{-\alpha} g^p(t) < \infty$. If $0 < m \leq \frac{f(\tau)}{g(\tau)} \leq M$, $\tau \in [0, t]$, then we have*

$$\left[{_H D_{1,t}^{-\alpha}} f^p(t)\right]^{\frac{2}{p}} + \left[{_H D_{1,t}^{-\alpha}} g^p(t)\right]^{\frac{2}{p}} \geq \left(\frac{(M+1)(m+1)}{M} - 2\right)$$
$$\left[{_H D_{1,t}^{-\alpha}} f^p(t)\right]^{\frac{1}{p}}\left[{_H D_{1,t}^{-\alpha}} g^p(t)\right]^{\frac{1}{p}}.$$

Definition 6.3 Let $\alpha > 0$, $\beta, \eta \in \mathbb{R}$, then the Saigo fractional integral $I_{0,x}^{\alpha,\beta,\eta}[f(x)]$ of order a for a real-valued continuous function $f(x)$ is defined by [9], see also ([10], [11, p. 19])

$$I_{0,x}^{\alpha,\beta,\eta}[f(x)] = \frac{x^{-\alpha-\beta}}{\Gamma(\alpha)} \int_0^x (x-t)^{\alpha-1}\, {_2F_1}\left(\alpha+\beta, -\eta; \alpha; 1-\frac{t}{x}\right) f(t)dt, \tag{6.2}$$

where the function $_2F_1(-)$ in the right-hand side of (6.2) is the Gaussian hypergeometric function defined by

$$_2F_1(a, b; c; x) = \sum_{n=0}^{\infty} \frac{(a)_n(b)_n}{(c)_n}\frac{x^n}{n!},$$

and $(a)_n$ is the Pochhammer symbol

$$(a)_n = a(a+1)\cdots(a+n-1) = \frac{\Gamma(\alpha+n)}{\Gamma(\alpha)}, \; (a)_0 = 1.$$

Theorem 6.7 [12] *Let $p \geq 1$ and let f, g be two positive functions on $[0, \infty)$, such that for all $x > 0$, $I_{0,x}^{\alpha,\beta,\eta}[f^p(x)] < \infty$, $I_{0,x}^{\alpha,\beta,\eta}[g^q(x)] < \infty$. If $0 < m \leq \frac{f(\tau)}{g(\tau)} \leq M$, $\tau \in (0, x)$, we have*

$$\left[I_{0,x}^{\alpha,\beta,\eta}[f^p(x)]\right]^{\frac{1}{p}} + \left[I_{0,x}^{\alpha,\beta,\eta}[g^q(x)]\right]^{\frac{1}{p}} \leq \frac{1+M(m+2)}{(m+1)(M+1)}\left[I_{0,x}^{\alpha,\beta,\eta}(f+g)^p(x)\right]^{\frac{1}{p}},$$

for any $\alpha > \max\{0, -\beta\}$, $\beta < 1$, $\beta - 1 < \eta < 0$.

Theorem 6.8 [12] *Let $p \geq 1$ and let f, g be two positive functions on $[0, \infty)$, such that for all $x > 0$, $I_{0,x}^{\alpha,\beta,\eta}[f^p(x)] < \infty$, $I_{0,x}^{\alpha,\beta,\eta}[g^q(x)] < \infty$. If $0 < m \leq \frac{f(\tau)}{g(\tau)} \leq M$, $\tau \in (0, x)$, we have*

$$\left[I_{0,x}^{\alpha,\beta,\eta}[f^p(x)]\right]^{\frac{2}{p}} + \left[I_{0,x}^{\alpha,\beta,\eta}[g^q(x)]\right]^{\frac{2}{p}}$$
$$\geq \left(\frac{(M+1)(m+1)}{M} - 2\right)\left[I_{0,x}^{\alpha,\beta,\eta}[f^p(x)]\right]^{\frac{1}{p}}\left[I_{0,x}^{\alpha,\beta,\eta}[g^q(x)]\right]^{\frac{1}{p}}.$$

Theorem 6.9 [12] *Let $p > 1$, $\frac{1}{p} + \frac{1}{q} = 1$ and let f, g be two positive functions on $[0, \infty)$, such that $I_{0,x}^{\alpha,\beta,\eta}[f(x)] < \infty$, $I_{0,x}^{\alpha,\beta,\eta}[g(x)] < \infty$. If $0 < m \leq \frac{f(\tau)}{g(\tau)} \leq M$, $\tau \in [0, x]$, we have*

$$\left[I_{0,x}^{\alpha,\beta,\eta}[f(x)]\right]^{\frac{1}{p}}\left[I_{0,x}^{\alpha,\beta,\eta}[g(x)]\right]^{\frac{1}{q}} \leq \left(\frac{M}{m}\right)^{\frac{1}{pq}}\left[I_{0,x}^{\alpha,\beta,\eta}[f(x)]^{\frac{1}{p}}[g(x)]^{\frac{1}{q}}\right].$$

For all $x > 0$, $\alpha > \max\{0, -\beta\}$, $\beta < 1$, $\beta - 1 < \eta < 0$.

Theorem 6.10 [12] *Let f and g be two positive function on $[0, \infty)$, such that $I_{0,x}^{\alpha,\beta,\eta}[f^p(x)] < \infty$, $I_{0,x}^{\alpha,\beta,\eta}[g^q(x)] < \infty$, $x > 0$. If $0 < m \leq \frac{f^p(\tau)}{g^q(\tau)} \leq M$, $\tau \in [0, x]$. Then we have*

$$\left[I_{0,x}^{\alpha,\beta,\eta}[f^p(x)]\right]^{\frac{1}{p}}\left[I_{0,x}^{\alpha,\beta,\eta}[g^q(x)]\right]^{\frac{1}{q}} \leq \left(\frac{M}{m}\right)^{\frac{1}{pq}}\left[I_{0,x}^{\alpha,\beta,\eta}[f(x)][g(x)]\right],$$

where $p > 1$, $\frac{1}{p} + \frac{1}{q} = 1$, for all $x > 0$, $\alpha > \max\{0, -\beta\}$, $\beta < 1$, $\beta - 1 < \eta < 0$.

Definition 6.4 [13] The (k, s)-fractional conformable integrals (left and right) of order $\alpha \in \mathbb{C}$, $Re(\alpha) > 0$ of a continuous function $f(x)$ on $[0, \infty)$ are given as follows:

$$\mathfrak{F}_{a^+,k}^{\alpha,s} f(t) = \frac{(s)^{1-\frac{\alpha}{k}}}{k\Gamma_k(\alpha)} \int_a^t ((t-a)^s - (x-a)^s)^{\frac{\alpha}{k}-1}(x-a)^{s-1} f(x)dx,$$
$$0 \leq a < t < b \leq \infty$$

and

$$\mathfrak{F}_{b^-,k}^{\alpha,s} f(t) = \frac{(s)^{1-\frac{\alpha}{k}}}{k\Gamma_k(\alpha)} \int_a^t ((b-x)^s - (b-t)^s)^{\frac{\alpha}{k}-1}(b-x)^{s-1} f(x)dx,$$
$$0 \leq a < t < b \leq \infty,$$

respectively, if integrals exist, where $k > 0$, $s \in \mathbb{R}\backslash 0$.

Theorem 6.11 [13] *For $k > 0$, $s \in \mathbb{R}\backslash\{0\}$, $\alpha > 0$, and $p \geq 1$. Let $f_1, f_2 \in L_{1,s}[a, t]$ be two positive functions on $[0, \infty)$, such that, for all $t > a$, $\mathfrak{F}_{a^+,k}^{\alpha,s} f_1^p(t) < \infty$, $\mathfrak{F}_{a^+,k}^{\alpha,s} f_2^p(t) < \infty$. If $0 < m \leq \frac{f_1(x)}{f_2(x)} \leq M$ for $m, M \in \mathbb{R}^+$ and for all $x \in [a, t]$, then*

$$\left[\mathfrak{F}_{a^+,k}^{\alpha,s} f_1^p(t)\right]^{\frac{1}{p}} + \left[\mathfrak{F}_{a^+,k}^{\alpha,s} f_2^p(t)\right]^{\frac{1}{p}} \leq c_1\left[\mathfrak{F}_{a^+,k}^{\alpha,s} (f_1 + f_2)^p(t)\right]^{\frac{1}{p}},$$

with $c_1 = \frac{M(m+1)+(M+1)}{(m+1)(M+1)}$.

Theorem 6.12 [13] *For* $k > 0, s \in \mathbb{R}\backslash\{0\}, \alpha > 0,$ *and* $p \geq 1$. *Let* $f_1, f_2 \in L_{1,s}[a, t]$ *be two positive functions on* $[0, \infty)$, *such that, for all* $t > a$, $\mathfrak{T}^{\alpha,s}_{a^+,k} f_1^p(t) < \infty$, $\mathfrak{T}^{\alpha,s}_{a^+,k} f_2^p(t) < \infty$. *If* $0 < m \leq \frac{f_1(x)}{f_2(x)} \leq M$ *for* $m, M \in \mathbb{R}^+$ *and for all* $x \in [a, t]$, *then*

$$\left[\mathfrak{T}^{\alpha,s}_{a^+,k} f_1^p(t)\right]^{\frac{2}{p}} + \left[\mathfrak{T}^{\alpha,s}_{a^+,k} f_2^p(t)\right]^{\frac{2}{p}} \geq c_2 \left[\mathfrak{T}^{\alpha,s}_{a^+,k} f_1^p(t)\right]^{\frac{1}{p}} \left[\mathfrak{T}^{\alpha,s}_{a^+,k} f_2^p(t)\right]^{\frac{1}{p}},$$

with $c_2 = \left(\frac{(M+1)(m+1)}{M} - 2\right)$.

Theorem 6.13 [13] *For* $k > 0, s \in \mathbb{R}\backslash\{0\}, \alpha > 0,$ *and* $p, q \geq 1$. *Let* $f_1, f_2 \in L_{1,s}[a, t]$ *be two positive functions on* $[0, \infty)$, *such that, for all* $t > a$, $\mathfrak{T}^{\alpha,s}_{a^+,k} f_1^p(t) < \infty$, $\mathfrak{T}^{\alpha,s}_{a^+,k} f_2^p(t) < \infty$. *If* $0 < m \leq \frac{f_1(x)}{f_2(x)} \leq M$ *for* $m, M \in \mathbb{R}^+$ *and for all* $x \in [a, t]$, *then*

$$\left[\mathfrak{T}^{\alpha,s}_{a^+,k} f_1^p(t)\right]^{\frac{1}{p}} \left[\mathfrak{T}^{\alpha,s}_{a^+,k} f_2^p(t)\right]^{\frac{1}{p}} \leq \left(\frac{M}{m}\right)^{\frac{1}{pq}} \left[\mathfrak{T}^{\alpha,s}_{a^+,k} f_1^{\frac{1}{p}}(t) f_2^{\frac{1}{q}}(t)\right].$$

Theorem 6.14 [13] *For* $k > 0$, $s \in \mathbb{R}\backslash\{0\}$, $\alpha > 0$, *and* $p, q \geq 1$ *and* $\frac{1}{p} + \frac{1}{q} = 1$. *Let* $f_1, f_2 \in L_{1,s}[a, t]$ *be two positive functions on* $[0, \infty)$, *such that, for all* $t > a$, $\mathfrak{T}^{\alpha,s}_{a^+,k} f_1^p(t) < \infty$, $\mathfrak{T}^{\alpha,s}_{a^+,k} f_2^p(t) < \infty$. *If* $0 < m \leq \frac{f_1(x)}{f_2(x)} \leq M$ *for* $m, M \in \mathbb{R}^+$ *and for all* $x \in [a, t]$, *then*

$$\left[\mathfrak{T}^{\alpha,s}_{a^+,k} f_1(t) f_2(t)\right] \leq c_3 \left[\mathfrak{T}^{\alpha,s}_{a^+,k} (f_1^p + f_2^p)(t)\right] + c_4 \left[\mathfrak{T}^{\alpha,s}_{a^+,k} (f_1^q + f_2^q)(t)\right],$$

with $c_3 = \frac{2^{p-1} M^p}{p(M+1)^p}$ *and* $c_4 = \frac{2^{q-1}}{q(m+1)^q}$.

Theorem 6.15 [13] *For* $k > 0, s \in \mathbb{R}\backslash\{0\}, \alpha > 0,$ *and* $p \geq 1$. *Let* $f_1, f_2 \in L_{1,s}[a, t]$ *be two positive functions on* $[0, \infty)$, *such that, for all* $t > a$, $\mathfrak{T}^{\alpha,s}_{a^+,k} f_1^p(t) < \infty$, $\mathfrak{T}^{\alpha,s}_{a^+,k} f_2^p(t) < \infty$. *If* $0 < c < m \leq \frac{f_1(x)}{f_2(x)} \leq M$ *for* $m, M \in \mathbb{R}^+$ *and for all* $x \in [a, t]$, *then*

$$\frac{M+1}{M-c} \left[\mathfrak{T}^{\alpha,s}_{a^+,k} (f_1(t) - c f_2(t))\right] \leq \left[\mathfrak{T}^{\alpha,s}_{a^+,k} f_1^p(t)\right]^{\frac{1}{p}}$$
$$+ \left[\mathfrak{T}^{\alpha,s}_{a^+,k} f_2^p(t)\right]^{\frac{1}{p}} \leq \frac{m+1}{m-c} \left[\mathfrak{T}^{\alpha,s}_{a^+,k} (f_1(t) - c f_2(t))\right]^{\frac{1}{p}}.$$

Theorem 6.16 [13] *For* $k > 0, s \in \mathbb{R}\backslash\{0\}, \alpha > 0,$ *and* $p \geq 1$. *Let* $f_1, f_2 \in L_{1,s}[a, t]$ *be two positive functions on* $[0, \infty)$, *such that, for all* $t > a$, $\mathfrak{T}^{\alpha,s}_{a^+,k} f_1^p(t) < \infty$, $\mathfrak{T}^{\alpha,s}_{a^+,k} f_2^p(t) < \infty$. *If* $0 \leq a \leq f_1(x) \leq A$ *and* $0 \leq b \leq f_2(x) \leq B$ *for* $m, M \in \mathbb{R}^+$ *and for all* $x \in [a, t]$, *then*

$$\left[\mathfrak{T}^{\alpha,s}_{a^+,k} f_1^p(t)\right]^{\frac{1}{p}} + \left[\mathfrak{T}^{\alpha,s}_{a^+,k} f_2^p(t)\right]^{\frac{1}{p}} \leq c_5 \left[\mathfrak{T}^{\alpha,s}_{a^+,k} (f_1 + f_2)^p(t)\right]^{\frac{1}{p}},$$

with $c_5 = \frac{A(a+B)+B(A+b)}{(A+b)(a+B)}$.

Theorem 6.17 [13] *For $k > 0$, $s \in \mathbb{R}\backslash\{0\}$, $\alpha > 0$. Let $f_1, f_2 \in L_{1,s}[a, t]$ be two positive functions on $[0, \infty)$, such that, for all $t > a$, $\mathfrak{F}^{\alpha,s}_{a^+,k} f_1^p(t) < \infty$, $\mathfrak{F}^{\alpha,s}_{a^+,k} f_2^p(t) < \infty$. If $0 < m \leq \frac{f_1(x)}{f_2(x)} \leq M$ for $m, M \in \mathbb{R}^+$ and for all $x \in [a, t]$, then*

$$\frac{1}{M}\left[\mathfrak{F}^{\alpha,s}_{a^+,k} f_1(t) f_2(t)\right] \leq \frac{1}{(m+1)(M+1)}\left[\mathfrak{F}^{\alpha,s}_{a^+,k}(f_1 + f_2)^2(t)\right] \leq \frac{1}{m}\left[\mathfrak{F}^{\alpha,s}_{a^+,k} f_1(t) f_2(t)\right].$$

Theorem 6.18 [13] *For $k > 0$, $s \in \mathbb{R}\backslash\{0\}$, $\alpha > 0$, and $p \geq 1$. Let $f_1, f_2 \in L_{1,s}[a, t]$ be two positive functions on $[0, \infty)$, such that, for all $t > a$, $\mathfrak{F}^{\alpha,s}_{a^+,k} f_1^p(t) < \infty$, $\mathfrak{F}^{\alpha,s}_{a^+,k} f_2^p(t) < \infty$. If $0 < m \leq \frac{f_1(x)}{f_2(x)} \leq M$ for $m, M \in \mathbb{R}^+$ and for all $x \in [a, t]$, then*

$$\left[\mathfrak{F}^{\alpha,s}_{a^+,k} f_1^p(t)\right]^{\frac{1}{p}} + \left[\mathfrak{F}^{\alpha,s}_{a^+,k} f_2^p(t)\right]^{\frac{1}{p}} \leq 2\left[\mathfrak{F}^{\alpha,s}_{a^+,k} h^p(f_1(t), f_2(t))\right]^{\frac{1}{p}},$$

where $h(f_1(x), f_2(x)) = \max\left\{M[\left(\frac{M}{m} + 1\right) f_1(t) - M f_2(t)], \frac{(m+M)f_2(t) - f_1(t)}{m}\right\}$.

Definition 6.5 [14] Let $\varphi \in X^p_c(a, b)$, $\alpha > 0$, and $\beta, \rho, \eta, \kappa \in \mathbb{R}$. Then, the fractional integrals of a function f, left and right, are given by

$${}^{\rho}\mathcal{I}^{\alpha,\beta}_{a+,\eta,\kappa}\varphi(x) := \frac{\rho^{1-\beta} x^{\kappa}}{\Gamma(\alpha)} \int_a^x \frac{\tau^{\rho(\eta+1)-1}}{(\tau^{\rho} - x^{\rho})^{1-\alpha}} \varphi(\tau) d\tau, \; 0 \leq a < x < b \leq \infty$$

and

$${}^{\rho}\mathcal{I}^{\alpha,\beta}_{b-,\eta,\kappa}\varphi(x) := \frac{\rho^{1-\beta} x^{\rho\eta}}{\Gamma(\alpha)} \int_x^b \frac{\tau^{\kappa+\rho-1}}{(\tau^{\rho} - x^{\rho})^{1-\alpha}} \varphi(\tau) d\tau, \; 0 \leq a < x < b \leq \infty,$$

respectively, if integrals exist.

Theorem 6.19 [15] *Let $\alpha > 0$, $\rho, \eta, \kappa, \beta \in \mathbb{R}$, and $p \geq 1$. Let $f, g \in X^p_c(a, x)$ be two positive functions in $[0, \infty)$, $\forall x > a$. If $0 < m \leq \frac{f(t)}{g(t)} \leq M$, for $m, M \in \mathbb{R}^*_+$ and $\forall t \in [a, x]$, then*

$$({}^{\rho}\mathcal{I}^{\alpha,\beta}_{a+,\eta,\kappa} f^p(x))^{\frac{1}{p}} + ({}^{\rho}\mathcal{I}^{\alpha,\beta}_{a+,\eta,\kappa} g^p(x))^{\frac{1}{p}} \leq c_1({}^{\rho}\mathcal{I}^{\alpha,\beta}_{a+,\eta,\kappa}(f + g)^p(x))^{\frac{1}{p}},$$

with $c_1 = \frac{M(m+1)+(M+1)}{(m+1)(M+1)}$.

Theorem 6.20 [15] *Let $\alpha > 0$, $\rho, \eta, \kappa, \beta \in \mathbb{R}$, and $p \geq 1$. Let $f, g \in X^p_c(a, x)$ be two positive functions in $[0, \infty)$, $\forall x > a$. If $0 < m \leq \frac{f(t)}{g(t)} \leq M$, for $m, M \in \mathbb{R}^*_+$ and $\forall t \in [a, x]$, then*

$$({}^{\rho}\mathcal{I}^{\alpha,\beta}_{a+,\eta,\kappa} f^p(x))^{\frac{2}{p}} + ({}^{\rho}\mathcal{I}^{\alpha,\beta}_{a+,\eta,\kappa} g^p(x))^{\frac{2}{p}} \leq c_2({}^{\rho}\mathcal{I}^{\alpha,\beta}_{a+,\eta,\kappa} f^p(x))^{\frac{1}{p}}({}^{\rho}\mathcal{I}^{\alpha,\beta}_{a+,\eta,\kappa} g^p(x))^{\frac{1}{p}},$$

with $c_2 = \frac{(M+1)(m+1)}{M} - 2$.

Theorem 6.21 [15] *Let $\alpha > 0$, $\rho, \eta, \kappa, \beta \in \mathbb{R}$, $p \geq 1$, and $\frac{1}{p} + \frac{1}{q} = 1$. Let $f, g \in X_c^p(a,x)$ be two positive functions in $[0, \infty)$, $\forall x > a$. If $0 < m \leq \frac{f(t)}{g(t)} \leq M$, for $m, M \in \mathbb{R}_+^*$ and $\forall t \in [a,x]$, then*

$$({}^{\rho}\mathcal{I}_{a+,\eta,\kappa}^{\alpha,\beta} f(x))^{\frac{1}{p}} ({}^{\rho}\mathcal{I}_{a+,\eta,\kappa}^{\alpha,\beta} g(x))^{\frac{1}{q}} \leq \left(\frac{M}{m}\right)^{\frac{1}{pq}} \left({}^{\rho}\mathcal{I}_{a+,\eta,\kappa}^{\alpha,\beta} f^{\frac{1}{p}}(x) g^{\frac{1}{q}}(x)\right).$$

Theorem 6.22 [15] *Let $\alpha > 0$, $\rho, \eta, \kappa, \beta \in \mathbb{R}$, $p \geq 1$, and $\frac{1}{p} + \frac{1}{q} = 1$. Let $f, g \in X_c^p(a,x)$ be two positive functions in $[0, \infty)$, $\forall x > a$. If $0 < m \leq \frac{f(t)}{g(t)} \leq M$, for $m, M \in \mathbb{R}_+^*$ and $\forall t \in [a,x]$, then*

$${}^{\rho}\mathcal{I}_{a+,\eta,\kappa}^{\alpha,\beta} f(x)g(x) \leq c_3({}^{\rho}\mathcal{I}_{a+,\eta,\kappa}^{\alpha,\beta}(f^p + g^p)(x)) + c_4({}^{\rho}\mathcal{I}_{a+,\eta,\kappa}^{\alpha,\beta}(f^q + g^q)(x)),$$

with $c_3 = \frac{2^{p-1}M^p}{p(M+1)^p}$ and $c_4 = \frac{2^{p-1}}{q(m+1)^q}$.

Theorem 6.23 [15] *Let $\alpha > 0$, $\rho, \eta, \kappa, \beta \in \mathbb{R}$, and $p \geq 1$. Let $f, g \in X_c^p(a,x)$ be two positive functions in $[0, \infty)$, $\forall x > a$. If $0 < m \leq \frac{f(t)}{g(t)} \leq M$, for $m, M \in \mathbb{R}_+^*$ and $\forall t \in [a,x]$, then*

$$\begin{aligned}\frac{M+1}{M-c}({}^{\rho}\mathcal{I}_{a+,\eta,\kappa}^{\alpha,\beta}(f(x) - cg(x)))^{\frac{1}{p}} &\leq ({}^{\rho}\mathcal{I}_{a+,\eta,\kappa}^{\alpha,\beta} f^p(x))^{\frac{1}{p}} + ({}^{\rho}\mathcal{I}_{a+,\eta,\kappa}^{\alpha,\beta} g^p(x))^{\frac{1}{p}} \\ &\leq \frac{m+1}{m-c}({}^{\rho}\mathcal{I}_{a+,\eta,\kappa}^{\alpha,\beta}(f(x) - cg(x)))^{\frac{1}{p}}.\end{aligned}$$

Theorem 6.24 [15] *Let $\alpha > 0$, $\rho, \eta, \kappa, \beta \in \mathbb{R}$, and $p \geq 1$. Let $f, g \in X_c^p(a,x)$ be two positive functions in $[0, \infty)$, $\forall x > a$. If $0 \leq a \leq f(t) \leq A$ and $0 \leq b \leq g(t) \leq B$, $\forall t \in [a,x]$, then*

$$({}^{\rho}\mathcal{I}_{a+,\eta,\kappa}^{\alpha,\beta} f^p(x))^{\frac{1}{p}} + ({}^{\rho}\mathcal{I}_{a+,\eta,\kappa}^{\alpha,\beta} g^p(x))^{\frac{1}{p}} \leq c_5({}^{\rho}\mathcal{I}_{a+,\eta,\kappa}^{\alpha,\beta}(f + g)^p(x))^{\frac{1}{p}},$$

with $c_5 = \frac{A(a+B)+B(A+b)}{(A+b)(a+B)}$.

Theorem 6.25 [15] *Let $\alpha > 0$, $\rho, \eta, \kappa, \beta \in \mathbb{R}$, and $p \geq 1$. Let $f, g \in X_c^p(a,x)$ be two positive functions in $[0, \infty)$, $\forall x > a$. If $0 < m \leq \frac{f(t)}{g(t)} \leq M$, for $m, M \in \mathbb{R}_+^*$ and $\forall t \in [a,x]$, then*

$$\begin{aligned}\frac{1}{M}({}^{\rho}\mathcal{I}_{a+,\eta,\kappa}^{\alpha,\beta} f(x)g(x)) &\leq \frac{1}{(m+1)(M+1)}({}^{\rho}\mathcal{I}_{a+,\eta,\kappa}^{\alpha,\beta}(f + g)^2(x)) \\ &\leq \frac{1}{m}({}^{\rho}\mathcal{I}_{a+,\eta,\kappa}^{\alpha,\beta} f(x)g(x)).\end{aligned}$$

Theorem 6.26 [15] *Let $\alpha > 0$, $\rho, \eta, \kappa, \beta \in \mathbb{R}$, and $p \geq 1$. Let $f, g \in X_c^p(a, x)$ be two positive functions in $[0, \infty)$, $\forall x > a$. If $0 < m \leq \frac{f(t)}{g(t)} \leq M$, for $m, M \in \mathbb{R}_+^*$ and $\forall t \in [a, x]$, then*

$$({}^{\rho}\mathcal{I}^{\alpha,\beta}_{a+,\eta,\kappa} f^p(x))^{\frac{1}{p}} ({}^{\rho}\mathcal{I}^{\alpha,\beta}_{a+,\eta,\kappa} g^p(x))^{\frac{1}{p}} \leq 2({}^{\rho}\mathcal{I}^{\alpha,\beta}_{a+,\eta,\kappa} h^p(f(x), g(x)))^{\frac{1}{p}},$$

with $h(f(x), g(x)) = \max\left\{ M\left[\left(\frac{M}{m}+1\right) f(x) - Mg(x)\right], \frac{(m+M)g(x)-f(x)}{m} \right\}$.

6.2 Reverse Minkowski Inequality Involving Mixed Conformable Fractional Integrals

Definition 6.6 [3] Let f be defined on $[a, b]$ and $\alpha \in \mathbb{C}$, $Re(\alpha) > 0$, $\rho > 0$. Then

i) The mixed left conformable fractional integral of f is defined by

$$ {}^{b}_{a}\mathfrak{J}^{\alpha,\rho} f(x) = \frac{1}{\Gamma(\alpha)} \int_a^x f(s) \left(\frac{(b-s)^{\rho} - (b-x)^{\rho}}{\rho} \right)^{\alpha-1} (b-s)^{\rho-1} ds \tag{6.3}$$

and

ii) The mixed right conformable fractional integral of f is defined by

$$ {}^{b}_{a}\mathfrak{J}^{\alpha,\rho} f(x) = \frac{1}{\Gamma(\alpha)} \int_x^b f(s) \left(\frac{(s-a)^{\rho} - (x-a)^{\rho}}{\rho} \right)^{\alpha-1} (s-a)^{\rho-1} ds. \tag{6.4}$$

For recent results related to operators in (6.3) and (6.4), we refer the reader [3].

Theorem 6.27 [16] *Let $\alpha, \rho > 0$, $p \geq 1$, and f, g be two positive functions on $[0, \infty)$, such that $\forall x > 0$, ${}^{b}\mathfrak{J}^{\alpha,\rho} f^p(x) < \infty$ and ${}^{b}\mathfrak{J}^{\alpha,\rho} g^p(x) < \infty$. If $0 < m \leq \frac{f(t)}{g(t)} \leq M$, for $\forall t \in [0, x]$, then*

$$\left({}^{b}\mathfrak{J}^{\alpha,\rho} f^p(x)\right)^{\frac{1}{p}} + \left({}^{b}\mathfrak{J}^{\alpha,\rho} g^p(x)\right)^{\frac{1}{p}} \leq c_1 \left({}^{b}\mathfrak{J}^{\alpha,\rho} (f+g)^p(x)\right)^{\frac{1}{p}} \tag{6.5}$$

with $c_1 = \dfrac{M(m+1) + (M+1)}{(m+1)(M+1)}$.

Proof: Using the condition $\frac{f(t)}{g(t)} \leq M$, $t \in [0, x]$, we can write

$$f(t) \leq M(f(t) - g(t)) - Mf(t),$$

which implies

$$(M+1)^p f^p(t) \leq M^p (f(t) + g(t))^p. \tag{6.6}$$

Multiplying by $\frac{1}{\Gamma(\alpha)}\left(\frac{(b-t)^\rho-(b-x)^\rho}{\rho}\right)^{\alpha-1}(b-t)^{\rho-1}$ both sides of (6.6) and integrating with respect to the variable t, we have

$$\frac{(M+1)^p}{\Gamma(\alpha)}\int_0^x\left(\frac{(b-t)^\rho-(b-x)^\rho}{\rho}\right)^{\alpha-1}(b-t)^{\rho-1}f^p(t)dt$$
$$\le\frac{M^p}{\Gamma(\alpha)}\int_0^x\left(\frac{(b-t)^\rho-(b-x)^\rho}{\rho}\right)^{\alpha-1}(b-t)^{\rho-1}(f+g)^p(t)dt. \tag{6.7}$$

Consequently, we can write

$$\left({}^b\mathfrak{J}^{\alpha,\rho}f^p(x)\right)^{\frac{1}{p}}\le\frac{M}{M+1}\left({}^b\mathfrak{J}^{\alpha,\rho}(f+g)^p(x)\right)^{\frac{1}{p}}. \tag{6.8}$$

On the other hand, as $mg(t)\le f(t)$, follow

$$\left(1+\frac{1}{m}\right)^p g^p(t)\le\left(\frac{1}{m}\right)^p(f(t)+g(t))^p. \tag{6.9}$$

Furthermore, multiplying by $\frac{1}{\Gamma(\alpha)}\left(\frac{(b-t)^\rho-(b-x)^\rho}{\rho}\right)^{\alpha-1}(b-t)^{\rho-1}$ both sides of (6.9) and integrating with respect to the variable t, we have

$$\left({}^b\mathfrak{J}^{\alpha,\rho}g^p(x)\right)^{\frac{1}{p}}\le\frac{1}{m+1}\left({}^b\mathfrak{J}^{\alpha,\rho}(f+g)^p(x)\right)^{\frac{1}{p}}. \tag{6.10}$$

From (6.8) and (6.10), the results follows. □

Equation (6.5) is the so-called reverse Minkowski's inequality associated with the mixed conformable fractional integral.

Theorem 6.28 [16] *Let $\alpha,\rho>0$, $p\ge1$, and f,g be two positive functions on $[0,\infty)$, such that $\forall x>0$, ${}^b\mathfrak{J}^{\alpha,\rho}f^p(x)<\infty$ and ${}^b\mathfrak{J}^{\alpha,\rho}g^p(x)<\infty$. If $0<m\le\frac{f(t)}{g(t)}\le M$, for $\forall t\in[0,x]$, then*

$$\left({}^b\mathfrak{J}^{\alpha,\rho}f^p(x)\right)^{\frac{2}{p}}+\left({}^b\mathfrak{J}^{\alpha,\rho}g^p(x)\right)^{\frac{2}{p}}\ge c_2\left({}^b\mathfrak{J}^{\alpha,\rho}f^p(x)\right)^{\frac{1}{p}}\left({}^b\mathfrak{J}^{\alpha,\rho}g^p(x)\right)^{\frac{1}{p}} \tag{6.11}$$

with $c_2=\dfrac{(M+1)(m+1)}{M}-2.$

Proof: Carrying out the product between (6.8) and (6.10), we have

$$\frac{(M+1)(m+1)}{M}\left({}^b\mathfrak{J}^{\alpha,\rho}f^p(x)\right)^{\frac{1}{p}}\left({}^b\mathfrak{J}^{\alpha,\rho}g^p(x)\right)^{\frac{1}{p}}\le\left({}^b\mathfrak{J}^{\alpha,\rho}(f+g)^p(x)\right)^{\frac{2}{p}}. \tag{6.12}$$

Using the Minkowski's inequality, on the right side of (6.12), we have

$$\frac{(M+1)(m+1)}{M}\left({}^b\mathfrak{J}^{\alpha,\rho}f^p(x)\right)^{\frac{1}{p}}\left({}^b\mathfrak{J}^{\alpha,\rho}g^p(x)\right)^{\frac{1}{p}}\le\left(\left({}^b\mathfrak{J}^{\alpha,\rho}f^p(x)\right)^{\frac{1}{p}}+\left({}^b\mathfrak{J}^{\alpha,\rho}g^p(x)\right)^{\frac{1}{p}}\right)^2. \tag{6.13}$$

So, from (6.13), we conclude that

$$\left(\frac{(M+1)(m+1)}{M}-2\right)\left({}^{b}\mathfrak{J}^{\alpha,\rho}f^{p}(x)\right)^{\frac{1}{p}}\left({}^{b}\mathfrak{J}^{\alpha,\rho}g^{p}(x)\right)^{\frac{1}{p}}$$
$$\leq\left({}^{b}\mathfrak{J}^{\alpha,\rho}f^{p}(x)\right)^{\frac{2}{p}}+\left({}^{b}\mathfrak{J}^{\alpha,\rho}g^{p}(x)\right)^{\frac{2}{p}}.$$

□

6.3 Related Inequalities

Theorem 6.29 *Let $\alpha,\rho>0$, $p\geq 1$, $\frac{1}{p}+\frac{1}{q}=1$, and f,g be two positive functions on $[0,\infty)$, such that $\forall x>0$, ${}^{b}\mathfrak{J}^{\alpha,\rho}f(x)<\infty$ and ${}^{b}\mathfrak{J}^{\alpha,\rho}g(x)<\infty$. If $0<m\leq\frac{f(t)}{g(t)}\leq M$, for $\forall t\in[0,x]$, then*

$$\left({}^{b}\mathfrak{J}^{\alpha,\rho}f(x)\right)^{\frac{1}{p}}\left({}^{b}\mathfrak{J}^{\alpha,\rho}g(x)\right)^{\frac{1}{p}}\leq\left(\frac{M}{m}\right)^{\frac{1}{pq}}\left({}^{b}\mathfrak{J}^{\alpha,\rho}f^{\frac{1}{p}}(x)g^{\frac{1}{p}}(x)\right). \tag{6.14}$$

Proof: Using the condition $\frac{f(t)}{g(t)}\leq M$, $t\in[0,x]$ with $x>0$, we have

$$f(t)\leq Mg(t)\Longrightarrow g^{\frac{1}{p}}(t)\geq M^{-\frac{1}{q}}f^{\frac{1}{q}}(t). \tag{6.15}$$

Multiplying by $f^{\frac{1}{p}}(t)$ both sides of (6.15), we can rewrite it as.follows

$$f^{\frac{1}{p}}(t)g^{\frac{1}{p}}(t)\geq M^{-\frac{1}{q}}f(t). \tag{6.16}$$

Now, multiplying by $\frac{1}{\Gamma(\alpha)}\left(\frac{(b-t)^{\rho}-(b-x)^{\rho}}{\rho}\right)^{\alpha-1}(b-t)^{\rho-1}$ both sides of (6.16) and integrating with respect to the variable t, we have

$$\frac{M^{-\frac{1}{q}}}{\Gamma(\alpha)}\int_{0}^{x}\left(\frac{(b-t)^{\rho}-(b-x)^{\rho}}{\rho}\right)^{\alpha-1}(b-t)^{\rho-1}f(t)dt$$
$$\leq\frac{1}{\Gamma(\alpha)}\int_{0}^{x}\left(\frac{(b-t)^{\rho}-(b-x)^{\rho}}{\rho}\right)^{\alpha-1}(b-t)^{\rho-1}f^{\frac{1}{p}}(t)g^{\frac{1}{q}}(t)dt. \tag{6.17}$$

Therefore, the inequality follows

$$M^{-\frac{1}{pq}}\left({}^{b}\mathfrak{J}^{\alpha,\rho}f(x)\right)^{\frac{1}{p}}\leq\left({}^{b}\mathfrak{J}^{\alpha,\rho}f^{\frac{1}{p}}(x)g^{\frac{1}{q}}(x)\right)^{\frac{1}{p}}. \tag{6.18}$$

On the other hand, we have

$$m^{\frac{1}{p}}g^{\frac{1}{p}}(t)\leq f^{\frac{1}{p}}(t),\ x>0. \tag{6.19}$$

Multiplying by $g^{\frac{1}{q}}(t)$ both sides of (6.19) and using the relation $\frac{1}{p}+\frac{1}{q}=1$, we have

$$m^{\frac{1}{p}}g(t)\leq f^{\frac{1}{p}}(t)g^{\frac{1}{q}}(t). \tag{6.20}$$

Multiplying by $\frac{1}{\Gamma(\alpha)}\left(\frac{(b-t)^\rho-(b-x)^\rho}{\rho}\right)^{\alpha-1}(b-t)^{\rho-1}$ both sides of (6.20) and integrating with respect to the variable t, we have

$$m^{\frac{1}{pq}}\left({}^b\mathfrak{J}^{\alpha,\rho}g(x)\right)^{\frac{1}{q}} \leq \left({}^b\mathfrak{J}^{\alpha,\rho}f^{\frac{1}{p}}(x)g^{\frac{1}{q}}(x)\right)^{\frac{1}{q}}. \tag{6.21}$$

Evaluating the product between (6.18) and (6.21) and using the relation $\frac{1}{p}+\frac{1}{q}=1$, we conclude that

$$\left({}^b\mathfrak{J}^{\alpha,\rho}f(x)\right)^{\frac{1}{p}}\left({}^b\mathfrak{J}^{\alpha,\rho}g(x)\right)^{\frac{1}{q}} \leq \left(\frac{M}{m}\right)^{\frac{1}{pq}}\left({}^b\mathfrak{J}^{\alpha,\rho}f^{\frac{1}{p}}(x)g^{\frac{1}{q}}(x)\right)^{\frac{1}{p}}.$$

□

Theorem 6.30 *Let $\alpha,\rho>0$, $p\geq 1$, $\frac{1}{p}+\frac{1}{q}=1$, and f,g be two positive functions on $[0,\infty)$, such that $\forall x>0$, ${}^b\mathfrak{J}^{\alpha,\rho}f^p(x)<\infty$, ${}^b\mathfrak{J}^{\alpha,\rho}f^q(x)<\infty$, ${}^b\mathfrak{J}^{\alpha,\rho}g^p(x)<\infty$, and ${}^b\mathfrak{J}^{\alpha,\rho}g^q(x)<\infty$. If $0<m\leq\frac{f(t)}{g(t)}\leq M$, for $\forall t\in[0,x]$, then*

$$ {}^b\mathfrak{J}^{\alpha,\rho}f(x)g(x) \leq c_3\left({}^b\mathfrak{J}^{\alpha,\rho}(f^p+g^p)(x)\right)+c_4\left({}^b\mathfrak{J}^{\alpha,\rho}(f^q+g^q)(x)\right), \tag{6.22}$$

with $c_3=\dfrac{2^{p-1}M^p}{p(M+1)^p}$ *and* $c_4=\dfrac{2^{p-1}}{q(m+1)^q}$.

Proof: Using the hypothesis, we have the following identity

$$(M+1)^p f^p(t)\leq M^p(f+g)^p(t). \tag{6.23}$$

Multiplying by $\frac{1}{\Gamma(\alpha)}\left(\frac{(b-t)^\rho-(b-x)^\rho}{\rho}\right)^{\alpha-1}(b-t)^{\rho-1}$ both sides of (6.23) and integrating with respect to the variable t, we get

$$\frac{(M+1)^p}{\Gamma(\alpha)}\int_0^x\left(\frac{(b-t)^\rho-(b-x)^\rho}{\rho}\right)^{\alpha-1}(b-t)^{\rho-1}f^p(t)$$
$$\leq\frac{M^p}{\Gamma(\alpha)}\int_0^x\left(\frac{(b-t)^\rho-(b-x)^\rho}{\rho}\right)^{\alpha-1}(b-t)^{\rho-1}(f+g)^p(t).$$

In this way, we have

$$ {}^b\mathfrak{J}^{\alpha,\rho}f^p(x)\leq\frac{M^p}{(M+1)^p}\,{}^b\mathfrak{J}^{\alpha,\rho}(f+g)^p(x). \tag{6.24}$$

On the other hand, as $0<m<\frac{f(t)}{g(t)}$, $t\in(0,x)$, we have

$$(m+1)^q g^q(t)\leq(f+g)^q(t). \tag{6.25}$$

Again, multiplying by $\frac{1}{\Gamma(\alpha)}\left(\frac{(b-t)^\rho-(b-x)^\rho}{\rho}\right)^{\alpha-1}(b-t)^{\rho-1}$ both sides of (6.25) and integrating with respect to the variable t, we get

$$ {}^b\mathfrak{J}^{\alpha,\rho}g^q(x)\leq\frac{1}{(m+1)^q}\,{}^b\mathfrak{J}^{\alpha,\rho}(f+g)^q(x). \tag{6.26}$$

Considering Young's inequality [17]

$$f(t)g(t)\leq\frac{f^p(t)}{p}+\frac{g^q(t)}{q}, \tag{6.27}$$

multiplying by $\frac{1}{\Gamma(\alpha)}\left(\frac{(b-t)^{\rho}-(b-x)^{\rho}}{\rho}\right)^{\alpha-1}(b-t)^{\rho-1}$ both sides of (6.27), and integrating with respect to the variable t, we have

$$^{b}\mathfrak{J}^{\alpha,\rho}(fg)(x) \leq \frac{1}{p}\left(^{b}\mathfrak{J}^{\alpha,\rho}f^{p}(x)\right)+\frac{1}{q}\left(^{b}\mathfrak{J}^{\alpha,\rho}g^{q}(x)\right). \tag{6.28}$$

Thus, using (6.24), (6.26), and (6.28), we get

$$\begin{aligned}^{b}\mathfrak{J}^{\alpha,\rho}(fg)(x) &\leq \frac{1}{p}\left(^{b}\mathfrak{J}^{\alpha,\rho}f^{p}(x)\right)+\frac{1}{q}\left(^{b}\mathfrak{J}^{\alpha,\rho}g^{q}(x)\right)\\ &\leq {}^{b}\mathfrak{J}^{\alpha,\rho}f^{p}(x)+{}^{b}\mathfrak{J}^{\alpha,\rho}g^{q}(x)\\ &\leq \frac{M^{p}}{p(M+1)^{p}}\left(^{b}\mathfrak{J}^{\alpha,\rho}(f+g)^{p}(x)\right)\\ &\quad+\frac{1}{q(m+1)^{q}}\left(^{b}\mathfrak{J}^{\alpha,\rho}(f+g)^{q}(x)\right).\end{aligned} \tag{6.29}$$

Using the following inequality, $(a+b)^{r} \leq 2^{p-1}(a^{r}+b^{r})$, $r>1$, $a,b\geq 0$, we get

$$^{b}\mathfrak{J}^{\alpha,\rho}(f+g)^{p}(x) \leq 2^{p-1}\,{}^{b}\mathfrak{J}^{\alpha,\rho}(f^{p}+g^{p})(x) \tag{6.30}$$

and

$$^{b}\mathfrak{J}^{\alpha,\rho}(f+g)q(x) \leq 2^{q-1}\,{}^{b}\mathfrak{J}^{\alpha,\rho}(f^{q}+g^{q})(x). \tag{6.31}$$

Thus, replacing (6.30) and (6.31) at (6.29), we conclude that

$$^{b}\mathfrak{J}^{\alpha,\rho}(fg)(x) \leq \frac{2^{p-1}M^{p}}{p(M+1)^{p}}\left(^{b}\mathfrak{J}^{\alpha,\rho}(f^{p}+g^{p})(x)\right)+\frac{2^{q-1}}{q(m+1)^{q}}\left(^{b}\mathfrak{J}^{\alpha,\rho}(f^{q}+g^{q})(x)\right).$$

□

Theorem 6.31 *Let $\alpha,\rho>0$, $p\geq 1$, and f,g be two positive functions on $[0,\infty)$, such that $\forall x>0$, ${}^{b}\mathfrak{J}^{\alpha,\rho}f^{p}(x)<\infty$ and ${}^{b}\mathfrak{J}^{\alpha,\rho}g^{p}(x)<\infty$. If $0<m\leq\frac{f(t)}{g(t)}\leq M$, for $\forall t\in[0,x]$, then*

$$\begin{aligned}\frac{M+1}{M-c}\left(^{b}\mathfrak{J}^{\alpha,\rho}(f(x)-cg(x))^{p}\right)^{\frac{1}{p}} &\leq \left(^{b}\mathfrak{J}^{\alpha,\rho}f^{p}(x)\right)^{\frac{1}{p}}+\left(^{b}\mathfrak{J}^{\alpha,\rho}g^{p}(x)\right)^{\frac{1}{p}}\\ &\leq \frac{m+1}{m-c}\left(^{b}\mathfrak{J}^{\alpha,\rho}(f(x)-cg(x))^{p}\right)^{\frac{1}{p}}.\end{aligned} \tag{6.32}$$

Proof: By hypothesis $0<c<m\leq M$, so

$$\begin{aligned}&mc \leq Mc \Longrightarrow mc+m \leq mc+M \leq Mc+M\\ &\Longrightarrow (M+1)(m-c) \leq (m+1)(M-c).\end{aligned}$$

Thus, we conclude that

$$\frac{M+1}{M-c} \leq \frac{m+1}{m-c}.$$

Also, we have

$$m - c \leq \frac{f(t) - cg(t)}{g(t)} \leq M - c,$$

which implies

$$\frac{(f(t) - cg(t))^p}{(M - c)^p} \leq g^p(t) \leq \frac{(f(t) - cg(t))^p}{(m - c)^p}. \tag{6.33}$$

Again, we have

$$\frac{1}{M} \leq \frac{g(t)}{f(t)} \leq \frac{1}{m} \Longrightarrow \frac{m - c}{cm} \leq \frac{f(t) - cg(t)}{cf(t)} \leq \frac{M - c}{cM},$$

which implies,

$$\left(\frac{M}{M - c}\right)^p (f(t) - cg(t))^p \leq f^p(t) \leq \left(\frac{m}{m - c}\right)^p (f(t) - cg(t))^p. \tag{6.34}$$

Multiplying by $\frac{1}{\Gamma(\alpha)}\left(\frac{(b-t)^\rho - (b-x)^\rho}{\rho}\right)^{\alpha-1}(b-t)^{\rho-1}f(t)$ both sides of (6.33) and integrating with respect to the variable t, we have

$$\int_0^x \frac{1}{\Gamma(\alpha)}\left(\frac{(b-t)^\rho - (b-x)^\rho}{\rho}\right)^{\alpha-1}(b-t)^{\rho-1}\frac{(f(t) - cg(t))^p}{(M - c)^p}dt$$
$$\leq \int_0^x \frac{1}{\Gamma(\alpha)}\left(\frac{(b-t)^\rho - (b-x)^\rho}{\rho}\right)^{\alpha-1}(b-t)^{\rho-1}g^p dt$$
$$\leq \int_0^x \frac{1}{\Gamma(\alpha)}\left(\frac{(b-t)^\rho - (b-x)^\rho}{\rho}\right)^{\alpha-1}(b-t)^{\rho-1}\frac{(f(t) - cg(t))^p}{(m - c)^p}dt.$$

In this way, we obtain

$$\frac{1}{M - c}\left({}^b\mathfrak{J}^{\alpha,\rho}(f(x) - cg(x))^p\right)^{\frac{1}{p}} \leq \left({}^b\mathfrak{J}^{\alpha,\rho}g^p(x)\right)^{\frac{1}{p}}$$
$$\leq \frac{1}{m - c}\left({}^b\mathfrak{J}^{\alpha,\rho}(f(x) - cg(x))^p\right)^{\frac{1}{p}}. \tag{6.35}$$

Realizing the same procedure as in (6.34), we have

$$\frac{M}{M - c}\left({}^b\mathfrak{J}^{\alpha,\rho}(f(x) - cg(x))^p\right)^{\frac{1}{p}} \leq \left({}^b\mathfrak{J}^{\alpha,\rho}f^p(x)\right)^{\frac{1}{p}}$$
$$\leq \frac{m}{m - c}\left({}^b\mathfrak{J}^{\alpha,\rho}(f(x) - cg(x))^p\right)^{\frac{1}{p}}. \tag{6.36}$$

Adding (6.35) and (6.36), we conclude that

$$\frac{M + 1}{M - c}\left({}^b\mathfrak{J}^{\alpha,\rho}(f(x) - cg(x))^p\right)^{\frac{1}{p}} \leq \left({}^b\mathfrak{J}^{\alpha,\rho}f^p(x)\right)^{\frac{1}{p}} + \left({}^b\mathfrak{J}^{\alpha,\rho}g^p(x)\right)^{\frac{1}{p}}$$
$$\leq \frac{m + 1}{m - c}\left({}^b\mathfrak{J}^{\alpha,\rho}(f(x) - cg(x))^p\right)^{\frac{1}{p}}.$$

□

Theorem 6.32 *Let $\alpha, \rho > 0$, $p \geq 1$, and f, g be two positive functions on $[0, \infty)$, such that $\forall x > 0$, ${}^{b}\mathfrak{J}^{\alpha,\rho} f^{p}(x) < \infty$, and ${}^{b}\mathfrak{J}^{\alpha,\rho} g^{p}(x) < \infty$. If $0 < m \leq \frac{f(t)}{g(t)} \leq M$, for $\forall t \in [0, x]$, then*

$$\left({}^{b}\mathfrak{J}^{\alpha,\rho} f^{p}(x)\right)^{\frac{1}{p}} + \left({}^{b}\mathfrak{J}^{\alpha,\rho} g^{p}(x)\right)^{\frac{1}{p}} \leq c_5 \left({}^{b}\mathfrak{J}^{\alpha,\rho} (f+g)^{p}(x)\right)^{\frac{1}{p}}, \tag{6.37}$$

with $c_5 = \dfrac{A(a+B) + B(A+b)}{(A+b)(a+B)}$.

Proof: By hypothesis, it follows that

$$\frac{1}{B} \leq \frac{1}{g(t)} \leq \frac{1}{b}. \tag{6.38}$$

Realizing the product between (6.38) and $0 < a \leq f(t) \leq A$, we have

$$\frac{a}{B} \leq \frac{f(t)}{g(t)} \leq \frac{A}{b}. \tag{6.39}$$

From (6.39), we get

$$g^{p}(t) \leq \left(\frac{B}{a+B}\right)^{p} (f(t) + g(t))^{p} \tag{6.40}$$

and

$$f^{p}(t) \leq \left(\frac{A}{b+A}\right)^{p} (f(t) + g(t))^{p}. \tag{6.41}$$

Multiplying by $\frac{1}{\Gamma(\alpha)} \left(\frac{(b-t)^{\rho} - (b-x)^{\rho}}{\rho}\right)^{\alpha-1} (b-t)^{\rho-1}$ both sides of (6.40) and integrating with respect to the variable t, we have

$$\frac{1}{\Gamma(\alpha)} \int_{0}^{x} \left(\frac{(b-t)^{\rho} - (b-x)^{\rho}}{\rho}\right)^{\alpha-1} (b-t)^{\rho-1} g^{p}(t) dt$$
$$\leq \left(\frac{B}{a+B}\right)^{p} \frac{1}{\Gamma(\alpha)} \int_{0}^{x} \left(\frac{(b-t)^{\rho} - (b-x)^{\rho}}{\rho}\right)^{\alpha-1} (b-t)^{\rho-1} (f(t) + g(t))^{p} dt.$$

Thus, it follows that

$$\left({}^{b}\mathfrak{J}^{\alpha,\rho} g^{p}(x)\right)^{\frac{1}{p}} \leq \frac{B}{a+B} \left({}^{b}\mathfrak{J}^{\alpha,\rho} (f+g)^{p}(x)\right)^{\frac{1}{p}}. \tag{6.42}$$

Similarly, we performed the calculations for (6.41), we get

$$\left({}^{b}\mathfrak{J}^{\alpha,\rho} f^{p}(x)\right)^{\frac{1}{p}} \leq \frac{A}{b+A} \left({}^{b}\mathfrak{J}^{\alpha,\rho} (f+g)^{p}(x)\right)^{\frac{1}{p}}. \tag{6.43}$$

Adding (6.42) and (6.43), we conclude that

$$\left({}^{b}\mathfrak{J}^{\alpha,\rho} f^{p}(x)\right)^{\frac{1}{p}} + \left({}^{b}\mathfrak{J}^{\alpha,\rho} g^{p}(x)\right)^{\frac{1}{p}} \leq \frac{A(a+B) + B(b+A)}{(a+B)(b+A)} \left({}^{b}\mathfrak{J}^{\alpha,\rho} (f+g)^{p}(x)\right)^{\frac{1}{p}}.$$

□

Theorem 6.33 *Let $\alpha, \rho > 0, p \geq 1$, and f, g be two positive functions on $[0, \infty)$, such that $\forall x > 0$, ${}^{b}\mathfrak{J}^{\alpha,\rho} f(x) < \infty$, and ${}^{b}\mathfrak{J}^{\alpha,\rho} g(x) < \infty$. If $0 < m \leq \frac{f(t)}{g(t)} \leq M$, for $\forall t \in [0, x]$, then*

$$\frac{1}{M}\left({}^{b}\mathfrak{J}^{\alpha,\rho} f(x)g(x)\right) \leq \frac{1}{(m+1)(M+1)}\left({}^{b}\mathfrak{J}^{\alpha,\rho}(f+g)^{2}(x)\right) \leq \frac{1}{m}\left({}^{b}\mathfrak{J}^{\alpha,\rho} f(x)g(x)\right). \tag{6.44}$$

Proof: Being $0 < m \leq \frac{f(t)}{g(t)} \leq M$, $\forall t \in [a, x]$, we have

$$g(t)(m+1) \leq g(t) + f(t) \leq g(t)(M+1). \tag{6.45}$$

Also, it follows that $\frac{1}{M} \leq \frac{g(t)}{f(t)} \leq \frac{1}{m}$, which implies

$$g(t)\left(\frac{M+1}{M}\right) \leq g(t) + f(t) \leq g(t)\left(\frac{m+1}{m}\right). \tag{6.46}$$

Evaluating the product between (6.45) and (6.46), we have

$$\frac{f(t)g(t)}{M} \leq \frac{(g(t)+f(t))^{2}}{(m+1)(M+1)} \leq \frac{f(t)g(t)}{m}. \tag{6.47}$$

Multiplying by $\frac{1}{\Gamma(\alpha)}\left(\frac{(b-t)^{\rho}-(b-x)^{\rho}}{\rho}\right)^{\alpha-1}(b-t)^{\rho-1}$ both sides of (6.47) and integrating with respect to the variable t, we have

$$\begin{aligned}
&\frac{1}{M}\frac{1}{\Gamma(\alpha)}\int_{0}^{x}\left(\frac{(b-t)^{\rho}-(b-x)^{\rho}}{\rho}\right)^{\alpha-1}(b-t)^{\rho-1} f(t)g(t)dt \\
&\leq \frac{1}{(m+1)(M+1)}\frac{1}{\Gamma(\alpha)}\int_{0}^{x}\left(\frac{(b-t)^{\rho}-(b-x)^{\rho}}{\rho}\right)^{\alpha-1}(b-t)^{\rho-1}(f(t)+g(t))^{2}dt \\
&\leq \frac{1}{m}\frac{1}{\Gamma(\alpha)}\int_{0}^{x}\left(\frac{(b-t)^{\rho}-(b-x)^{\rho}}{\rho}\right)^{\alpha-1}(b-t)^{\rho-1} f(t)g(t)dt.
\end{aligned}$$

Thus, we conclude that

$$\frac{1}{M}\left({}^{b}\mathfrak{J}^{\alpha,\rho} f(x)g(x)\right) \leq \frac{1}{(m+1)(M+1)}\left({}^{b}\mathfrak{J}^{\alpha,\rho}(f+g)^{2}(x)\right) \leq \frac{1}{m}\left({}^{b}\mathfrak{J}^{\alpha,\rho} f(x)g(x)\right).$$

□

Theorem 6.34 *Let $\alpha, \rho > 0$, $p \geq 1$, and f, g be two positive functions on $[0, \infty)$, such that $\forall x > 0$, ${}^{b}\mathfrak{J}^{\alpha,\rho} f^{p}(x) < \infty$, and ${}^{b}\mathfrak{J}^{\alpha,\rho} g^{p}(x) < \infty$. If $0 < m \leq \frac{f(t)}{g(t)} \leq M$, for $\forall t \in [0, x]$, then*

$$\left({}^{b}\mathfrak{J}^{\alpha,\rho} f^{p}(x)\right)^{\frac{1}{p}} + \left({}^{b}\mathfrak{J}^{\alpha,\rho} g^{p}(x)\right)^{\frac{1}{p}} \leq 2\left({}^{b}\mathfrak{J}^{\alpha,\rho} h^{p}(f(x), g(x))\right)^{\frac{1}{p}}.$$

Proof: From the hypothesis $0 < m \leq \frac{f(t)}{g(t)} \leq M$, $\forall t \in [a, x]$, we have

$$0 < m \leq M + m - \frac{f(t)}{g(t)} \tag{6.48}$$

and

$$M + m - \frac{f(t)}{g(t)} \leq M. \tag{6.49}$$

Thus, using (6.48) and (6.49), we get

$$g(t) < \frac{(M+m)g(t) - f(t)}{m} \leq h(f(t), g(t)), \tag{6.50}$$

where $h(f(t), g(t)) = \max\left\{M\left[\left(\frac{M}{m}+1\right)f(t) - Mg(t)\right], \frac{(M+m)g(t)-f(t)}{m}\right\}$. Using the hypothesis, it follows that $0 < \frac{1}{M} \leq \frac{g(t)}{f(t)} \leq \frac{1}{m}$. In this way, we obtain

$$\frac{1}{M} \leq \frac{1}{M} + \frac{1}{m} - \frac{g(t)}{f(t)} \tag{6.51}$$

and

$$\frac{1}{M} + \frac{1}{m} - \frac{g(t)}{f(t)} \leq \frac{1}{m}. \tag{6.52}$$

Then, from (6.51) and (6.52), we have

$$\frac{1}{M} \leq \frac{\left(\frac{1}{m} + \frac{1}{M}\right)f(t) - g(t)}{f(t)} \leq \frac{1}{m},$$

which can be rewrite as

$$\begin{aligned}
f(t) &\leq M\left(\frac{1}{m} + \frac{1}{M}\right)f(t) - Mg(t) \\
&= \frac{M(M+m)f(t) - M^2 mg(t)}{Mm} \\
&= \left(\frac{M}{m}+1\right)f(t) - Mg(t) \\
&\leq M\left[\left(\frac{M}{m}+1\right)f(t) - Mg(t)\right] \\
&\leq h(f(t), g(t)).
\end{aligned} \tag{6.53}$$

Thus, using (6.50) and (6.53), we can write

$$f^p(t) \leq h^p(f(t), g(t)) \tag{6.54}$$

and

$$g^p(t) \leq h^p(f(t), g(t)). \tag{6.55}$$

Multiplying by $\frac{1}{\Gamma(\alpha)}\left(\frac{(b-t)^\rho-(b-x)^\rho}{\rho}\right)^{\alpha-1}(b-t)^{\rho-1}$ both sides of (6.54) and integrating with respect to the variable t, we have

$$\begin{aligned}
&\frac{1}{\Gamma(\alpha)}\int_0^x \left(\frac{(b-t)^\rho - (b-x)^\rho}{\rho}\right)^{\alpha-1}(b-t)^{\rho-1} f^p(t)dt \\
&\leq \frac{1}{\Gamma(\alpha)}\int_0^x \left(\frac{(b-t)^\rho - (b-x)^\rho}{\rho}\right)^{\alpha-1}(b-t)^{\rho-1} h^p(f(t), g(t))dt.
\end{aligned}$$

In this way, we obtain

$$\left({}^{b}\mathfrak{J}^{\alpha,\rho}f^{p}(x)\right)^{\frac{1}{p}} \leq \left({}^{b}\mathfrak{J}^{\alpha,\rho}h^{p}(f(x),g(x))\right)^{\frac{1}{p}}. \tag{6.56}$$

Using the same procedure as above, for (6.55), we have

$$\left({}^{b}\mathfrak{J}^{\alpha,\rho}g^{p}(x)\right)^{\frac{1}{p}} \leq \left({}^{b}\mathfrak{J}^{\alpha,\rho}h^{p}(f(x),g(x))\right)^{\frac{1}{p}}. \tag{6.57}$$

Thus, using (6.56) and (6.57), we conclude that

$$\left({}^{b}\mathfrak{J}^{\alpha,\rho}f^{p}(x)\right)^{\frac{1}{p}} + \left({}^{b}\mathfrak{J}^{\alpha,\rho}g^{p}(x)\right)^{\frac{1}{p}} \leq 2\left({}^{b}\mathfrak{J}^{\alpha,\rho}h^{p}(f(x),g(x))\right)^{\frac{1}{p}}.$$

□

Bibliography

1 Hardy, G.H., Littlewood, J.E., and Pólya, G. (1934). *Inequalities*, Cambridge University Press.
2 Bougoffa, L. (2006). On Minkowski and Hardy integral inequalities. *J. Inequal. Pure and Appl. Math.* **7**(2): 1–3.
3 Abdeljawad, T. (2015). On conformable fractional calculus. *J. Comput. Appl. Math.* **279**: 57–66.
4 Abdeljawad, T. (2020). Fractional operators with boundary points dependent kernels and integration by parts. *Discrete Contin. Dyn. Syst., Ser. S* **13**(3): 351–375. https://doi.org/10.3934/dcdss.2020020.
5 Dahmani, Z. (2010). On Minkowski and Hermite–Hadamard integral inequalities via fractional integration. *Ann. Funct. Anal.* **1**: 51–58.
6 Kilbas, A.A. (2001). Hadamard type fractional calculus. *J. Korean Math. Soc.* **38**: 1191–1204.
7 Chinchane, V.L. and Pachpatte, D.B. (2013). New fractional inequalities via Hadamard fractional integral. *Int. J. Funct. Anal.* **5**: 165–176.
8 Taf, S. and Brahim, K. (2015). Some new results using Hadamard fractional integral. *Int. J. Nonlinear Anal. Appl.* **2**: 24–42.
9 Saigo, M. (1978). A remark on integral operators involving the Gauss hypergeometric functions. *Math. Rep. Kyushu Univ.* **11**: 135–143.
10 Raina, R.K. (2010). Solution of Abel-type integral equation involving the Appell hypergeometric function. *Integral Transforms Spec. Funct.* **21** (7): 515–522.
11 Kiryakova, V. (1994). *Generalized Fractional Calculus and Applications, Pitman Research Notes in Mathematics Series*, vol. **301**. Harlow: Longman Scientific and Technical.
12 Chinchane, V.L. and Pachpatte, D.B. (2014). New fractional inequalities involving Saigo fractional integral operator. *Math. Sci. Lett.* **3**: 133–139.

13 Mubeen, S., Habib, S., and Naemm, M.N. (2019). The Minkowkski inequality involving generalized k-fractional conformable integral. *J. Inequal. Appl.* **2019** (81): 1–18.

14 Katugampola, U.N. (2016). New fractional integral unifying six existing fractional integrals, arxiv.org/abs/1612.08596.

15 Vanterler da C. Sousa, J. and Capelas de Oliveira, E. (2018). The Minkowski's inequality by means of a generalized fractional integral. *AIMS Math.* **3** (1): 131–147.

16 Nisar, K.S., Tassaddiq, A., Rahman, G., and Khan, A. (2019). Some inequalities via fractional conformable integral operators. *J. Inequal. Appl.* **2019** (217): 1–8.

17 Kreyszig, E. (1989). *Introductory Functional Analysis with Applications*, vol. **1**. New York: Wiley.

7

New Estimations for Different Kinds of Convex Functions via Conformable Integrals and Riemann–Liouville Fractional Integral Operators

Ahmet Ocak Akdemir[1] *and Hemen Dutta*[2]

[1]*Department of Mathematics, Faculty of Science and Letters, Ağrı İbrahim Çeçen University, Ağrı, Turkey*
[2]*Department of Mathematics, Gauhati University, Guwahati, Assam, India*

7.1 Introduction

Let us start with some well-known definitions to refresh our memories.

In [1], Niculescu mentioned the following considerable definitions:

The *AG*-convex functions (usually known as log-convex functions) are those functions $f : I \to (0, \infty)$ for which

$$x, y \in I \text{ and } \lambda \in [0, 1] \Longrightarrow f(\lambda x + (1-\lambda)y) \leq f(x)^{1-\lambda} f(y)^{\lambda}, \tag{7.1}$$

i.e. for which $\log f$ is convex.

The *GG*-convex functions (called in what follows multiplicatively convex functions) are those functions $f : I \to J$ (acting on subintervals of $(0, \infty)$) such that

$$x, y \in I \text{ and } \lambda \in [0, 1] \Longrightarrow f(x^{1-\lambda} y^{\lambda}) \leq f(x)^{1-\lambda} f(y)^{\lambda}. \tag{7.2}$$

The class of all *GA*-convex functions is constituted by all functions $f : I \to \mathbb{R}$ (defined on subintervals of $(0, \infty)$) for which

$$x, y \in I \text{ and } \lambda \in [0, 1] \Longrightarrow f(x^{1-\lambda} y^{\lambda}) \leq \lambda f(x) + (1-\lambda) f(y). \tag{7.3}$$

In [2], Anderson et al. give the conditions and properties of some classes of convex functions. For example, recall that the condition of *GA*- convexity is $x^2 f'' + x f' \geq 0$. Based on these classes of functions, several researchers have been established new integral inequalities by using different techniques, some of them can be found in [1–11].

Let $f : I \subseteq \mathbb{R} \to \mathbb{R}$ be a convex function defined on the interval I of real numbers and $a < b$. The following double inequality:

$$f\left(\frac{a+b}{2}\right) \leq \frac{1}{b-a} \int_a^b f(x)dx \leq \frac{f(a)+f(b)}{2}$$

Fractional Order Analysis: Theory, Methods and Applications, First Edition.
Edited by Hemen Dutta, Ahmet Ocak Akdemir, and Abdon Atangana.

is well known in the literature as Hadamard's inequality. Both inequalities hold in the reversed direction if f is concave.

In [12], Dragomir defined convex functions on the co-ordinates as following.

Definition 7.1 Let us consider the bidimensional interval $\Delta = [a,b] \times [c,d]$ in $\mathbb{R}^2$ with $a < b$, $c < d$. A function $f : \Delta \to \mathbb{R}$ will be called convex on the co-ordinates if the partial mappings $f_y : [a,b] \to \mathbb{R}, f_y(u) = f(u,y)$ and $f_x : [c,d] \to \mathbb{R}, f_x(v) = f(x,v)$ are convex where defined for all $y \in [c,d]$ and $x \in [a,b]$. Recall that the mapping $f : \Delta \to \mathbb{R}$ is convex on Δ if the following inequality holds,

$$f(\lambda x + (1-\lambda)z, \lambda y + (1-\lambda)w) \leq \lambda f(x,y) + (1-\lambda) f(z,w),$$

for all $(x,y), (z,w) \in \Delta$, and $\lambda \in [0,1]$.

In [12], Dragomir established the following inequalities of Hadamard's type for co-ordinated convex functions on a rectangle from the plane $\mathbb{R}^2$.

Theorem 7.1 *Suppose that $f : \Delta = [a,b] \times [c,d] \to \mathbb{R}$ is convex on the co-ordinates on Δ. Then one has the inequalities:*

$$\begin{aligned}
& f\left(\frac{a+b}{2}, \frac{c+d}{2}\right) \\
& \leq \frac{1}{2}\left[\frac{1}{b-a}\int_a^b f\left(x, \frac{c+d}{2}\right) dx + \frac{1}{d-c}\int_c^d f\left(\frac{a+b}{2}, y\right) dy\right] \\
& \leq \frac{1}{(b-a)(d-c)}\int_a^b \int_c^d f(x,y) dx\, dy \\
& \leq \frac{1}{4}\left[\frac{1}{(b-a)}\int_a^b f(x,c) dx + \frac{1}{(b-a)}\int_a^b f(x,d) dx \right. \\
& \left. + \frac{1}{(d-c)}\int_c^d f(a,y) dy + \frac{1}{(d-c)}\int_c^d f(b,y) dy\right] \\
& \leq \frac{f(a,c) + f(a,d) + f(b,c) + f(b,d)}{4}.
\end{aligned} \tag{7.4}$$

The above inequalities are sharp.

Similar results can be found in [12–18].

An important definition called Riemann–Liouville fractional integrals, which is milestone in the theory of fractional calculus:

Definition 7.2 **(See [19])** Let $f \in L_1[a,b]$. The Riemann–Liouville integrals $J_{a+}^{\alpha} f$ and $J_{b-}^{\alpha} f$ of order $\alpha > 0$ are defined by

$$J_{a+}^{\alpha} f(t) = \frac{1}{\Gamma(\alpha)} \int_a^t (t-x)^{\alpha-1} f(x) dx, \quad t > a$$

and

$$J_{b-}^{\alpha}f(t) = \frac{1}{\Gamma(\alpha)} \int_t^b (x-t)^{\alpha-1} f(x)dx, \quad t < b$$

respectively, where $\Gamma(\alpha) = \int_0^{\infty} e^{-t}t^{\alpha-1}dt$. Here $J_{a+}^0 f(t) = J_{b-}^0 f(t) = f(t)$.

In the case of $\alpha = 1$, the fractional integral reduces to classical integral. Several researchers have focused on new integral inequalities involving Riemann–Liouville fractional integrals in recent years, see Refs. [19–24] and [25–33].

Recently, in [34], Khalil et al. gave a new definition that is called conformable derivative and its properties. This definition includes a local fractional derivative, so a huge amount of mathematicians called conformable derivative instead of fractional. The conformable derivative attracts attention with conformity to classical derivative. Khalil et al. have introduced the conformable derivative by the equation, which has a limit form similar to the classical derivative. Khalil et al. have proved that this definition provides multiplication and division rules. They also express the Rolle theorem and the mean value theorem for functions, which are differentiable with conformable local fractional order. Many researchers have used integral and derivatives with arbitrary order to provide generalizations of many dynamic equations and integral inequalities. The definitions of local and nonlocal differentiation and integration have been discussed in the community of fractional calculus. Both concepts allow us to integrate or differentiate with respect to arbitrary order. Moreover, some researchers have defined a new modification of conformable derivative that is called local fractional proportional derivatives.

In [34], Khalil et al. gave the following definition.

Definition 7.3 (Conformable integral) Let $\alpha \in (0, 1]$, $0 \le \kappa_1 < \kappa_2$. A function $h : [\kappa_1, \kappa_2] \to \mathbb{R}$ is α-fractional integrable on $[\kappa_1, \kappa_2]$ if the integral

$$\int_{\kappa_1}^{\kappa_2} h(x)d_{\alpha}x = \int_{\kappa_1}^{\kappa_2} h(x)x^{\alpha-1}dx$$

exists and is finite. All α-fractional integrable functions on $[\kappa_1, \kappa_2]$ is indicated by $L_{\alpha}([\kappa_1, \kappa_2])$.

Remark 7.1 (See [34])

$$I_{\alpha}^{\kappa_1}(h_1)(s) = I_1^{\kappa_1}(s^{\alpha-1}h_1) = \int_{\kappa_1}^{s} \frac{h_1(x)}{x^{1-\alpha}}dx,$$

where the integral is the usual Riemann improper integral, and $\alpha \in (0, 1]$.

In [34] and [35], authors have pointed that the Riemann–Liouville derivatives are not valid for product of two functions. In this case, the inequalities that have been proved by Riemann–Liouville integrals are not valid. The results that are obtained by using the conformable integrals have a wide range of validity. (Let us consider the function f defined as $f : \mathbb{R}^+ \to \mathbb{R}, f = x^2 e^x$, which is convex.) For related papers, see Refs. [34–39].

7.2 Some Generalizations for Geometrically Convex Functions

The main purpose of this section is to prove more general integral inequalities for functions whose derivatives of absolute values are *GA*- and *GG*-convex functions via α-fractional integrals. In this sense, we have established a new integral identity by using conformable integrals that generalizes some previous results. Several special cases of our results have also been considered.

An useful integral identity has been embodied the following lemma.

Lemma 7.1 *Let $f : I \subset \mathbb{R}_+ = (0, \infty) \to \mathbb{R}$ be an α-fractional twice differentiable function on the interior I^0 of I, where $a, b \in I$ with $a < b$ and $f'' \in L_\alpha[a, b]$. Then the inequality:*

$$\begin{aligned} &\Upsilon_f(a, b; \alpha; n) \\ &= \frac{\ln a - \ln x}{[(n+1)\alpha][(n+1)\alpha - 1]} \int_0^1 (a^{1-t}x^t)^{(n+1)\alpha+1} f''(a^{1-t}x^t)dt \\ &+ \frac{\ln x - \ln b}{[(n+1)\alpha][(n+1)\alpha - 1]} \int_0^1 (b^{1-t}x^t)^{(n+1)\alpha+1} f''(b^{1-t}x^t)dt, \end{aligned} \tag{7.5}$$

where

$$\begin{aligned} \Upsilon_f(a, b; \alpha; n) = {} & \frac{a^{(n+1)\alpha} f'(a) - b^{(n+1)\alpha} f'(b)}{[(n+1)\alpha][(n+1)\alpha - 1]} - \left(\frac{a^{(n+1)\alpha-1} f(a) - b^{(n+1)\alpha-1} f(b)}{(n+1)\alpha} \right) \\ & - \int_a^b u^{(n+1)\alpha-2} f(u)du \end{aligned}$$

holds for all $x \in [a, b]$, $n \geq 1$.

Proof: Suppose that

$$\begin{aligned} I = {} & \frac{\ln a - \ln x}{[(n+1)\alpha][(n+1)\alpha - 1]} \int_0^1 (a^t x^{1-t})^{(n+1)\alpha+1} f''(a^{1-t}x^t)dt \\ & + \frac{\ln x - \ln b}{[(n+1)\alpha][(n+1)\alpha - 1]} \int_0^1 (b^{1-t}x^t)^{(n+1)\alpha+1} f''(b^{1-t}x^t)dt \\ = {} & \frac{\ln a - \ln x}{[(n+1)\alpha][(n+1)\alpha - 1]} I_1 + \frac{\ln x - \ln b}{[(n+1)\alpha][(n+1)\alpha - 1]} I_2. \end{aligned} \tag{7.6}$$

By using integration by parts, we get

$$
\begin{aligned}
I_1 &= (\ln a - \ln x)\int_0^1 (a^t x^{1-t})^{(n+2)\alpha} D_\alpha(f')(a^t x^{1-t}) t^{1-\alpha} d_\alpha t \\
&= \int_0^1 (a^t x^{1-t})^{(n+1)\alpha+1} f''(a^t x^{1-t}) dt. \qquad (7.7)
\end{aligned}
$$

By taking into account the change of the variable $u = a^t x^{1-t}$ and integration by parts, we have

$$
\begin{aligned}
I_1 &= \frac{1}{\ln x - \ln a}\int_a^x u^{(n+1)\alpha} f''(u)du \\
&= \frac{1}{\ln x - \ln a}\left[u^{(n+1)\alpha} f'(u)|_a^x - (n+1)\alpha \int_a^x u^{(n+1)\alpha-1} f'(u)du\right] \\
&= \frac{1}{\ln x - \ln a}[x^{(n+1)\alpha} f'(x) - a^{(n+1)\alpha} f'(a) - (n+1)\alpha[x^{(n+1)\alpha-1} f(x) \\
&\quad - a^{(n+1)\alpha-1} f(a)] + [(n+1)\alpha][(n+1)\alpha - 1]\int_a^x u^{(n+1)\alpha-2} f(u)du\Bigg] \\
&= \frac{1}{\ln x - \ln a}[(x^{(n+1)\alpha} f'(x) - a^{(n+1)\alpha} f'(a)) - (n+1)\alpha(x^{(n+1)\alpha} f(x) \\
&\quad - a^{(n+1)\alpha} f(a)) + [(n+1)\alpha][(n+1)\alpha - 1]\int_a^x u^{(n+1)\alpha-2} f(u)du\Bigg]. \qquad (7.8)
\end{aligned}
$$

Now multiplying the resulting inequality by $\frac{\ln x - \ln a}{[(n+1)\alpha][(n+1)\alpha-1]}$, we obtain

$$
\begin{aligned}
&\frac{\ln x - \ln a}{[(n+1)\alpha][(n+1)\alpha - 1]} I_1 \\
&= \frac{x^{(n+1)\alpha} f'(x) - a^{(n+1)\alpha} f'(a)}{[(n+1)\alpha][(n+1)\alpha - 1]} - \frac{x^{(n+1)\alpha-1} f(x) - a^{(n+1)\alpha-1} f(a)}{[(n+1)\alpha][(n+1)\alpha - 1]} \\
&\quad + \int_a^x u^{(n+1)\alpha-2} f(u)du. \qquad (7.9)
\end{aligned}
$$

Similarly, one can have

$$
\begin{aligned}
&\frac{\ln b - \ln x}{[(n+1)\alpha][(n+1)\alpha - 1]} I_2 \\
&= \frac{b^{(n+1)\alpha} f'(b) - x^{(n+1)\alpha} f'(x)}{[(n+1)\alpha][(n+1)\alpha - 1]} - \frac{b^{(n+1)\alpha-1} f(b) - x^{(n+1)\alpha-1} f(x)}{[(n+1)\alpha][(n+1)\alpha - 1]} \\
&\quad + \int_x^b u^{(n+1)\alpha-2} f(u)du. \qquad (7.10)
\end{aligned}
$$

By adding (7.9) and (7.10), we get

$$
\begin{aligned}
I &= \frac{b^{(n+1)\alpha} f'(b) - a^{(n+1)\alpha} f'(a)}{[(n+1)\alpha][(n+1)\alpha - 1]} - \left(\frac{b^{(n+1)\alpha-1} f(b) - a^{(n+1)\alpha-1} f(a)}{(n+1)\alpha - 1}\right) \\
&\quad + \int_a^b u^{(n+1)\alpha-2} f(u)du
\end{aligned}
$$

$$= \frac{a^{(n+1)\alpha}f'(a) - b^{(n+1)\alpha}f'(b)}{[(n+1)\alpha][(n+1)\alpha - 1]} - \left(\frac{a^{(n+1)\alpha-1}f(a) - b^{(n+1)\alpha-1}f(b)}{(n+1)\alpha - 1}\right)$$
$$- \int_a^b u^{(n+1)\alpha-2}f(u)du,$$

which is the required result. □

Remark 7.2 Several new identities can be derived from Lemma 7.1, by selecting of the special cases of n and α.

Remark 7.3 Let $\alpha = 1$, then Lemma 7.1 reduces to Lemma 2.1 of [3].

Theorem 7.2 *Let $f : I \subset \mathbb{R}_+ = (0, \infty) \to \mathbb{R}$ be an α-fractional twice differentiable function on the interior I^0 of I, where $a, b \in I$ with $a < b$ and $f'' \in L_\alpha[a, b]$. If $|f''|$ is GG-convex on I, then the inequality:*

$$|\Upsilon_f(a, b; \alpha; n)| \leq \frac{1}{[(n+1)\alpha][(n+1)\alpha - 1]}$$
$$\times [(\ln x - \ln a)L(x^{(n+1)\alpha+1}|f''(x)|, a^{(n+1)\alpha+1}|f''(a)|)$$
$$+ (\ln b - \ln x)L(b^{(n+1)\alpha+1}|f''(b)|, x^{(n+1)\alpha+1}|f''(x)|)]$$

holds for all $x \in [a, b]$ and $n \geq 1$.

Proof: From Lemma 7.1 and by using the GG-convexity of $|f''|$, we can write

$$|\Upsilon_f(a, b; \alpha; n)|$$
$$\leq \frac{\ln x - \ln a}{[(n+1)\alpha][(n+1)\alpha - 1]} \int_0^1 (a^t x^{1-t})^{(n+1)\alpha+1}|f''(a^t x^{1-t})|dt$$
$$+ \frac{\ln b - \ln x}{[(n+1)\alpha][(n+1)\alpha - 1]} \int_0^1 (b^{1-t}x^t)^{(n+1)\alpha+1}|f''(b^{1-t}x^t)|dt$$
$$\leq \frac{\ln x - \ln a}{[(n+1)\alpha][(n+1)\alpha - 1]} \int_0^1 (a^t x^{1-t})^{(n+1)\alpha+1}|f''(a)|^t|f''(x)|^{1-t})|dt$$
$$+ \frac{\ln b - \ln x}{[(n+1)\alpha][(n+1)\alpha - 1]} \int_0^1 (b^{1-t}x^t)^{(n+1)\alpha+1}|f''(b)|^{1-t}|f''(x)|^t dt.$$

By computing the above integrals and simplifying, we deduce

$$|\Upsilon_f(a, b; \alpha; n)| \leq \frac{1}{[(n+1)\alpha][(n+1)\alpha - 1]}$$
$$\times [(\ln x - \ln a)L(x^{(n+1)\alpha+1}|f''(x)|, a^{(n+1)\alpha+1}|f''(a)|)$$
$$+ (\ln b - \ln x)L(b^{(n+1)\alpha+1}|f''(b)|, x^{(n+1)\alpha+1}|f''(x)|)],$$

which is the desired result. □

Corollary 7.1 *If we choose $\alpha = 1$, then under the assumption of Theorem 7.2, one have the following new result:*

$$\left|\frac{a^{n+1}f'(a) - b^{n+1}f'(b)}{n(n+1)} - \frac{a^n f(a) - b^n f(b)}{n+1} - \int_a^b f(u)du\right|$$
$$\le \frac{1}{n(n+1)} \times [(\ln x - \ln a)L(x^{n+2}|f''(x)|, a^{n+2}|f''(a)|)$$
$$+ (\ln b - \ln x)L(b^{n+2}|f''(b)|, x^{n+2}|f''(x)|)].$$

Corollary 7.2 *If we choose $n = 1$ and $\alpha = 1$, then under the assumption of Theorem 7.2, one have a new result as:*

$$\left|\frac{a^2 f'(a) - b^2 f'(b)}{2} - \frac{af(a) - bf(b)}{2} - \int_a^b f(u)du\right|$$
$$\le \frac{1}{2} \times [(\ln x - \ln a)L(x^3|f''(x)|, a^3|f''(a)|)$$
$$+ (\ln b - \ln x)L(b^3|f''(b)|, x^3|f''(x)|)].$$

Theorem 7.3 *Let $f : I \subset \mathbb{R}_+ = (0, \infty) \to \mathbb{R}$ be an α-fractional twice differentiable function on the interior I° of I, where $a, b \in I$ with $a < b$ and $f'' \in L_\alpha[a, b]$. If $|f''|^q$ is GG-convex on I for $q \ge 1$, then the inequality:*

$$|\Upsilon_f(a, b; \alpha; n)|$$
$$\le \frac{\ln x - \ln a}{[(n+1)\alpha][(n+1)\alpha - 1]} L^{1-\frac{1}{q}}(x^{(n+1)\alpha+1}, a^{(n+1)\alpha+1})$$
$$\times L^{\frac{1}{q}}(x^{(n+1)\alpha+1}|f''(x)|^q, a^{(n+1)\alpha+1}|f''(a)|^q) + \frac{\ln b - \ln x}{[(n+1)\alpha][(n+1)\alpha - 1]}$$
$$\times L^{1-\frac{1}{q}}(b^{(n+1)\alpha+1}, x^{(n+1)\alpha+1}) L^{\frac{1}{q}}(b^{(n+1)\alpha+1}|f''(b)|^q, x^{(n+1)\alpha+1}|f''(x)|^q)$$

holds for all $x \in [a, b]$ and $n \ge 1$.

Proof: From Lemma 7.1, by using the GG-convexity of $|f''|^q$ and the Hölder integral inequality, we have

$$|\Upsilon_f(a, b; \alpha; n)|$$
$$\le \frac{\ln x - \ln a}{[(n+1)\alpha][(n+1)\alpha - 1]} \int_0^1 (a^t x^{1-t})^{(n+1)\alpha+1}|f''(a^t x^{1-t})|dt$$
$$+ \frac{\ln b - \ln x}{[(n+1)\alpha][(n+1)\alpha - 1]} \int_0^1 (b^{1-t}x^t)^{(n+1)\alpha+1}|f''(b^{1-t}x^t)|dt$$
$$\le \frac{\ln x - \ln a}{[(n+1)\alpha][(n+1)\alpha - 1]} \left(\int_0^1 (a^t x^{1-t})^{(n+1)\alpha+1} dt\right)^{1-\frac{1}{q}}$$

$$\times\left(\int_0^1 (a^tx^{1-t})^{(n+1)\alpha+1}|f''(a^tx^{1-t})|^q dt\right)^{\frac{1}{q}}$$
$$+\frac{\ln b-\ln x}{[(n+1)\alpha][(n+1)\alpha-1]}$$
$$\left(\int_0^1 (b^{1-t}x^t)^{(n+1)\alpha+1} dt\right)^{1-\frac{1}{q}}$$
$$\times\left(\int_0^1 (b^{1-t}x^t)^{(n+1)\alpha+1}|f''(b^tx^t)|^q dt\right)^{\frac{1}{q}}.$$

Namely,

$$|\Upsilon_f(a,b;\alpha;n)|$$
$$\leq \frac{\ln x-\ln a}{[(n+1)\alpha][(n+1)\alpha-1]}\left(\int_0^1 (a^tx^{1-t})^{(n+1)\alpha+1} dt\right)^{1-\frac{1}{q}}$$
$$\times\left(\int_0^1 (a^tx^{1-t})^{(n+1)\alpha+1}|f''(a)|^{qt}|f''(x)|^{q(1-t)})|dt\right)^{\frac{1}{q}}$$
$$\frac{\ln b-\ln x}{[(n+1)\alpha][(n+1)\alpha-1]}\left(\int_0^1 (x^tb^{1-t})^{(n+1)\alpha+1} dt\right)^{1-\frac{1}{q}}$$
$$\times\left(\int_0^1 (x^tb^{1-t})^{(n+1)\alpha+1}|f''(b)|^{q(1-t)}|f''(x)|^{qt})|^q dt\right)^{\frac{1}{q}}.$$

By a simple computation, we obtain

$$|\Upsilon_f(a,b;\alpha;n)|$$
$$\leq \frac{\ln x-\ln a}{[(n+1)\alpha][(n+1)\alpha-1]}L^{1-\frac{1}{q}}(x^{(n+1)\alpha+1},a^{(n+1)\alpha+1})$$
$$\times L^{\frac{1}{q}}(x^{(n+1)\alpha+1}|f''(x)|^q,a^{(n+1)\alpha+1}|f''(a)|^q)+\frac{\ln b-\ln x}{[(n+1)\alpha][(n+1)\alpha-1]}$$
$$\times L^{1-\frac{1}{q}}(b^{(n+1)\alpha+1},x^{(n+1)\alpha+1})L^{\frac{1}{q}}(b^{(n+1)\alpha+1}|f''(b)|^q,x^{(n+1)\alpha+1}|f''(x)|^q),$$

which is the required result. □

Corollary 7.3 *If we choose $\alpha=1$, then under the assumption of Theorem 7.3, we have the following inequality:*

$$\left|\frac{a^{n+1}f'(a)-b^{n+1}f'(b)}{n(n+1)}-\frac{a^nf(a)-b^nf(b)}{n+1}-\int_a^b f(u)du\right|$$
$$\leq \frac{\ln x-\ln a}{n(n+1)}L^{1-\frac{1}{q}}(x^{(n+2)},a^{(n+2)})\times L^{\frac{1}{q}}(x^{(n+2)}|f''(x)|^q,a^{(n+2)}|f''(a)|^q)$$
$$+\frac{\ln b-\ln x}{n(n+1)}\times L^{1-\frac{1}{q}}(b^{(n+2)},x^{(n+2)})L^{\frac{1}{q}}(b^{(n+2)}|f''(b)|^q,x^{(n+2)}|f''(x)|^q).$$

Corollary 7.4 *If we choose $n = 1$ and $\alpha = 1$, then under the assumption of Theorem 7.3, we get the following inequality:*

$$\left|\frac{a^2 f'(a) - b^2 f'(b)}{2} - \frac{af(a) - bf(b)}{2} - \int_a^b f(u)du\right|$$
$$\leq \frac{\ln x - \ln a}{2} L^{1-\frac{1}{q}}(x^3, a^3) \times L^{\frac{1}{q}}(x^3|f''(x)|^q, a^3|f''(a)|^q)$$
$$+ \frac{\ln b - \ln x}{3} \times L^{1-\frac{1}{q}}(b^3, x^3) L^{\frac{1}{q}}(b^3|f''(b)|^q, x^3|f''(x)|^q).$$

Theorem 7.4 *Let $f : I \subset \mathbb{R}_+ = (0, \infty) \to \mathbb{R}$ be an α-fractional twice differentiable function on the interior I° of I, where $a, b \in I$ with $a < b$ and $f'' \in L_\alpha[a, b]$. If $|f''|^q$ is GG-convex on I for $q \geq 1$, then the inequality:*

$$|\Upsilon_f(a, b; \alpha; n)|$$
$$\leq \frac{(\ln x - \ln a)\left[L^{\frac{1}{q}}(|f''(x)|^q, |f''(a)|^q)\right]}{[(n+1)\alpha][(n+1)\alpha - 1]}\left(L^{1-\frac{1}{q}}\left(x^{[(n+1)\alpha+1]\frac{q}{q-1}}, a^{[(n+1)\alpha+1]\frac{q}{q-1}}\right)\right)$$
$$+\frac{(\ln b - \ln x)\left[L^{\frac{1}{q}}(|f''(b)|^q, |f''(x)|^q)\right]}{[(n+1)\alpha][(n+1)\alpha - 1]}\left(L^{1-\frac{1}{q}}\left(b^{[(n+1)\alpha+1]\frac{q}{q-1}}, x^{[(n+1)\alpha+1]\frac{q}{q-1}}\right)\right)$$

holds for all $x \in [a, b]$ and $n \geq 1$.

Proof: From Lemma 7.1, by using the *GG*-convexity of $|f''|^q$ and by using the Hölder integral inequality, we have

$$|\Upsilon_f(a, b; \alpha; n)|$$
$$\leq \frac{\ln x - \ln a}{[(n+1)\alpha][(n+1)\alpha - 1]}\int_0^1 (a^t x^{1-t})^{(n+1)\alpha+1}|f''(a^t x^{1-t})|dt$$
$$+\frac{\ln b - \ln x}{[(n+1)\alpha][(n+1)\alpha - 1]}\int_0^1 (b^{1-t} x^t)^{(n+1)\alpha+1}|f''(b^{1-t} x^t)|dt$$
$$\leq \frac{(\ln x - \ln a)x^{(n+1)\alpha+1}}{[(n+1)\alpha][(n+1)\alpha - 1]}\left(\int_0^1 \left(\frac{a}{x}\right)^{[(n+1)\alpha+1]\frac{qt}{q-1}} dt\right)^{1-\frac{1}{q}}$$
$$\times\left(\int_0^1 |f''(a)|^{qt}|f''(x)|^{q(1-t)}\right)^{\frac{1}{q}}$$
$$+\frac{(\ln b - \ln x)b^{(n+1)\alpha+1}}{[(n+1)\alpha][(n+1)\alpha - 1]}\left(\int_0^1 \left(\frac{x}{b}\right)^{[(n+1)\alpha+1]\frac{qt}{q-1}} dt\right)^{1-\frac{1}{q}}$$
$$\times\left(\int_0^1 |f''(x)|^{qt}|f''(b)|^{q(1-t)}\right)^{\frac{1}{q}}$$

that is,

$$
\begin{aligned}
&|\Upsilon_f(a,b;\alpha;n)| \\
&\leq \frac{(\ln x - \ln a)\left[L^{\frac{1}{q}}(|f''(x)|^q, |f''(a)|^q)\right]}{[(n+1)\alpha][(n+1)\alpha-1]}\left(L^{1-\frac{1}{q}}\left(x^{[(n+1)\alpha+1]\frac{q}{q-1}}, a^{[(n+1)\alpha+1]\frac{q}{q-1}}\right)\right) \\
&+\frac{(\ln b - \ln x)\left[L^{\frac{1}{q}}(|f''(b)|^q, |f''(x)|^q)\right]}{[(n+1)\alpha][(n+1)\alpha-1]}\left(L^{1-\frac{1}{q}}\left(b^{[(n+1)\alpha+1]\frac{q}{q-1}}, x^{[(n+1)\alpha+1]\frac{q}{q-1}}\right)\right),
\end{aligned}
$$

which is the required result. □

Corollary 7.5 *If we choose $\alpha = 1$, then under the assumption of Theorem 7.4, the following inequality holds*

$$
\begin{aligned}
&\left|\frac{a^{n+1}f'(a) - b^{n+1}f'(b)}{n(n+1)} - \frac{a^n f(a) - b^n f(b)}{n+1} - \int_a^b f(u)du\right| \\
&\leq \frac{(\ln x - \ln a)\left[L^{\frac{1}{q}}(|f''(x)|^q, |f''(a)|^q)\right]}{n(n+1)}\left(L^{1-\frac{1}{q}}\left(x^{(n+2)\frac{q}{q-1}}, a^{(n+2)\frac{q}{q-1}}\right)\right) \\
&+\frac{(\ln b - \ln x)\left[L^{\frac{1}{q}}(|f''(b)|^q, |f''(x)|^q)\right]}{n(n+1)}\left(L^{1-\frac{1}{q}}\left(b^{(n+2)\frac{q}{q-1}}, x^{(n+2)\frac{q}{q-1}}\right)\right).
\end{aligned}
$$

Corollary 7.6 *If we choose $n = 1$ and $\alpha = 1$, then under the assumption of Theorem 7.4, the following inequality holds*

$$
\begin{aligned}
&\left|\frac{a^2 f'(a) - b^2 f'(b)}{2} - \frac{af(a) - bf(b)}{2} - \int_a^b f(u)du\right| \\
&\leq \frac{(\ln x - \ln a)\left[L^{\frac{1}{q}}(|f''(x)|^q, |f''(a)|^q)\right]}{2}\left(L^{1-\frac{1}{q}}\left(x^{\frac{3q}{q-1}}, a^{\frac{3q}{q-1}}\right)\right) \\
&+\frac{(\ln b - \ln x)\left[L^{\frac{1}{q}}(|f''(b)|^q, |f''(x)|^q)\right]}{2}\left(L^{1-\frac{1}{q}}\left(b^{\frac{3q}{q-1}}, x^{\frac{3q}{q-1}}\right)\right).
\end{aligned}
$$

Theorem 7.5 *Let $f : I \subset \mathbb{R}_+ = (0, \infty) \to \mathbb{R}$ be an α-fractional twice differentiable function on the interior I° of I, where $a, b \in I$ with $a < b$ and $f'' \in L_\alpha[a,b]$. If $|f''|^q$ is GG-convex on I for $q \geq 1$, then the inequality:*

$$
\begin{aligned}
&|\Upsilon_f(a,b;\alpha;n)| \\
&\leq \frac{\ln x - \ln a}{[(n+1)\alpha][(n+1)\alpha-1]}\left[L^{\frac{1}{q}}\left(x^{[(n+1)\alpha+1]q}|f''(x)|^q, a^{[(n+1)\alpha+1]q}|f''(a)|^q\right)\right] \\
&+\frac{\ln b - \ln x}{[(n+1)\alpha][(n+1)\alpha-1]}\left[L^{\frac{1}{q}}\left(b^{[(n+1)\alpha+1]q}|f''(b)|^q, x^{[(n+1)\alpha+1]q}|f''(x)|^q\right)\right]
\end{aligned}
$$

holds for all $x \in [a,b]$ and $n \geq 1$.

Proof: From Lemma 7.1, by using the *GG*-convexity of $|f''|^q$ and by using a variant of the Hölder integral inequality, we have

$$
\begin{aligned}
&|\Upsilon_f(a,b;\alpha;n)| \\
&\leq \frac{\ln x - \ln a}{[(n+1)\alpha][(n+1)\alpha - 1]} \int_0^1 (a^t x^{1-t})^{(n+1)\alpha+1} |f''(a^t x^{1-t})| dt \\
&+ \frac{\ln b - \ln x}{[(n+1)\alpha][(n+1)\alpha - 1]} \int_0^1 (b^{1-t} x^t)^{(n+1)\alpha+1} |f''(b^{1-t} x^t)| dt \\
&\leq \frac{\ln x - \ln a}{[(n+1)\alpha][(n+1)\alpha - 1]} \left(\int_0^1 1 dt \right)^{1-\frac{1}{q}} \\
&\left[x^{(n+1)\alpha+1} |f''(x)|^q \int_0^1 \left(\frac{a^{[(n+1)\alpha+1]q} |f''(a)|^q}{x^{[(n+1)\alpha+1]q} |f''(x)|^q} \right)^t dt \right]^{\frac{1}{q}} \\
&+ \frac{\ln b - \ln x}{[(n+1)\alpha][(n+1)\alpha - 1]} \left(\int_0^1 1 dt \right)^{1-\frac{1}{q}} \\
&\left[b^{(n+1)\alpha+1} |f''(b)|^q \int_0^1 \left(\frac{b^{[(n+1)\alpha+1]q} |f''(b)|^q}{x^{[(n+1)\alpha+1]q} |f''(x)|^q} \right)^t dt \right]^{\frac{1}{q}}.
\end{aligned}
$$

By making use of necessary computations for the above integrals, we deduce

$$
\begin{aligned}
&|\Upsilon_f(a,b;\alpha;n)| \\
&\leq \frac{\ln x - \ln a}{[(n+1)\alpha][(n+1)\alpha - 1]} \left[L^{\frac{1}{q}} \left(x^{[(n+1)\alpha+1]q} |f''(x)|^q, a^{[(n+1)\alpha+1]q} |f''(a)|^q \right) \right] \\
&+ \frac{\ln b - \ln x}{[(n+1)\alpha][(n+1)\alpha - 1]} \left[L^{\frac{1}{q}} \left(b^{[(n+1)\alpha+1]q} |f''(b)|^q, x^{[(n+1)\alpha+1]q} |f''(x)|^q \right) \right],
\end{aligned}
$$

which is the required result. □

Corollary 7.7 *If we choose $n = 1$ and $\alpha = 1$, then under the assumption of Theorem 7.5, we have the following inequality:*

$$
\begin{aligned}
&\left| \frac{a^2 f'(a) - b^2 f'(b)}{2} - \frac{af(a) - bf(b)}{2} - \int_a^b f(u) du \right| \\
&\leq \frac{\ln x - \ln a}{2} \left[L^{\frac{1}{q}} \left(x^{3q} |f''(x)|^q, a^{3q} |f''(a)|^q \right) \right] \\
&+ \frac{\ln b - \ln x}{2} \left[L^{\frac{1}{q}} \left(b^{3q} |f''(b)|^q, x^{3q} |f''(x)|^q \right) \right].
\end{aligned}
$$

7.3 New Inequalities for Co-ordinated Convex Functions

The main purpose of this section is to prove some new inequalities of Hadamard-type for convex functions on the co-ordinates by using Riemann–Liouville fractional integrals. Throughout of this chapter, we will use the following notation:

$$C = \frac{(x-a)^\alpha (y-c)^\beta f(a,c) + (x-a)^\alpha (d-y)^\beta f(a,d)}{(b-a)(d-c)}$$
$$+ \frac{(b-x)^\alpha (y-c)^\beta f(b,c) + (b-x)^\alpha (d-y)^\beta f(b,d)}{(b-a)(d-c)}$$
$$- \frac{\Gamma(\beta+1)}{(b-a)(d-c)} [(x-a)^\alpha J^\beta_{y+} f(a,d) + (x-a)^\alpha J^\beta_{y-} f(a,c)$$
$$+ (b-x)^\alpha J^\beta_{y-} f(b,c) + (b-x)^\alpha J^\beta_{y+} f(b,d)]$$
$$- \frac{\Gamma(\alpha+1)}{(b-a)(d-c)} [(d-y)^\beta J^\alpha_{x-} f(a,d) + (y-c)^\beta J^\alpha_{x-} f(b,c)$$
$$+ (d-y)^\beta J^\alpha_{x+} f(b,d) + (y-c)^\beta J^\alpha_{x+} f(a,d)]$$
$$+ \frac{\Gamma(\alpha+1)\Gamma(\beta+1)}{(b-a)(d-c)} [J^{\alpha,\beta}_{x^- y^-} f(u,v) + J^{\alpha,\beta}_{x^+ y^-} f(u,v) + J^{\alpha,\beta}_{x^- y^+} f(u,v) + J^{\alpha,\beta}_{x^+ y^+} f(u,v)],$$

where

$$J^{\alpha,\beta}_{x^- y^-} f(u,v) = \frac{1}{\Gamma(\alpha)\Gamma(\beta)} \int_x^b \int_c^y (b-u)^{\alpha-1} (v-c)^{\beta-1} f(u,v) du\, dv,$$
$$J^{\alpha,\beta}_{x^+ y^-} f(u,v) = \frac{1}{\Gamma(\alpha)\Gamma(\beta)} \int_x^b \int_y^d (b-u)^{\alpha-1} (d-v)^{\beta-1} f(u,v) du\, dv,$$

$$J^{\alpha,\beta}_{x^- y^+} f(u,v) = \frac{1}{\Gamma(\alpha)\Gamma(\beta)} \int_a^x \int_c^y (u-a)^{\alpha-1} (v-c)^{\beta-1} f(u,v) du\, dv,$$
$$J^{\alpha,\beta}_{x^+ y^+} f(u,v) = \frac{1}{\Gamma(\alpha)\Gamma(\beta)} \int_a^x \int_y^d (u-a)^{\alpha-1} (d-v)^{\beta-1} f(u,v) du\, dv.$$

To prove our main result, we need the following Lemma.

Lemma 7.2 *Let $f : \Delta = [a,b] \times [c,d] \to \mathbb{R}$ be a twice partial differentiable mapping on $\Delta = [a,b] \times [c,d]$. If $\frac{\partial^2 f}{\partial t\, \partial s} \in L(\Delta)$ and $\alpha, \beta > 0$, $a, c \geq 0$, then the following equality holds for Riemann–Liouville fractional integrals:*

$$C = \frac{(x-a)^{\alpha+1} (y-c)^{\beta+1}}{(b-a)(d-c)} \int_0^1 \int_0^1 (t^\alpha - 1)(s^\beta - 1) \frac{\partial^2 f}{\partial t\, \partial s}$$
$$\times (tx + (1-t)a, sy + (1-s)c) ds\, dt$$
$$+ \frac{(x-a)^{\alpha+1} (d-y)^{\beta+1}}{(b-a)(d-c)} \int_0^1 \int_0^1 (t^\alpha - 1)(1 - s^\beta) \frac{\partial^2 f}{\partial t\, \partial s}$$
$$\times (tx + (1-t)a, sy + (1-s)d) ds\, dt$$
$$+ \frac{(b-x)^{\alpha+1} (y-c)^{\beta+1}}{(b-a)(d-c)} \int_0^1 \int_0^1 (1 - t^\alpha)(s^\beta - 1) \frac{\partial^2 f}{\partial t\, \partial s}$$
$$\times (tx + (1-t)b, sy + (1-s)c) ds\, dt$$

$$+\frac{(b-x)^{\alpha+1}(d-y)^{\beta+1}}{(b-a)(d-c)}\int_0^1\int_0^1(1-t^\alpha)(1-s^\beta)\frac{\partial^2 f}{\partial t\,\partial s}$$
$$\times(tx+(1-t)b,sy+(1-s)d)ds\,dt.$$

Proof: It suffices to note that

$$I=\underbrace{\frac{(x-a)^{\alpha+1}(y-c)^{\beta+1}}{(b-a)(d-c)}\int_0^1\int_0^1(t^\alpha-1)(s^\beta-1)\frac{\partial^2 f}{\partial t\,\partial s}(tx+(1-t)a,sy+(1-s)c)ds\,dt}_{I_1}$$
$$+\underbrace{\frac{(x-a)^{\alpha+1}(d-y)^{\beta+1}}{(b-a)(d-c)}\int_0^1\int_0^1(t^\alpha-1)(1-s^\beta)\frac{\partial^2 f}{\partial t\,\partial s}(tx+(1-t)a,sy+(1-s)d)ds\,dt}_{I_2}$$
$$+\underbrace{\frac{(b-x)^{\alpha+1}(y-c)^{\beta+1}}{(b-a)(d-c)}\int_0^1\int_0^1(1-t^\alpha)(s^\beta-1)\frac{\partial^2 f}{\partial t\,\partial s}(tx+(1-t)b,sy+(1-s)c)ds\,dt}_{I_3}$$
$$+\underbrace{\frac{(b-x)^{\alpha+1}(d-y)^{\beta+1}}{(b-a)(d-c)}\int_0^1\int_0^1(1-t^\alpha)(1-s^\beta)\frac{\partial^2 f}{\partial t\,\partial s}(tx+(1-t)b,sy+(1-s)d)ds\,dt.}_{I_4}$$

Integrating by parts, we get

$$I_1=\frac{(x-a)^{\alpha+1}(y-c)^{\beta+1}}{(b-a)(d-c)}\int_0^1(s^\beta-1)\left[\frac{t^\alpha-1}{x-a}\frac{\partial f}{\partial s}(tx+(1-t)a,sy+(1-s)c)\Big|_0^1\right.$$
$$\left.-\frac{\alpha}{x-a}\int_0^1 t^{\alpha-1}\frac{\partial f}{\partial s}(tx+(1-t)a,sy+(1-s)c)dt\right]ds$$
$$=\frac{(x-a)^{\alpha+1}(y-c)^{\beta+1}}{(b-a)(d-c)}\int_0^1(s^\beta-1)\left[\frac{1}{x-a}\frac{\partial f}{\partial s}(a,sy+(1-s)c)\right.$$
$$\left.-\frac{\alpha}{x-a}\int_0^1 t^{\alpha-1}\frac{\partial f}{\partial s}(tx+(1-t)a,sy+(1-s)c)dt\right]ds.$$

By integrating again, we obtain

$$I_1=\frac{(x-a)^{\alpha+1}(y-c)^{\beta+1}}{(b-a)(d-c)}\int_0^1(s^\beta-1)\left[\frac{t^\alpha-1}{x-a}\frac{\partial f}{\partial s}(tx+(1-t)a,sy+(1-s)c)\Big|_0^1\right.$$
$$\left.-\frac{\alpha}{x-a}\int_0^1 t^{\alpha-1}\frac{\partial f}{\partial s}(tx+(1-t)a,sy+(1-s)c)dt\right]ds$$
$$=\frac{(x-a)^\alpha(y-c)^\beta}{(b-a)(d-c)}f(a,c)-\frac{\beta(x-a)^\alpha(y-c)^\beta}{(b-a)(d-c)}\int_0^1 s^{\beta-1}f(a,sy+(1-s)c)ds$$
$$-\frac{\alpha(x-a)^\alpha(y-c)^\beta}{(b-a)(d-c)}\int_0^1 t^{\alpha-1}f(tx+(1-t)a,c)dt$$

$$+\frac{\alpha\beta(x-a)^{\alpha}(y-c)^{\beta}}{(b-a)(d-c)}\int_0^1\int_0^1 t^{\alpha-1}s^{\beta-1}\frac{\partial f}{\partial s}(tx+(1-t)a,sy+(1-s)c)dt\,ds.$$

By a similar argument, we have

$$I_2=\frac{(x-a)^{\alpha}(d-y)^{\beta}}{(b-a)(d-c)}f(a,d)-\frac{\beta(x-a)^{\alpha}(d-y)^{\beta}}{(b-a)(d-c)}\int_0^1 s^{\beta-1}f(a,sy+(1-s)d)ds$$
$$-\frac{\alpha(x-a)^{\alpha}(d-y)^{\beta}}{(b-a)(d-c)}\int_0^1 t^{\alpha-1}f(tx+(1-t)a,d)dt$$
$$+\frac{\alpha\beta(x-a)^{\alpha}(d-y)^{\beta}}{(b-a)(d-c)}\int_0^1\int_0^1 t^{\alpha-1}s^{\beta-1}\frac{\partial f}{\partial s}(tx+(1-t)a,sy+(1-s)d)dt\,ds,$$

$$I_3=\frac{(b-x)^{\alpha}(y-c)^{\beta}}{(b-a)(d-c)}f(b,c)-\frac{\beta(b-x)^{\alpha}(y-c)^{\beta}}{(b-a)(d-c)}\int_0^1 s^{\beta-1}f(b,sy+(1-s)c)ds$$
$$-\frac{\alpha(b-x)^{\alpha}(y-c)^{\beta}}{(b-a)(d-c)}\int_0^1 t^{\alpha-1}f(tx+(1-t)b,c)dt$$
$$+\frac{\alpha\beta(b-x)^{\alpha}(y-c)^{\beta}}{(b-a)(d-c)}\int_0^1\int_0^1 t^{\alpha-1}s^{\beta-1}\frac{\partial f}{\partial s}(tx+(1-t)b,sy+(1-s)c)dt\,ds,$$

and

$$I_4=\frac{(b-x)^{\alpha}(d-y)^{\beta}}{(b-a)(d-c)}f(b,d)-\frac{\beta(b-x)^{\alpha}(d-y)^{\beta}}{(b-a)(d-c)}\int_0^1 s^{\beta-1}f(b,sy+(1-s)d)ds$$
$$-\frac{\alpha(b-x)^{\alpha}(d-y)^{\beta}}{(b-a)(d-c)}\int_0^1 t^{\alpha-1}f(tx+(1-t)b,d)dt$$
$$+\frac{\alpha\beta(b-x)^{\alpha}(d-y)^{\beta}}{(b-a)(d-c)}\int_0^1\int_0^1 t^{\alpha-1}s^{\beta-1}\frac{\partial f}{\partial s}(tx+(1-t)b,sy+(1-s)d)dt\,ds.$$

By using the change of the variables, we obtain

$$I_1+I_2+I_3+I_4$$
$$=\frac{1}{(b-a)(d-c)}\Bigg[A-\beta(x-a)^{\alpha}\int_y^d (d-v)^{\beta-1}f(a,v)dv-\beta(x-a)^{\alpha}$$
$$\times\int_c^y (v-c)^{\beta-1}f(a,v)dv$$
$$-\beta(b-x)^{\alpha}\int_c^y (v-c)^{\beta-1}f(b,v)dv-\beta(b-x)^{\alpha}\int_y^d (d-v)^{\beta-1}f(b,v)dv$$
$$-\alpha(d-y)^{\beta}\int_a^x (u-a)^{\alpha-1}f(u,d)du-\alpha(y-c)^{\beta}\int_a^x (u-a)^{\alpha-1}f(u,c)du$$
$$-\alpha(d-y)^{\beta}\int_x^b (b-u)^{\alpha-1}f(u,d)du-\alpha(y-c)^{\beta}\int_x^b (b-u)^{\alpha-1}f(u,c)du$$
$$+\alpha\beta\int_x^b\int_c^y (b-u)^{\alpha-1}(v-c)^{\beta-1}f(u,v)du\,dv+\alpha\beta\int_x^b\int_y^d (b-u)^{\alpha-1}$$

$$\times(d-v)^{\beta-1}f(u,v)du\,dv$$
$$+\alpha\beta\int_a^x\int_c^y (u-a)^{\alpha-1}(v-c)^{\beta-1}f(u,v)du\,dv+\alpha\beta\int_a^x\int_y^d (u-a)^{\alpha-1}$$
$$\left.\times(d-v)^{\beta-1}f(u,v)du\,dv\right].$$

By using the Riemann–Liouville fractional integrals, the proof is completed. □

Theorem 7.6 *Let $f:\Delta=[a,b]\times[c,d]\to\mathbb{R}$ be a partial differentiable mapping on $\Delta=[a,b]\times[c,d]$ and $\frac{\partial^2 f}{\partial t\,\partial s}\in L(\Delta)$, $\alpha,\beta>0$. If $\left|\frac{\partial^2 f}{\partial t\,\partial s}\right|$ is a convex function on the co-ordinates on Δ, then the following inequality holds for Riemann–Liouville fractional integrals:*

$$|C|$$
$$\le T_1\left(\frac{((x-a)^{\alpha+1}+(b-x)^{\alpha+1})((y-c)^{\beta+1}+(d-y)^{\beta+1})}{(b-a)(d-c)}\right)\left|\frac{\partial^2 f}{\partial t\,\partial s}(x,y)\right|$$
$$+T_2\left[\left(\frac{(x-a)^{\alpha+1}((y-c)^{\beta+1}+(d-y)^{\beta+1})}{(b-a)(d-c)}\right)\left|\frac{\partial^2 f}{\partial t\,\partial s}(a,y)\right|\right.$$
$$\left.+\left(\frac{(b-x)^{\alpha+1}((y-c)^{\beta+1}+(d-y)^{\beta+1})}{(b-a)(d-c)}\right)\left|\frac{\partial^2 f}{\partial t\,\partial s}(b,y)\right|\right]$$
$$+T_3\left[\left(\frac{(x-a)^{\alpha+1}((y-c)^{\beta+1}+(d-y)^{\beta+1})}{(b-a)(d-c)}\right)\left|\frac{\partial^2 f}{\partial t\,\partial s}(x,c)\right|\right.$$
$$\left.+\left(\frac{(b-x)^{\alpha+1}((d-y)^{\beta+1}+(y-c)^{\beta+1})}{(b-a)(d-c)}\right)\left|\frac{\partial^2 f}{\partial t\,\partial s}(x,d)\right|\right]$$
$$+T_4\left[\frac{(x-a)^{\alpha+1}(y-c)^{\beta+1}}{(b-a)(d-c)}\left|\frac{\partial^2 f}{\partial t\,\partial s}(a,c)\right|+\frac{(x-a)^{\alpha+1}(d-y)^{\beta+1}}{(b-a)(d-c)}\left|\frac{\partial^2 f}{\partial t\,\partial s}(a,d)\right|,\right.$$

where

$$T_1=\frac{\alpha\beta}{4(\alpha+2)(\beta+2)},\quad T_2=\frac{\alpha\beta(\beta+3)}{4(\alpha+2)(\beta^2+3\beta+2)}$$
$$T_3=\frac{\alpha\beta(\alpha+3)}{4(\beta+2)(\alpha^2+3\alpha+2)},\quad T_4=\frac{\alpha\beta(\alpha+3)(\beta+3)}{4(\alpha^2+3\alpha+2)(\beta^2+3\beta+2)}.$$

Proof: From Lemma 7.2 and using the property of modulus, we have

$$|C|$$
$$\le\frac{(x-a)^{\alpha+1}(y-c)^{\beta+1}}{(b-a)(d-c)}$$

$$\times \int_0^1 \int_0^1 |(t^\alpha - 1)(s^\beta - 1)| \left| \frac{\partial^2 f}{\partial t\, \partial s}(tx + (1-t)a, sy + (1-s)c) \right| ds\, dt$$
$$+ \frac{(x-a)^{\alpha+1}(d-y)^{\beta+1}}{(b-a)(d-c)}$$
$$\times \int_0^1 \int_0^1 |(t^\alpha - 1)(1 - s^\beta)| \left| \frac{\partial^2 f}{\partial t\, \partial s}(tx + (1-t)a, sy + (1-s)d) \right| ds\, dt$$
$$+ \frac{(b-x)^{\alpha+1}(y-c)^{\beta+1}}{(b-a)(d-c)}$$
$$\times \int_0^1 \int_0^1 |(1 - t^\alpha)(s^\beta - 1)| \left| \frac{\partial^2 f}{\partial t\, \partial s}(tx + (1-t)b, sy + (1-s)c) \right| ds\, dt$$
$$+ \frac{(b-x)^{\alpha+1}(d-y)^{\beta+1}}{(b-a)(d-c)}$$
$$\times \int_0^1 \int_0^1 |(1 - t^\alpha)(1 - s^\beta)| \left| \frac{\partial^2 f}{\partial t\, \partial s}(tx + (1-t)b, sy + (1-s)d) \right| ds\, dt.$$

Since $\left| \frac{\partial^2 f}{\partial t\, \partial s} \right|$ is co-ordinated convex, we can write

$$|C|$$
$$\leq \frac{(x-a)^{\alpha+1}(y-c)^{\beta+1}}{(b-a)(d-c)} \int_0^1 |(s^\beta - 1)| \left[\int_0^1 (t^\alpha - 1)t \left| \frac{\partial^2 f}{\partial t\, \partial s}(x, sy + (1-s)c) \right| dt \right.$$
$$\left. + \int_0^1 (t^\alpha - 1)(1-t) \left| \frac{\partial^2 f}{\partial t\, \partial s}(a, sy + (1-s)c) \right| dt \right] ds$$
$$+ \frac{(x-a)^{\alpha+1}(d-y)^{\beta+1}}{(b-a)(d-c)} \int_0^1 |(1 - s^\beta)| \left[\int_0^1 (t^\alpha - 1)t \left| \frac{\partial^2 f}{\partial t\, \partial s}(x, sy + (1-s)d) \right| dt \right.$$
$$\left. + \int_0^1 (t^\alpha - 1)(1-t) \left| \frac{\partial^2 f}{\partial t\, \partial s}(a, sy + (1-s)d) \right| dt \right] ds$$
$$+ \frac{(b-x)^{\alpha+1}(y-c)^{\beta+1}}{(b-a)(d-c)} \int_0^1 |(s^\beta - 1)| \left[\int_0^1 (t^\alpha - 1)t \left| \frac{\partial^2 f}{\partial t\, \partial s}(x, sy + (1-s)c) \right| dt \right.$$
$$\left. + \int_0^1 (t^\alpha - 1)(1-t) \left| \frac{\partial^2 f}{\partial t\, \partial s}(b, sy + (1-s)c) \right| dt \right] ds$$
$$+ \frac{(b-x)^{\alpha+1}(d-y)^{\beta+1}}{(b-a)(d-c)} \int_0^1 |(1 - s^\beta)| \left[\int_0^1 (1 - t^\alpha)t \left| \frac{\partial^2 f}{\partial t\, \partial s}(x, sy + (1-s)d) \right| dt \right.$$
$$\left. + \int_0^1 (1 - t^\alpha)(1-t) \left| \frac{\partial^2 f}{\partial t\, \partial s}(b, sy + (1-s)d) \right| dt \right] ds.$$

By computing these integrals, we obtain

$$\begin{aligned}
&|C| \\
&\leq \frac{(x-a)^{\alpha+1}(y-c)^{\beta+1}}{(b-a)(d-c)} \int_0^1 |(s^\beta - 1)| \left[\left(\frac{1}{\alpha+2} - \frac{1}{2} \right) \left| \frac{\partial^2 f}{\partial t\, \partial s}(x, sy+(1-s)c) \right| \right. \\
&\left. + \left(\frac{1}{(\alpha+1)(\alpha+2)} - \frac{1}{2} \right) \left| \frac{\partial^2 f}{\partial t\, \partial s}(a, sy+(1-s)c) \right| \right] ds \\
&+ \frac{(x-a)^{\alpha+1}(d-y)^{\beta+1}}{(b-a)(d-c)} \int_0^1 |(1-s^\beta)| \left[\left(\frac{1}{\alpha+2} - \frac{1}{2} \right) \left| \frac{\partial^2 f}{\partial t\, \partial s}(x, sy+(1-s)d) \right| \right. \\
&\left. + \left(\frac{1}{(\alpha+1)(\alpha+2)} - \frac{1}{2} \right) \left| \frac{\partial^2 f}{\partial t\, \partial s}(a, sy+(1-s)d) \right| \right] ds \\
&+ \frac{(b-x)^{\alpha+1}(y-c)^{\beta+1}}{(b-a)(d-c)} \int_0^1 |(s^\beta - 1)| \left[\left(\frac{1}{\alpha+2} - \frac{1}{2} \right) \left| \frac{\partial^2 f}{\partial t\, \partial s}(x, sy+(1-s)c) \right| \right. \\
&\left. + \left(\frac{1}{(\alpha+1)(\alpha+2)} - \frac{1}{2} \right) \left| \frac{\partial^2 f}{\partial t\, \partial s}(b, sy+(1-s)c) \right| \right] ds \\
&+ \frac{(b-x)^{\alpha+1}(d-y)^{\beta+1}}{(b-a)(d-c)} \int_0^1 |(1-s^\beta)| \left[\left(\frac{1}{\alpha+2} - \frac{1}{2} \right) \left| \frac{\partial^2 f}{\partial t\, \partial s}(x, sy+(1-s)d) \right| \right. \\
&\left. + \left(\frac{1}{(\alpha+1)(\alpha+2)} - \frac{1}{2} \right) \left| \frac{\partial^2 f}{\partial t\, \partial s}(b, sy+(1-s)d) \right| \right] ds.
\end{aligned}$$

Using co-ordinated convexity of $\left| \frac{\partial^2 f}{\partial t\, \partial s} \right|$ again and computing all integrals, we obtain

$$\begin{aligned}
&|C| \\
&\leq T_1 \left(\frac{((x-a)^{\alpha+1} + (b-x)^{\alpha+1})((y-c)^{\beta+1} + (d-y)^{\beta+1})}{(b-a)(d-c)} \right) \left| \frac{\partial^2 f}{\partial t\, \partial s}(x, y) \right| \\
&+ T_2 \left[\left(\frac{(x-a)^{\alpha+1}((y-c)^{\beta+1} + (d-y)^{\beta+1})}{(b-a)(d-c)} \right) \left| \frac{\partial^2 f}{\partial t\, \partial s}(a, y) \right| \right. \\
&\left. + \left(\frac{(b-x)^{\alpha+1}((y-c)^{\beta+1} + (d-y)^{\beta+1})}{(b-a)(d-c)} \right) \left| \frac{\partial^2 f}{\partial t\, \partial s}(b, y) \right| \right] \\
&+ T_3 \left[\left(\frac{(x-a)^{\alpha+1}(y-c)^{\beta+1} + (b-x)^{\alpha+1}(d-y)^{\beta+1}}{(b-a)(d-c)} \right) \left| \frac{\partial^2 f}{\partial t\, \partial s}(x, c) \right| \right. \\
&\left. + \left(\frac{(x-a)^{\alpha+1}(d-y)^{\beta+1} + (b-x)^{\alpha+1}(y-c)^{\beta+1}}{(b-a)(d-c)} \right) \left| \frac{\partial^2 f}{\partial t\, \partial s}(x, d) \right| \right]
\end{aligned}$$

$$+T_4\left[\frac{(x-a)^{\alpha+1}(y-c)^{\beta+1}}{(b-a)(d-c)}\left|\frac{\partial^2 f}{\partial t\,\partial s}(a,c)\right|+\frac{(x-a)^{\alpha+1}(d-y)^{\beta+1}}{(b-a)(d-c)}\left|\frac{\partial^2 f}{\partial t\,\partial s}(a,d)\right|\right.$$

$$\left.+\frac{(b-x)^{\alpha+1}(y-c)^{\beta+1}}{(b-a)(d-c)}\left|\frac{\partial^2 f}{\partial t\,\partial s}(b,c)\right|+\frac{(b-x)^{\alpha+1}(d-y)^{\beta+1}}{(b-a)(d-c)}\left|\frac{\partial^2 f}{\partial t\,\partial s}(b,d)\right|\right],$$

which completes the proof. □

Theorem 7.7 *Let $f : \Delta = [a,b]\times[c,d] \to \mathbb{R}$ be a partial differentiable mapping on $\Delta = [a,b]\times[c,d]$ and $\frac{\partial^2 f}{\partial t\,\partial s} \in L(\Delta)$, $\alpha,\beta \in (0,1]$. If $\left|\frac{\partial^2 f}{\partial t\,\partial s}\right|^q$, $q > 1$, is a convex function on the co-ordinates on Δ, then the following inequality holds for Riemann–Liouville fractional integrals:*

$$\begin{aligned}
&|C| \\
&\leq \left(\frac{1}{4^{\frac{1}{q}}(b-a)(d-c)(\alpha p+1)^{\frac{1}{p}}(\beta p+1)^{\frac{1}{p}}}\right) \\
&\times(x-a)^{\alpha+1}(y-c)^{\beta+1}\left(\left|\frac{\partial^2 f}{\partial t\,\partial s}(x,y)\right|^q+\left|\frac{\partial^2 f}{\partial t\,\partial s}(x,c)\right|^q\right. \\
&\left.+\left|\frac{\partial^2 f}{\partial t\,\partial s}(a,y)\right|^q+\left|\frac{\partial^2 f}{\partial t\,\partial s}(a,c)\right|^q\right)^{\frac{1}{q}} \\
&+(x-a)^{\alpha+1}(d-y)^{\beta+1}\left(\left|\frac{\partial^2 f}{\partial t\,\partial s}(x,y)\right|^q+\left|\frac{\partial^2 f}{\partial t\,\partial s}(x,d)\right|^q\right. \\
&\left.+\left|\frac{\partial^2 f}{\partial t\,\partial s}(a,y)\right|^q+\left|\frac{\partial^2 f}{\partial t\,\partial s}(a,d)\right|^q\right)^{\frac{1}{q}} \\
&+(b-x)^{\alpha+1}(y-c)^{\beta+1}\left(\left|\frac{\partial^2 f}{\partial t\,\partial s}(x,y)\right|^q+\left|\frac{\partial^2 f}{\partial t\,\partial s}(x,c)\right|^q\right. \\
&\left.+\left|\frac{\partial^2 f}{\partial t\,\partial s}(b,y)\right|^q+\left|\frac{\partial^2 f}{\partial t\,\partial s}(b,c)\right|^q\right)^{\frac{1}{q}} \\
&+(b-x)^{\alpha+1}(d-y)^{\beta+1}\left(\left|\frac{\partial^2 f}{\partial t\,\partial s}(x,y)\right|^q+\left|\frac{\partial^2 f}{\partial t\,\partial s}(x,d)\right|^q\right. \\
&\left.+\left|\frac{\partial^2 f}{\partial t\,\partial s}(b,y)\right|^q ds\,dt+\left|\frac{\partial^2 f}{\partial t\,\partial s}(b,d)\right|^q\right)^{\frac{1}{q}},
\end{aligned}$$

where $p^{-1}+q^{-1}=1$.

Proof: From Lemma 7.2 and by applying the well-known Hölder inequality for double integrals, then one has

$$
\begin{aligned}
&|C| \\
&\leq \frac{(x-a)^{\alpha+1}(y-c)^{\beta+1}}{(b-a)(d-c)}\left(\int_0^1\int_0^1 |(t^\alpha-1)(s^\beta-1)|^p ds\, dt\right)^{\frac{1}{p}} \\
&\times\left(\int_0^1\int_0^1 \left|\frac{\partial^2 f}{\partial t\,\partial s}(tx+(1-t)a, sy+(1-s)c)\right|^q ds\, dt\right)^{\frac{1}{q}} \\
&+\frac{(x-a)^{\alpha+1}(d-y)^{\beta+1}}{(b-a)(d-c)}\left(\int_0^1\int_0^1 |(t^\alpha-1)(1-s^\beta)|^p ds\, dt\right)^{\frac{1}{p}} \\
&\times\left(\int_0^1\int_0^1 \left|\frac{\partial^2 f}{\partial t\,\partial s}(tx+(1-t)a, sy+(1-s)d)\right|^q ds\, dt\right)^{\frac{1}{q}} \\
&+\frac{(b-x)^{\alpha+1}(y-c)^{\beta+1}}{(b-a)(d-c)}\left(\int_0^1\int_0^1 |(1-t^\alpha)(s^\beta-1)|^p ds\, dt\right)^{\frac{1}{p}} \\
&\times\left(\int_0^1\int_0^1 \left|\frac{\partial^2 f}{\partial t\,\partial s}(tx+(1-t)b, sy+(1-s)c)\right|^q ds\, dt\right)^{\frac{1}{q}} \\
&+\frac{(b-x)^{\alpha+1}(d-y)^{\beta+1}}{(b-a)(d-c)}\left(\int_0^1\int_0^1 |(1-t^\alpha)(1-s^\beta)|^p ds\, dt\right)^{\frac{1}{p}} \\
&\times\left(\int_0^1\int_0^1 \left|\frac{\partial^2 f}{\partial t\,\partial s}(tx+(1-t)b, sy+(1-s)d)\right|^q ds\, dt\right)^{\frac{1}{q}}.
\end{aligned}
$$

By using the fact that

$$|t_1^\alpha - t_2^\alpha| \leq |t_1 - t_2|^\alpha$$

for $\alpha \in (0,1]$ and $t_1, t_2 \in [0,1]$, we get

$$\int_0^1 |1-t^\alpha|^p dt \leq \int_0^1 |1-t|^{\alpha p} dt = \frac{1}{\alpha p+1}$$

and

$$\int_0^1 |1-s^\beta|^p dt \leq \int_0^1 |1-s|^{\beta p} ds = \frac{1}{\beta p+1}.$$

Since $\left|\frac{\partial^2 f}{\partial t\,\partial s}\right|$ is co-ordinated convex, we can write

$$
\begin{aligned}
&|C| \\
&\leq \left(\frac{1}{(b-a)(d-c)(\alpha p+1)^{\frac{1}{p}}(\beta p+1)^{\frac{1}{p}}}\right)
\end{aligned}
$$

$$
\begin{aligned}
&\times(x-a)^{\alpha+1}(y-c)^{\beta+1}\left(\int_0^1\int_0^1 ts\left|\frac{\partial^2 f}{\partial t\,\partial s}(x,y)\right|^q ds\,dt\right.\\
&+\int_0^1\int_0^1 t(1-s)\left|\frac{\partial^2 f}{\partial t\,\partial s}(x,c)\right|^q ds\,dt\\
&\left.+\int_0^1\int_0^1 (1-t)s\left|\frac{\partial^2 f}{\partial t\,\partial s}(a,y)\right|^q ds\,dt+\int_0^1\int_0^1 (1-t)(1-s)\left|\frac{\partial^2 f}{\partial t\,\partial s}(a,c)\right|^q ds\,dt\right)^{\frac{1}{q}}\\
&+(x-a)^{\alpha+1}(d-y)^{\beta+1}\left(\int_0^1\int_0^1 ts\left|\frac{\partial^2 f}{\partial t\,\partial s}(x,y)\right|^q ds\,dt\right.\\
&+\int_0^1\int_0^1 t(1-s)\left|\frac{\partial^2 f}{\partial t\,\partial s}(x,d)\right|^q ds\,dt\\
&\left.+\int_0^1\int_0^1 (1-t)s\left|\frac{\partial^2 f}{\partial t\,\partial s}(a,y)\right|^q ds\,dt+\int_0^1\int_0^1 (1-t)(1-s)\left|\frac{\partial^2 f}{\partial t\,\partial s}(a,d)\right|^q ds\,dt\right)^{\frac{1}{q}}\\
&+(b-x)^{\alpha+1}(y-c)^{\beta+1}\left(\int_0^1\int_0^1 ts\left|\frac{\partial^2 f}{\partial t\,\partial s}(x,y)\right|^q ds\,dt\right.\\
&+\int_0^1\int_0^1 t(1-s)\left|\frac{\partial^2 f}{\partial t\,\partial s}(x,c)\right|^q ds\,dt\\
&\left.+\int_0^1\int_0^1 (1-t)s\left|\frac{\partial^2 f}{\partial t\,\partial s}(b,y)\right|^q ds\,dt+\int_0^1\int_0^1 (1-t)(1-s)\left|\frac{\partial^2 f}{\partial t\,\partial s}(b,c)\right|^q ds\,dt\right)^{\frac{1}{q}}\\
&+(b-x)^{\alpha+1}(d-y)^{\beta+1}\left(\int_0^1\int_0^1 ts\left|\frac{\partial^2 f}{\partial t\,\partial s}(x,y)\right|^q ds\,dt\right.\\
&+\int_0^1\int_0^1 t(1-s)\left|\frac{\partial^2 f}{\partial t\,\partial s}(x,d)\right|^q ds\,dt\\
&\left.+\int_0^1\int_0^1 (1-t)s\left|\frac{\partial^2 f}{\partial t\,\partial s}(b,y)\right|^q ds\,dt+\int_0^1\int_0^1 (1-t)(1-s)\left|\frac{\partial^2 f}{\partial t\,\partial s}(b,d)\right|^q ds\,dt\right)^{\frac{1}{q}}.
\end{aligned}
$$

By computing these integrals, we obtain

$$
\begin{aligned}
|C|\leq &\left(\frac{1}{4^{\frac{1}{q}}(b-a)(d-c)(\alpha p+1)^{\frac{1}{p}}(\beta p+1)^{\frac{1}{p}}}\right)\\
&\times(x-a)^{\alpha+1}(y-c)^{\beta+1}\left(\left|\frac{\partial^2 f}{\partial t\,\partial s}(x,y)\right|^q+\left|\frac{\partial^2 f}{\partial t\,\partial s}(x,c)\right|^q\right.
\end{aligned}
$$

$$+\left|\frac{\partial^2 f}{\partial t\,\partial s}(a,y)\right|^q+\left|\frac{\partial^2 f}{\partial t\,\partial s}(a,c)\right|^q\Bigg)^{\frac{1}{q}}+(x-a)^{\alpha+1}(d-y)^{\beta+1}$$

$$+\left|\frac{\partial^2 f}{\partial t\,\partial s}(x,d)\right|^q+\left|\frac{\partial^2 f}{\partial t\,\partial s}(a,y)\right|^q+\left|\frac{\partial^2 f}{\partial t\,\partial s}(a,d)\right|^q\Bigg)^{\frac{1}{q}}$$

$$+(b-x)^{\alpha+1}(y-c)^{\beta+1}\Bigg(\left|\frac{\partial^2 f}{\partial t\,\partial s}(x,y)\right|^q+\left|\frac{\partial^2 f}{\partial t\,\partial s}(x,c)\right|^q$$

$$+\left|\frac{\partial^2 f}{\partial t\,\partial s}(b,y)\right|^q+\left|\frac{\partial^2 f}{\partial t\,\partial s}(b,c)\right|^q\Bigg)^{\frac{1}{q}}+(b-x)^{\alpha+1}(d-y)^{\beta+1}\Bigg(\left|\frac{\partial^2 f}{\partial t\,\partial s}(x,y)\right|^q$$

$$+\left|\frac{\partial^2 f}{\partial t\,\partial s}(x,d)\right|^q+\left|\frac{\partial^2 f}{\partial t\,\partial s}(b,y)\right|^q ds\,dt+\left|\frac{\partial^2 f}{\partial t\,\partial s}(b,d)\right|^q\Bigg)^{\frac{1}{q}},$$

which completes the proof. □

Theorem 7.8 *Let $f:\Delta=[a,b]\times[c,d]\to\mathbb{R}$ be a partial differentiable mapping on $\Delta=[a,b]\times[c,d]$ and $\frac{\partial^2 f}{\partial t\,\partial s}\in L(\Delta)$, $\alpha,\beta\in(0,1]$. If $\left|\frac{\partial^2 f}{\partial t\,\partial s}\right|^q$, $q\geq 1$, is a convex function on the co-ordinates on Δ, then the following inequality holds for Riemann–Liouville fractional integrals:*

$$|C|$$

$$\leq\left(\frac{\alpha\beta}{(\alpha+1)(\beta+1)}\right)^{1-\frac{1}{q}}\left[\frac{(x-a)^{\alpha+1}(y-c)^{\beta+1}}{(b-a)(d-c)}\left(K\left|\frac{\partial^2 f}{\partial t\,\partial s}(x,y)\right|^q+L\left|\frac{\partial^2 f}{\partial t\,\partial s}(a,y)\right|^q\right.\right.$$

$$\left.+M\left|\frac{\partial^2 f}{\partial t\,\partial s}(x,c)\right|^q+N\left|\frac{\partial^2 f}{\partial t\,\partial s}(a,c)\right|^q\right)^{\frac{1}{q}}$$

$$+\left[\frac{(x-a)^{\alpha+1}(d-y)^{\beta+1}}{(b-a)(d-c)}\left(K\left|\frac{\partial^2 f}{\partial t\,\partial s}(x,y)\right|^q+L\left|\frac{\partial^2 f}{\partial t\,\partial s}(a,y)\right|^q\right.\right.$$

$$\left.+M\left|\frac{\partial^2 f}{\partial t\,\partial s}(x,d)\right|^q+N\left|\frac{\partial^2 f}{\partial t\,\partial s}(a,d)\right|^q\right)^{\frac{1}{q}}$$

$$+\left[\frac{(b-x)^{\alpha+1}(y-c)^{\beta+1}}{(b-a)(d-c)}\left(K\left|\frac{\partial^2 f}{\partial t\,\partial s}(x,y)\right|^q+L\left|\frac{\partial^2 f}{\partial t\,\partial s}(b,y)\right|^q\right.\right.$$

$$\left.+M\left|\frac{\partial^2 f}{\partial t\,\partial s}(x,c)\right|^q+N\left|\frac{\partial^2 f}{\partial t\,\partial s}(b,c)\right|^q\right)^{\frac{1}{q}}$$

$$+\left[\frac{(b-x)^{\alpha+1}(d-y)^{\beta+1}}{(b-a)(d-c)}\left(K\left|\frac{\partial^2 f}{\partial t\,\partial s}(x,y)\right|^q+L\left|\frac{\partial^2 f}{\partial t\,\partial s}(b,y)\right|^q\right.\right.$$

$$\left.+M\left|\frac{\partial^2 f}{\partial t\,\partial s}(x,d)\right|^q+N\left|\frac{\partial^2 f}{\partial t\,\partial s}(b,d)\right|^q\right)^{\frac{1}{q}},$$

where

$$K=\left(\frac{1}{\alpha+2}-\frac{1}{2}\right)\left(\frac{1}{\beta+2}-\frac{1}{2}\right),\quad L=\left(\frac{1}{(\alpha+1)(\alpha+2)}-\frac{1}{2}\right)\left(\frac{1}{\beta+2}-\frac{1}{2}\right)$$

$$M=\left(\frac{1}{\alpha+2}-\frac{1}{2}\right)\left(\frac{1}{(\beta+1)(\beta+2)}-\frac{1}{2}\right),$$

$$N=\left(\frac{1}{(\alpha+1)(\alpha+2)}-\frac{1}{2}\right)\left(\frac{1}{(\beta+1)(\beta+2)}-\frac{1}{2}\right).$$

Proof: From Lemma 7.2 and by applying the well-known power-mean inequality for double integrals, then one has

$$\begin{aligned}
&|C|\\
&\leq\frac{(x-a)^{\alpha+1}(y-c)^{\beta+1}}{(b-a)(d-c)}\left(\int_0^1\int_0^1|(t^\alpha-1)(s^\beta-1)|ds\,dt\right)^{1-\frac{1}{q}}\\
&\times\left(\int_0^1\int_0^1|(t^\alpha-1)(s^\beta-1)|\left|\frac{\partial^2 f}{\partial t\,\partial s}(tx+(1-t)a,sy+(1-s)c)\right|^q ds\,dt\right)^{\frac{1}{q}}\\
&+\frac{(x-a)^{\alpha+1}(d-y)^{\beta+1}}{(b-a)(d-c)}\left(\int_0^1\int_0^1|(t^\alpha-1)(1-s^\beta)|ds\,dt\right)^{1-\frac{1}{q}}\\
&\times\left(\int_0^1\int_0^1|(t^\alpha-1)(1-s^\beta)|\left|\frac{\partial^2 f}{\partial t\,\partial s}(tx+(1-t)a,sy+(1-s)d)\right|^q ds\,dt\right)^{\frac{1}{q}}\\
&+\frac{(b-x)^{\alpha+1}(y-c)^{\beta+1}}{(b-a)(d-c)}\left(\int_0^1\int_0^1|(1-t^\alpha)(s^\beta-1)|ds\,dt\right)^{\frac{1}{p}}\\
&\times\left(\int_0^1\int_0^1|(1-t^\alpha)(s^\beta-1)|\left|\frac{\partial^2 f}{\partial t\,\partial s}(tx+(1-t)b,sy+(1-s)c)\right|^q ds\,dt\right)^{\frac{1}{q}}\\
&+\frac{(b-x)^{\alpha+1}(d-y)^{\beta+1}}{(b-a)(d-c)}\left(\int_0^1\int_0^1|(1-t^\alpha)(1-s^\beta)|ds\,dt\right)^{1-\frac{1}{q}}\\
&\times\left(\int_0^1\int_0^1|(1-t^\alpha)(1-s^\beta)|\left|\frac{\partial^2 f}{\partial t\,\partial s}(tx+(1-t)b,sy+(1-s)d)\right|^q ds\,dt\right)^{\frac{1}{q}}.
\end{aligned}$$

By computing these integrals, we obtain

$$
\begin{aligned}
&|C| \\
&\leq \left(\frac{\alpha\beta}{(\alpha+1)(\beta+1)}\right)^{1-\frac{1}{q}} \left[\frac{(x-a)^{\alpha+1}(y-c)^{\beta+1}}{(b-a)(d-c)} \left(K\left|\frac{\partial^2 f}{\partial t\, \partial s}(x,y)\right|^q + L\left|\frac{\partial^2 f}{\partial t\, \partial s}(a,y)\right|^q\right.\right. \\
&\left.\left. + M\left|\frac{\partial^2 f}{\partial t\, \partial s}(x,c)\right|^q + N\left|\frac{\partial^2 f}{\partial t\, \partial s}(a,c)\right|^q\right)^{\frac{1}{q}}\right. \\
&+ \left[\frac{(x-a)^{\alpha+1}(d-y)^{\beta+1}}{(b-a)(d-c)} \left(K\left|\frac{\partial^2 f}{\partial t\, \partial s}(x,y)\right|^q + L\left|\frac{\partial^2 f}{\partial t\, \partial s}(a,y)\right|^q\right.\right. \\
&\left.\left. + M\left|\frac{\partial^2 f}{\partial t\, \partial s}(x,d)\right|^q + N\left|\frac{\partial^2 f}{\partial t\, \partial s}(a,d)\right|^q\right)^{\frac{1}{q}}\right. \\
&+ \left[\frac{(b-x)^{\alpha+1}(y-c)^{\beta+1}}{(b-a)(d-c)} \left(K\left|\frac{\partial^2 f}{\partial t\, \partial s}(x,y)\right|^q + L\left|\frac{\partial^2 f}{\partial t\, \partial s}(b,y)\right|^q\right.\right. \\
&\left.\left. + M\left|\frac{\partial^2 f}{\partial t\, \partial s}(x,c)\right|^q + N\left|\frac{\partial^2 f}{\partial t\, \partial s}(b,c)\right|^q\right)^{\frac{1}{q}}\right. \\
&+ \left[\frac{(b-x)^{\alpha+1}(d-y)^{\beta+1}}{(b-a)(d-c)} \left(K\left|\frac{\partial^2 f}{\partial t\, \partial s}(x,y)\right|^q + L\left|\frac{\partial^2 f}{\partial t\, \partial s}(b,y)\right|^q\right.\right. \\
&\left.\left. + M\left|\frac{\partial^2 f}{\partial t\, \partial s}(x,d)\right|^q + N\left|\frac{\partial^2 f}{\partial t\, \partial s}(b,d)\right|^q\right)^{\frac{1}{q}}\right.,
\end{aligned}
$$

which completes the proof. □

Bibliography

1 Niculescu, C.P. (2000). Convexity according to the geometric mean. *Math. Inequal. Appl.* **3** (2): 155–167. https://doi.org/doi.org/10.7153/mia-03-19.

2 Anderson, G.D., Vamanamurthy, M.K., and Vuorinen, M. (2007). Generalized convexity and inequalities. *J. Math. Anal. Appl* **335**: 1294–1308.

3 Akdemir, A.O., Özdemir, M.E., Ardıç, M.A., and Yalçın, A. (2017). Some new generalizations for GA-convex functions. *Filomat* **31** (4): 1009–1016.

4 Akdemir, A.O., Set, E., Özdemir, M.E., and Yalçın, A. (2018). New generalizations for functions whose second derivatives are GG-convex. *J. Uzbek Math.* **4**: 22–34. https://doi.org/10.29229/uzmj.2018-4-3.

5 Latif, M.A. (2014). New Hermite–Hadamard type integral inequalities for GA-convex functions with applications. *Analysis* **34** (4): 379–389. https://doi.org/10.1515/anly-2012-1235.

6 Niculescu, C.P. (2003). Convexity according to means. *Math. Inequal. Appl* **6** (4): 571–579. https://doi.org/doi.org/10.7153/mia-06-53.

7 Rashid, S., Akdemir, A.O., Noor, M.A., and Noor, K.I. New inequalities of the Hermite–Hadamard type for geometrically convex functions via conformable fractional integral operators(Preprint).

8 Noor, M.A., Noor, K.I., and Rashid, S. Fractional integral inequalities of the exponential GA-convex functions(Preprint).

9 Satnoianu, R.A. (2002). Improved GA-convexity inequalities. *J. Inequal. Pure Appl. Math.* **3** (5): Article ID 82.

10 Zhang, T.-Y., Ji, A.-P., and Qi, F. (2013). Some inequalities of Hermite–Hadamard type for GA-convex functions with applications to means. *Le Matematiche* **18**: 229–239. https://doi.org/10.4418/2013.68.1.17.

11 Zhang, X.-M., Chu, Y.-M., and Zhang, X.-H. (2010). The Hermite–Hadamard type inequality of GA-convex functions and its application. *J. Inequal. Appl.* 507–560. https://doi.org/10.1155/2010/507560.

12 Dragomir, S.S. (2001). On Hadamard's inequality for convex functions on the co-ordinates in a rectangle from the plane. *Taiwanese J. Math.* **5**: 775–788.

13 Alomari, M. and Darus, M. (2009). On the Hadamard's inequality for log -convex functions on the coordinates. *J. Inequal. Appl.* **2009**: Article ID 283147.

14 Bakula, M.K. and Pečarić, J. (2006). On the Jensen's inequality for convex functions on the co-ordinates in a rectangle from the plane. *Taiwanese J. Math.* **5**: 1271–1292.

15 Özdemir, M.E., Set, E., and Sarıkaya, M.Z. (2011). Some new Hadamard's type inequalities for co-ordinated m-convex and (α,m)-convex functions. *Hacettepe J. Math. Stat.* **40**: 219–229.

16 Sarıkaya, M.Z., Set, E., Özdemir, M.E., and Dragomir, S.S. 2012 New some Hadamard's type inequalities for co-ordinated convex functions. *Tamsui Oxf. J. Inf. Math. Sci.* **28** (2): 137–152.

17 Özdemir, M.E., Kavurmacı, H., Akdemir, A.O., and Avcı, M. (2012). Inequalities for convex and s-convex functions on $\Delta = [a,b] \times [c,d]$. *J. Inequal. Appl.* 20. https://doi.org/10.1186/1029-242X-2012-20.

18 Özdemir, M.E., Latif, M.A., and Akdemir, A.O. (2012). On some Hadamard-type inequalities for product of two s-convex functions on the co-ordinates. *J. Inequal. Appl.* 21. https://doi.org/10.1186/1029-242X-2012-21.

19 Samko, S.G., Kilbas, A.A., and Marichev, O.I.(1993). *Fractional Integral and Derivatives, Theory and Applications.* Gordon and Breach, Yverdon et alibi.

20 Dahmani, Z. (2010). On Minkowski and Hermite–Hadamard integral inequalities via fractional integration. *Ann. Funct. Anal.* **1** (1): 51–58.

21 Set, E. (2012). New inequalities of Ostrowski type for mappings whose derivatives are s-convex in the second sense via fractional integrals. *Comput. Math. Appl.* **63**: 1147–1154.

22 Sarıkaya, M.Z., Set, E., Yaldız, H., and Başak, N. (2013). Hermite–Hadamard's inequalities for fractional integrals and related fractional inequalities. *Math. Comput. Model.* **57**: 2403–2407.

23 Tariboon, J., Ntouyas, S.K., and Sudsutad, W. (2014). Some new Riemann–Liouville fractional integral inequalities. *Int. J. Math. Math. Sci.* **2014** Article ID 869434, 6.

24 Set, E., Akdemir, A.O., and Özdemir, M.E. (2017). Simpson type integral inequalities for convex functions via Riemann–Liouville integrals. *Filomat* **31** (14): 4415–4420.

25 Dahmani, Z. (2010). New inequalities in fractional integrals. *Int. J. Nonlinear Sci.* **9** (4): 493–497.

26 Dahmani, Z., Tabharit, L., and Taf, S. (2010). Some fractional integral inequalities. *Nonlinear Sci. Lett. A* **1** (2): 155–160.

27 Dahmani, Z., Tabharit, L., and Taf, S. (2010). New generalizations of Grüss inequality using Riemann–Liouville fractional integrals. *Bull. Math. Anal. Appl.* **2** (3): 93–99.

28 Sarıkaya, M.Z. and Öğünmez, H. 2012 On new inequalities via Riemann–Liouville fractional integration. *Abstr. Appl. Anal.* **2012**: Article ID 428983.

29 Gorenflo, R. and Mainardi, F. (1997). *Fractional Calculus: Integral and Differential Equations of Fractional Order*, 223–276.Wien: Springer-Verlag.

30 Miller, S. and Ross, B. (1993). *An Introduction to the Fractional Calculus and Fractional Differential Equations*, 2. USA: Wiley.

31 Podlubny, I. (1999). *Fractional Differential Equations*. San Diego, CA: Academic Press.

32 Ekinci, A. and Özdemir, M.E. (2019) Some New Integral Inequalities Via Riemann–Liouville Integral Operators. *Appl. and Comput. Math.* **18** (3): 288–295.

33 Gradshteyn, I.S. and Ryzhik, I.M. (2007). *Table of Integrals, Series, and Products*, 7e. Academic Press, Elsevier Inc.

34 Khalil, R., Al Horani, M., Yousef, A., and Sababheh, M. (2014). A new definition of fractional derivative. *J. Comput. Appl. Math.* **264**: 65–70.

35 Abdeljawad, T. (2015). On conformable fractional calculus. *J. Comput. Appl. Math.* **279**: 57–66.

36 Set, E. and Çelik, B. (2017). Certain Hermite–Hadamard type inequalities associated with conformable fractional integral operators. *Creative Math. Inf.* **26** (3): 321–330.

37 Set, E., Gözpınar, A., and Ekinci, A. (2017). Hermite–Hadamard type inequalities via conformable fractional integrals. *Acta Math. Univ. Comenianae* **LXXXVI** (2): 309–320.

38 Awan, M.U., Noor, M.A., Mihai, M.V., and Noor, K.I. (2017). Conformable fractional Hermite–Hadamard inequalities via preinvex functions. *Tbilisi Math. J.* **10** (4): 129–141.

39 Akdemir, A.O., Ekinci, A., and Set, E. (2017). Conformable fractional integrals and related new integral inequalities. *J. Nonlinear Convex Anal.* **18** (4): 661–674.

8

Legendre-Spectral Algorithms for Solving Some Fractional Differential Equations

Youssri H. Youssri and Waleed M. Abd-Elhameed

Department of Mathematics, Faculty of Science, Cairo University, Giza, Egypt

8.1 Introduction

Spectral methods are crucial for obtaining solutions of ordinary, Partial, and fractional differential equations (FDEs). These methods have many advantages if compared with other methods such as finite element and finite difference methods. In many physical and chemical applications, solutions with many decimal places of accuracy are needed, so it is very useful to employ various spectral methods because these methods can provide exponential convergence of the solutions. For applications of spectral methods in different disciplines, one can consult [1–5]. The philosophy of the application of various spectral methods is built on writing the solution of a certain problem as a suitable combination of certain polynomials, which are often orthogonal.

It is well-known that there are three popular types of spectral methods, they are tau, collocation, and Galerkin methods. The philosophy of applying Galerkin method is based on constructing suitable combinations of orthogonal polynomials satisfying the underlying initial/boundary conditions, and after that enforcing the residual to be orthogonal with the selected basis functions. This method is fruitfully applied to linear differential equations (see, for example, [6, 7]). The tau method has the advantage that it avoids some of the problems of the Galerkin methods since we can choose any set of orthogonal polynomials as basis functions and the boundary (initial) conditions are set as constraints (see, for example, [8]). The philosophy of applying the collocation method is to select certain collocation points such that the differential equation is satisfied exactly at these points. This method is convenient for treating nonlinear problems (see, for example, [9–13]).

Fractional Order Analysis: Theory, Methods and Applications, First Edition.
Edited by Hemen Dutta, Ahmet Ocak Akdemir, and Abdon Atangana.

Fractional calculus is a very important branch of mathematical analysis. There are extensive studies concerning fractional calculus from both theoretical and practical points of view. It is well-known that the fractional calculus deals with derivatives and integrals to an arbitrary order (real or complex). The applications of fractional calculus are numerous in many fields. For example, several problems in mechanics (theory of viscoelasticity and viscoplasticity), (bio-)chemistry (modeling of polymers and proteins), medicine (modeling of human tissue under mechanical loads) electrical engineering (transmission of ultrasound waves), and other problems can be modeled by FDEs. Analytical solutions of FDEs are not always available; and therefore, it is an important matter to obtain numerical solutions for such equations via several techniques. In this respect, a great number of researchers have considerable interests in investigating numerically various types of FDEs. For example, Taylor collocation method is followed in [14], Adomian's decomposition method is followed in [15, 16], finite difference method is followed in [17], variational iteration method is followed in [18], homotopy analysis method and homotopy perturbation methods are followed, respectively, in [19], [20], wavelets methods are used in [21, 22], operational matrix methods are employed in [23–25], and the collocation method is applied in [26]. Further, recently, Abd-Elhameed and Youssri [27, 28] have developed some new algorithms based on the two kinds of Chebyshev polynomials, namely, Chebyshev polynomials of fifth- and sixth-kinds to treat some kinds of multiterm FDEs.

Telegraph equations are commonly used in the study of wave propagation of electric signals in a cable transmission line and also in wave phenomena. In recent years, much attention has been given in the literature to the development of numerical schemes for the telegraph type equations (see, for example, [29–35]).

Solving the fractional partial differential equations (FPDEs) analytically is a challenging topic. This is due to the nonlocal properties of fractional operators, so obtaining numerical techniques for these types of equations is very important. There are different techniques for treating numerically such kinds of problems, see, for example, [36–40]. Among the important FPDEs are the FDEs. These equations are used to model problems in many branches of science such as physics, hydrology, and finance. Fractional space derivatives may be used to formulate anomalous dispersion models, where a particle plume spreads at a rate that is different than the classical Brownian motion model. When a fractional derivative of order ν where $1 < \nu < 2$ replaces the second derivative in a diffusion or dispersion model, it leads to a superdiffusive flow model. Fractional diffusion equation plays a pivotal role in modeling anomalous diffusion and subdiffusion systems (see [41]). Different numerical techniques are employed for solving space fractional-order diffusion equation. For instance, numerical solutions based on the application of second-order difference methods are proposed for solving space fractional diffusion equations [40]. Also, some numerical approximations for the same equations are obtained via Chebyshev finite difference method in [42]. Some other numerical solutions using several techniques can be found in [43–47].

The main objective of this work is to present and analyze some new algorithms for treating some kinds of FDEs. More precisely, we will focus our attention to the following items:

- Approximating the time-fractional telegraph equation based on the application of spectral Legendre–Galerkin method.
- Applying collocation and Petrov–Galerkin methods in order to solve numerically the space fractional diffusion equation. We develop two efficient spectral algorithms based on a truncated series of shifted Legendre polynomials.

The contents of the chapter are as follows. In the next section, some properties and new formulae concerned with Legendre polynomials and their shifted ones are given. A Galerkin approach for treating fractional telegraph type equation is presented in detail in Section 8.3. Section 8.4 is devoted to displaying some test problems for fractional telegraph equation. Discussion of the convergence and error analysis of the suggested double expansion is given in Section 8.4. In Section 8.5, we are interested in analyzing and presenting two numerical algorithms for treating the space fractional diffusion problem. Error analysis of the suggested expansion in Section 8.5 is investigated in Section 8.6. Some numerical experiments and comparisons are given in Section 8.7 to ensure the applicability and efficiency of the presented algorithms in Section 8.5. Finally, some conclusions are reported in Section 8.8.

8.2 Some Properties and Relations Concerned with Shifted Legendre Polynomials

It is well known that the standard Legendre polynomials $L_k(z);\ z \in [-1,1]$ form a complete orthogonal system for $L^2(-1,1)$. Moreover, these polynomials satisfy the following orthogonality relation

$$\int_{-1}^{1} L_j(z)\, L_k(z)\, dz = \frac{2}{2k+1}\, \delta_{j,k},$$

where $\delta_{j,k}$ is the well-known Kronecker delta function.

We denote by $L_k^*(z)$ the shifted Legendre polynomials defined on $[0,1]$ as:

$$L_k^*(z) = L_k(2z-1).$$

The shifted Legendre polynomials form a complete orthogonal system for $L^2[0,1]$. They have the following orthogonality relation

$$\int_{0}^{1} L_j^*(z)\, L_k^*(z)\, dz = \frac{1}{2k+1}\, \delta_{j,k}. \tag{8.1}$$

$L_k^*(z)$ has following the analytic form (see [48])

$$L_k^*(z) = \sum_{i=0}^{k} \frac{(-1)^{k+i}\,(k+i)!}{(k-i)!\,(i!)^2} z^i. \tag{8.2}$$

Also, the following recurrence relation is satisfied by $L_k^*(z)$:

$$(k+1)L_{k+1}^*(z) = (2k+1)(2z-1)L_k^*(z) - kL_{k-1}^*(z), \quad L_0^*(z) = 1,$$
$$L_1^*(z) = 2z - 1, \quad k \geq 1. \tag{8.3}$$

Now, assume that for any function $f(z) \in L^2[0,1]$, $f(z)$ can be expanded as

$$f(z) = \sum_{k=0}^{\infty} c_k\, L_k^*(z),$$

where

$$c_k = (2k+1) \int_0^1 f(z)\, L_k^*(z)\, dz. \tag{8.4}$$

The following theorem and lemma are needed in the sequel.

Theorem 8.1 *If $I_i^{(r)}(x)$ denotes the r times repeated integration of $L_i^*(x)$:*

$$I_i^{(r)}(x) = \underbrace{\int\int \cdots \int}_{r\ \text{times}} L_i^*(x)\, \underbrace{dx\, dx \cdots dx}_{r\ \text{times}}.$$

The following relation holds [49]:

$$I_i^{(r)}(x) = \frac{(b-a)^r}{2^{2r}} \sum_{j=0}^{r} \binom{r}{j} (-1)^j \frac{\left(i+r-2j+\frac{1}{2}\right)\Gamma(i-j+\frac{1}{2})}{\Gamma(i+r-j+\frac{3}{2})} L_{i+r-2j}^*(x)$$
$$+ \pi_{r-1}(x),$$

where $\pi_{r-1}(x)$ is a polynomial of degree does not exceed $(r-1)$.

Lemma 8.1 *Let $\phi_k(t) = t(1-t)\, L_k^*(t)$ and let μ be any real number. The following integral formula is valid:*

$$\int_0^1 t^\mu \phi_k(t)\, dt = \frac{(2+\mu-k-k^2)(\Gamma(2+\mu))^2}{\Gamma(3+\mu-k)\,\Gamma(4+\mu+k)}. \tag{8.5}$$

Proof: First, we will show that, for any real number μ, one has

$$\int_0^1 t^\mu\, L_k^*(t)\, dt = \frac{(\Gamma(\mu+1))^2}{\Gamma(\mu-k+1)\,\Gamma(\mu+k+2)}. \tag{8.6}$$

The power form representation (8.2) for $L_k^*(t)$ enables one to write

$$\int_0^1 t^\mu\, L_k^*(t)\, dt = \sum_{i=0}^{k} \frac{(-1)^{i+k}(i+k)!}{(i!)^2(k-i)!} \int_0^1 t^{i+\mu}\, dt$$
$$= \sum_{i=0}^{k} \frac{(-1)^{i+k}(i+k)!}{(i!)^2(i+\mu+1)(k-i)!}.$$

Now, if we set

$$B_{\mu,k} = \sum_{i=0}^{k} \frac{(-1)^{i+k}(i+k)!}{(i!)^2(i+\mu+1)(k-i)!},$$

then, it can be shown by means of Zeilberger's algorithm [50] that $B_{\mu,k}$ satisfies the following difference equation:

$$(k-\mu)\, B_{\mu,k} + (k+\mu+2)\, B_{\mu,k+1} = 0, \quad B_{\mu,0} = \frac{1}{\mu+1},$$

which can be immediately solved to give

$$B_{\mu,k} = \frac{(\Gamma(\mu+1))^2}{\Gamma(\mu-k+1)\,\Gamma(\mu+k+2)}.$$

Now, since

$$\int_0^1 t^\mu \phi_k(t)\, dt = \int_0^1 (t^{\mu+1} - t^{\mu+2})\, L_k^*(t)\, dt,$$

then making use of (8.6), formula (8.5) can be obtained. This proves Lemma 8.1. □

Now, we denote by $L_{b,k}(z)$, the shifted Legendre polynomials defined on (0,b) as:

$$L_{b,k}(z) = L_k\left(\frac{2z}{b} - 1\right). \tag{8.7}$$

All relations and formula of shifted Legendre polynomials on $[0,1]$ can be easily generalized to give their counterparts on $[0,b]$. For example, the polynomials $L_{b,k}(z)$ have the following orthogonality relation

$$\int_0^b L_{b,i}(z)\, L_{b,k}(z)\, dz = \frac{b}{2k+1}\, \delta_{k,j}, \tag{8.8}$$

and they have the analytic form (see [48])

$$L_{b,k}(z) = \sum_{i=0}^{k} \frac{(-1)^{k+i}\,(k+i)!}{(k-i)!\,(i!)^2\, b^i} z^i. \tag{8.9}$$

For any function $f(z) \in L^2[0,b]$, $f(z)$ may be expanded as:

$$f(z) = \sum_{k=0}^{\infty} c_k\, L_{b,k}(z),$$

where

$$c_k = \frac{2k+1}{b} \int_0^b f(z)\, L_{b,k}(z)\, dz. \tag{8.10}$$

For more properties of Legendre polynomials and their shifted ones, see, for example, [48].

8.3 Galerkin Approach for Treating Fractional Telegraph Type Equation

This section focuses on presenting and analyzing in detail how to handle fractional telegraph type equation (see [51]):

$$(D_t^{\alpha} + D_t^{\alpha-1} - D_x^2 + 1)\, u(x,t) = f(x,t), \tag{8.11}$$

with the initial and boundary conditions:

$$u(x,0) = \xi(x), \quad u(x,1) = \eta(x), \tag{8.12}$$

$$u(0,t) = \rho(t), \quad u(1,t) = \sigma(t), \tag{8.13}$$

where $\alpha \in (1,2]$, $f(x,t)$, $\xi(x)$, $\eta(x)$, $\rho(t)$, and $\sigma(t)$ are continuous functions. We will employ the spectral Galerkin method for solving (8.11)–(8.13). The philosophy of the application of this method is basically built on choosing suitable basis functions satisfying the underlying conditions governed by the differential equation. Now, we make use of the transformation:

$$v(x,t) = u(x,t) - \mu(x,t),$$

where

$$\mu(x,t) = (1-x)\,\rho(t) + x\,\sigma(t) + (1-t)(\xi(x) - (1-x)\,\xi(0) - x\,\xi(1)) + t(\eta(x) - (1-x)\,\rho(1) - x\,\sigma(1)),$$

to turn Eq. (8.11) governed by (8.12) and (8.13) into the following modified one:

$$(D_t^{\alpha} + D_t^{\alpha-1} - D_x^2 + 1)v(x,t) = g(x,t), \tag{8.14}$$

governed by the homogeneous initial and boundary conditions:

$$v(x,0) = v(x,1) = v(0,t) = v(1,t) = 0 \tag{8.15}$$

and

$$g(x,t) = f(x,t) - (D_t^{\alpha} + D_t^{\alpha-1} - D_x^2 + 1)\mu(x,t).$$

To seek for a numerical solution for (8.14) and (8.15), we propose the following approximate solution:

$$v(x,t) \approx v_n(x,t) = \sum_{i=0}^{n}\sum_{j=0}^{n} c_{i,j}\ \phi_i(x)\ \phi_j(t), \tag{8.16}$$

where the basis functions $\phi_i(z)$ are chosen to be

$$\phi_k(x) = x(1-x)\ L_k^*(x).$$

It is clear that each member of the basis functions $\{\phi_k(x)\}_{k\geq 0}$ fulfill the boundary conditions in (8.45). Moreover, we observe that the set of polynomials $\{\phi_k(x)\}_{k\geq 0}$ is linearly independent on $[0,1]$. In addition, the polynomials $\{\phi_k(x)\}_{k\geq 0}$ are orthogonal on $[0,1]$ with respect to the weight function $w(x) = \dfrac{1}{x^2(\ell - x)^2}$, in the sense that

$$\int_0^1 \frac{\phi_m(x)\ \phi_n(x)\ dx}{x^2(1-x)^2} = \begin{cases} \frac{1}{2n+1}, & m = n, \\ 0, & m \neq n. \end{cases}$$

The expansion coefficients in (8.16) are determined by the formula

$$c_{i,j} = (2i+1)(2j+1)\int_0^1\int_0^1 v(x,t)\ \phi_i(x)\ \phi_j(t)\ w(x,t)\ dx\ dt, \tag{8.17}$$

where $w(x,t) = \dfrac{1}{x^2(1-x)^2\ t^2(1-t)^2}$..

Now, if we define the following spaces:

$$\begin{aligned} S_n &= \text{span}\{L_i^*(x)L_j^*(t) : \quad i,j = 0,1,\ldots,n\}, \\ Z_n &= \{z(x,t) \in S_n : \ z(x,0) = z(x,1) = z(0,t) = z(1,t) = 0, \quad 0 < x,t < 1\}, \end{aligned} \tag{8.18}$$

then the Legendre–Galerkin approximation to (8.14) and (8.15) is to find $v_n(x,t) \in Z_n$ such that $\forall v(x,t) \in Z_n$

$$((D_t^\alpha + D_t^{\alpha-1} - D_x^2 + 1)\ v_n(x,t), v(x,t)) = (g(x,t), v(x,t)), \tag{8.19}$$

where $(u,v) = \int\int_\Omega uv\ dx\ dt$ is the scalar product in the Sobolev space $L^2(\Omega)$, $\Omega = (0,1)\times(0,1)$. The Galerkin variational formulation (8.19) is equivalent to

$$\begin{aligned} ((D_t^\alpha + D_t^{\alpha-1} - D_x^2 + 1)\ v_n(x,t), \phi_r(x)\ \phi_s(t)) &= (g(x,t), \phi_r(x)\ \phi_s(t)), \\ 0 \leq r,s \leq n. \end{aligned} \tag{8.20}$$

If we denote

$$v_n(x,t) = \sum_{i=0}^{n}\sum_{j=0}^{n} c_{ij}\ \phi_i(x)\phi_j(t), \qquad C = (c_{ij})_{0\leq i,j\leq n}, \tag{8.21}$$

$$a_{r,s} = (\phi_r(x), \phi_s(x)), \qquad A = (a_{r,s})_{0\le r,s\le n}, \tag{8.22}$$

$$b^{(\alpha)}_{r,s} = (D^{\alpha}_t \phi_r(t), \phi_s(t)), \qquad B^{(\alpha)} = (b^{(\alpha)}_{r,s})_{0\le r,s\le n}, \tag{8.23}$$

$$d_{r,s} = (D^2_x \phi_r(x), \phi_s(x)), \qquad D = (d_{r,s})_{0\le r,s\le n}, \tag{8.24}$$

$$g_{r,s} = (g(x,t), \phi_r(x)\phi_s(t)), \qquad G = (g_{r,s})_{0\le r,s\le n}, \tag{8.25}$$

then Eq. (8.20) can be written alternatively in the following matrix form:

$$(B^{(\alpha)} + B^{(\alpha-1)} - D + A)C\,A^T - G = 0, \tag{8.26}$$

where the nonzero elements of the matrices $A, B^{(\alpha)}$, and D are given explicitly in the following theorem.

Theorem 8.2 *The nonzero elements of the matrices $A, B^{(\alpha)}$, and D are given explicitly as follows:*

$$b^{(\alpha)}_{r,s} = \sum_{j=0}^{r} \frac{(-1)^{j+r}(j+r)!(\kappa^{(\alpha)}_{j,s} - \kappa^{(\alpha)}_{j+1,s})}{(j!)^2(r-j)!},$$

$$\kappa^{(\alpha)}_{j,s} = \frac{(j+1)!\Gamma(j-\alpha+3)^2(-\alpha+j-s(s+1)+3)}{\Gamma(j-\alpha+2)\Gamma(j-s-\alpha+4)\Gamma(j+s-\alpha+5)},$$

$$d_{r,r} = -\frac{r^4+2r^3+3r^2+2r-2}{2(2r-1)(2r+1)(2r+3)},$$

$$d_{r,r+2} = d_{r+2,r} = \frac{(r+1)^2(r+2)^2}{4(2r+1)(2r+3)(2r+5)},$$

$$a_{r,r} = \frac{3r^4+6r^3-11r^2-14r+12}{8(2r-3)(2r-1)(2r+1)(2r+3)(2r+5)},$$

$$a_{r,r+2} = a_{r+2,r} = -\frac{(r+1)(r+2)(r^2+3r-2)}{4(2r-1)(2r+1)(2r+3)(2r+5)(2r+7)},$$

$$a_{r,r+4} = a_{r+4,r} = \frac{(r+1)(r+2)(r+3)(r+4)}{16(2r+1)(2r+3)(2r+5)(2r+7)(2r+9)}.$$

Proof: First, to compute the elements of the matrix $A = (a_{r,s})$, we have

$$a_{r,s} = (x(1-x)L^*_r(x), x(1-x)L^*_s(x)). \tag{8.27}$$

If we write the recurrence relation (8.3) as

$$x\,L^*_r(x) = \gamma_r\,L^*_{r-1}(x) + \beta_r\,L^*_r(x) + \alpha_r\,L^*_{r+1}(x), \tag{8.28}$$

where

$$\gamma_r = \frac{r}{2(2r+1)}, \quad \beta_r = \frac{1}{2}, \quad \alpha_r = \frac{r+1}{2(2r+1)},$$

then it is easy to see that

$$x(1-x)\,L_r^*(x) = \sum_{\ell=0}^{4} \theta_{\ell,r}\, L_{r+\ell-2}^*(x) \tag{8.29}$$

and

$$\theta_{1,r} = -\gamma_r\,\gamma_{r-1},\quad \theta_{2,r} = \gamma_r(1-\beta_r-\beta_{r-1}),\quad \theta_{3,r} = \beta_r - \alpha_r\,\gamma_{r+1} - (\beta_r)^2 - \gamma_r\,\alpha_{r-1},$$
$$\theta_{4,r} = \alpha_r(1-\beta_{r+1}-\beta_r),\quad \theta_{5,r} = -\alpha_r\,\alpha_{r+1}.$$

Now, in virtue of relation (8.29), the elements $a_{r,s}$ are given by the formula

$$a_{r,s} = \left(\sum_{\ell=0}^{4} \theta_{\ell,r}\, L_{r+\ell-2}^*(x), \sum_{\ell=0}^{4} \theta_{\ell,s}\, L_{s+\ell-2}^*(x)\right). \tag{8.30}$$

With the aid of (8.30) together with the orthogonality relation (8.1), the elements $a_{r,s}$ can be easily obtained. Now, to compute the nonzero elements of the matrix $D=(d_{r,s})$. We have

$$d_{r,s} = (D^2\phi_r(x), \phi_s(x)) = \int_0^1 D^2(x(1-x)\,L_r^*(x))x(1-x)L_s^*(x)\,dx,$$

which can be–after performing some computations–written as

$$d_{r,s} = \sum_{i=0}^{r} \frac{(-1)^{i+r}(i+r)!}{(i!)^2(r-i)!}\,\{i(i+1)\xi_{s,i-1} - (i+1)(i+2)\xi_{s,i}\}$$

and

$$\xi_{s,i} = \int_0^1 x^i\,\phi_s(x).$$

Lemma 8.1 enables one to write the elements $d_{r,s}$ in the following explicit form

$$a_{r,s} = \sum_{i=0}^{r} \frac{(-1)^{i+r}(i+r)!}{(r-i)!}\left(\frac{i(i+1)(i-s(s+1)+1)}{(i-s+1)!(i+s+2)!}\right.$$
$$\left. - \frac{(i+1)^3(i+2)(i-s(s+1)+2)}{(i-s+2)!(i+s+3)!}\right). \tag{8.31}$$

The summation in the right hand side of (8.31) can be computed in a closed form for all values of r and s. Hence all nonzero elements of the matrix $a_{r,s}$ can be obtained. For $s=r+2$, it can be shown with the aid Zeilberger's algorithm that $\xi_r = a_{r,r+2}$ satisfies the following difference equation:

$$(2r+7)(r+1)^2\xi_{r+1} - (2r+1)(r+3)^2\xi_r = 0,\quad \xi_0 = \frac{1}{15}. \tag{8.32}$$

The exact solution of (8.32) is

$$d_{r,r+2} = \frac{(r+1)^2(r+2)^2}{4(2r+1)(2r+3)(2r+5)}.$$

Also, for $s = r$, $\eta_r = d_{r,r}$ satisfies the following difference equation:

$$(1-2r)(r^4+6r^3+15r^2+18r+6)\eta_r + (2r+5)(r^4+2r^3+3r^2+2r-2)\eta_{r+1} = 0,$$
$$\eta_0 = -\frac{1}{3}. \tag{8.33}$$

The exact solution of (8.33) is

$$d_{r,r} = -\frac{r^4+2r^3+3r^2+2r-2}{2(2r-1)(2r+1)(2r+3)}.$$

Moreover, it can be shown that

$$d_{r+2,r} = d_{r,r+2},$$

and all the other elements of the matrix D are zeros. Finally, we compute the elements of the matrix $B^{(\alpha)}$. The elements $b_{r,s}^{(\alpha)}$ can be calculated by the formula

$$b_{r,s}^{(\alpha)} = (D^\alpha \phi_r(x), \phi_s(x)).$$

Relation (8.2) along with Lemma 8.1 enables one to write

$$b_{r,s}^{(\alpha)} = \sum_{i=0}^{r} \frac{(-1)^{r+i}\,(r+i)!}{(r-i)!\,(i!)^2}\left[\frac{(i+1)!}{\Gamma(i+2-\alpha)}\zeta_{s,i+1-\alpha} - \frac{(i+2)!}{\Gamma(i+3-\alpha)}\zeta_{s,i+2-\alpha}\right],$$

where

$$\zeta_{k,\mu} = \frac{(2+\mu-k-k^2)(\Gamma(2+\mu))^2}{\Gamma(3+\mu-k)\,\Gamma(4+\mu+k)},$$

and hence

$$b_{r,s}^{(\alpha)} = \sum_{i=0}^{r} \frac{(-1)^{r+i}\,(r+i)!}{(r-i)!\,(i!)^2}[\kappa_{i,s}^{(\alpha)} - \kappa_{i+1,s}^{(\alpha)}].$$

This completes the proof of the theorem. □

8.4 Discussion of the Convergence and Error Analysis of the Suggested Double Expansion

This section is devoted to investigating the convergence and error analysis of the proposed approximate solution. We give a priori estimate of the resulting global error. In this respect, we state and prove two theorems. First, the following lemma is needed.

Lemma 8.2 *For every positive integer p, one has*

$$\sum_{r=0}^{p} \frac{\binom{p}{r}\left(i+p-2r+\frac{1}{2}\right)\Gamma\left(i-r+\frac{1}{2}\right)}{\Gamma\left(i+p-r+\frac{3}{2}\right)} = \frac{\Gamma\left(\frac{2i-2p+3}{4}\right)}{\Gamma\left(\frac{2i+2p+3}{4}\right)}. \tag{8.34}$$

Proof: If we set

$$A_p = \sum_{r=0}^{p} \frac{\binom{p}{r}\left(i+p-2r+\frac{1}{2}\right)\Gamma\left(i-r+\frac{1}{2}\right)}{\Gamma\left(i+p-r+\frac{3}{2}\right)}$$

and make use of the celebrated algorithm of Zeilberger, then it can be shown that the following difference equation of order one is satisfied by A_p

$$(2i-2p-1)(2i+2p+3)A_{p+2} - 16\,A_p = 0, \quad A_0 = 1.$$

The above recurrence relation can be immediately solved to give

$$A_p = \frac{\Gamma\left(\frac{2i-2p+3}{4}\right)}{\Gamma\left(\frac{2i+2p+3}{4}\right)}.$$

This proves the lemma. □

Theorem 8.3 *Let $v(x,t)$ be the exact solution of Eqs.* (8.14) *and* (8.15). *Suppose that $v(x,t) = v_1(x)\,v_2(t)$, and assume that there exist positive constants M_1, M_2, such that, $\left|\frac{d^p\,v_1}{d\,x^p}\right| < M_1$, $\left|\frac{d^q\,v_2}{d\,t^q}\right| < M_2$, for some positive integers p, q. If $v(x,t)$ is approximated as: $v(x,t) \approx \sum_{i=0}^{n}\sum_{j=0}^{n} c_{i,j}\,\phi_i(x)\,\phi_j(t)$, then the expansion coefficients c_{ij} satisfy the following estimate*

$$|c_{ij}| = \mathcal{O}(i^{\frac{1}{2}-p}\,j^{\frac{1}{2}-q}), \; \forall i > p > \frac{3}{2},\; j > q > \frac{3}{2}.$$

Proof: Starting from Eq. (8.17), we have

$$c_{i,j} = (2i+1)(2j+1)\int_0^1\int_0^1 v(x,t)\,\phi_i(x)\,\phi_j(t)\,w(x,t)\,dx\,dt.$$

The choice of the polynomials $\phi_i(x)$ implies that

$$c_{ij} = (2i+1)(2j+1)\int_0^1 v_1(x)\,L_i(2x-1)\,dx \int_0^1 v_2(t)\,L_j(2t-1)\,dt. \tag{8.35}$$

If integration by parts is applied p-times to the first integral in the right-hand side of (8.35) and q-times to the second integral, then after taking Theorem 8.3 into consideration, the coefficients c_{ij} are given by

$$c_{ij} = (2i+1)(2j+1)(-1)^{p+q}\int_0^1 v_1^{(p)}(x)\,I_i^{(p)}(x)\,dx \int_0^1 v_2^{(q)}(t)\,I_j^{(q)}(t)\,dt,$$

where

$$I_i^{(p)}(x) = 4^{-p} \sum_{r=0}^{p} \frac{(-1)^r \binom{p}{r} \left(i+p-2r+\frac{1}{2}\right) \Gamma\left(i-r+\frac{1}{2}\right)}{\Gamma\left(i+p-r+\frac{3}{2}\right)} L_{i+p-2r}(2x-1).$$

Now, the hypothesis of the theorem along with Bernstein type inequality

$$|L_i(2x-1)| \leq (2\,\pi\, i\,(x-x^2))^{-\frac{1}{2}}$$

yield

$$\begin{aligned} |c_{ij}| &< 2^{-(1+2p+2q)}\,\pi\; M_1\, M_2\,(2i+1)(2j+1) \sum_{r=0}^{p} \\ &\times \frac{\binom{p}{r}\left(i+p-2r+\frac{1}{2}\right)}{\Gamma\left(i-r+\frac{1}{2}\right)\Gamma(i+p-r+\frac{3}{2})\sqrt{i-p}} \\ &\times \sum_{r=0}^{q} \frac{\binom{q}{r}\left(j+q-2r+\frac{1}{2}\right)\Gamma\left(j-r+\frac{1}{2}\right)}{\Gamma\left(j+q-r+\frac{3}{2}\right)\sqrt{j-q}} \\ &< 2^{-(1+2p+2q)}\,\pi\; M_1\, M_2\,(2i+1)(2j+1) \sum_{r=0}^{p} \frac{\binom{p}{r}\left(i+p+\frac{1}{2}\right)\Gamma\left(i-r+\frac{1}{2}\right)}{\Gamma\left(i+p-r+\frac{3}{2}\right)\sqrt{i-p}} \\ &\times \sum_{r=0}^{q} \frac{\binom{q}{r}(j+q+\frac{1}{2})\Gamma\left(j-r+\frac{1}{2}\right)}{\Gamma\left(j+q-r+\frac{3}{2}\right)\sqrt{j-q}}. \end{aligned}$$

The application of Lemma 8.2 leads to

$$|c_{ij}| < \frac{2^{-(1+2p+2q)}\,\pi\; M_1\, M_2\,(2i+1)(2j+1)\,\Gamma\left(\frac{3}{4}+\frac{i}{2}-\frac{p}{2}\right)\Gamma\left(\frac{3}{4}+\frac{j}{2}-\frac{q}{2}\right)}{\sqrt{i-p}\,\sqrt{j-q}\,\Gamma\left(\frac{3}{4}+\frac{i}{2}+\frac{p}{2}\right)\Gamma\left(\frac{3}{4}+\frac{j}{2}+\frac{q}{2}\right)}.$$

Finally, if we apply the Stirling asymptotic approximation of Gamma function (see [52])

$$\Gamma(z) = \mathcal{O}(z^{z-\frac{1}{2}}\, e^{-z}),$$

we get the desired result. □

Theorem 8.4 *Under the hypothesis of Theorem 8.3, we have the following error estimate*

$$e_n = |u(x,t) - u_n(x,t)| = \mathcal{O}(n^{3-p-q}).$$

Proof: We have

$$
\begin{aligned}
e_n &= |u(x,t) - u_n(x,t)| \\
&= \left| \sum_{i=n+1}^{\infty} \sum_{j=n+1}^{\infty} c_{i,j}\, L_i(2x-1)\, L_j(2t-1) \right| \\
&\leq \sum_{i=n+1}^{\infty} \sum_{j=n+1}^{\infty} |c_{i,j}\, L_i(2x-1)\, L_j(2t-1)|.
\end{aligned}
$$

If we use the well-known identity $|L_r(2z-1)| \leq 1$ and apply Theorem 8.3, then we conclude that there exists a generic constant Υ, such that

$$
|c_{ij}| \leq \Upsilon\, i^{\frac{1}{2}-p}\, j^{\frac{1}{2}-q},
$$

and accordingly, we have

$$
e_n \leq \Upsilon \sum_{i=n+1}^{\infty} \sum_{j=n+1}^{\infty} i^{1-p}\, j^{1-q}. \tag{8.36}
$$

Since the reminder of the convergent series $\sum_{i=0}^{\infty} f(i)$ satisfies the following estimate (see [53])

$$
\sum_{i=n+1}^{\infty} f(i) \leq \int_n^{\infty} f(x)\, dx - \frac{1}{2} f(n+1) < \int_n^{\infty} f(x)\, dx,
$$

then we have

$$
e_n \leq \Upsilon \int_n^{\infty} x^{\frac{1}{2}-p}\, dx \int_n^{\infty} y^{\frac{1}{2}-q}\, dy = \frac{\Upsilon}{\left(p-\frac{3}{2}\right)\left(q-\frac{3}{2}\right)}\, n^{3-p-q},
$$

which completes the proof of the theorem. □

8.5 Some Test Problems for Fractional Telegraph Equation

In this section, in order to illustrate the efficiency of the method described in this chapter, which we call Legendre–Galerkin matrix method (LGMM), we present some numerical results to find approximations $u_n(x,t)$ to the solutions $u(x,t)$ of some test problems. We consider the following error norm,

$$
E = \max_{(x,t)\in(0,1)\times(0,1)} |u(x,t) - u_n(x,t)|.
$$

All calculations are carried out using Mathematica 11.

Example 8.1 Consider the following telegraph problem [51]:

$$\frac{\partial^\alpha u}{\partial t^\alpha} + \frac{\partial^{\alpha-1} u}{\partial t^{\alpha-1}} + u - \frac{\partial^2 u}{\partial x^2} = (2 - 2t + t^2)(x - x^2)e^{-t} + 2t^2 e^{-t},$$
$$(x, t) \in (0, 1) \times (0, 1), \tag{8.37}$$

subject to the initial and boundary conditions

$$u(0, t) = u(1, t) = 0, \quad t \in [0, 1],$$
$$u(x, 0) = 0, \ u(x, 1) = (x - x^2)e^{-1}, \quad x \in [0, 1].$$

The exact solution of Eq. (8.69) in case $\alpha = 2$ is given by $u(x, t) = e^{-t}\, x^{1+\alpha}$. We apply LGMM, the maximum pointwise errors are listed in Table 8.1, and we compare our results with those obtained by the method developed in [51] in Table 8.2.

Example 8.2 Consider the following telegraph problem:

$$(D_t^\alpha + D_t^{\alpha-1} - D_x^2 + 1)\, u(x, t) = \frac{t^{-\alpha} \sinh x}{\Gamma(1-\alpha)}, \quad (x, t) \in (0, 1) \times (0, 1), \tag{8.38}$$

subject to the initial and boundary conditions

$$u(0, t) = 0 \quad u(1, t) = e^{-1} \sinh 1, \quad t \in [0, 1],$$
$$u(x, 0) = \sinh x \quad u(x, 1) = e^{-1} \sinh x, \quad x \in [0, 1],$$

with the exact solution $u(x, t) = e^{-t} \sinh x$. We apply LGMM, the maximum pointwise errors are listed in Table 8.3.

Table 8.1 Maximum pointwise error of Example 8.1.

n	α	E	α	E	α	E
4		2.54×10^{-5}		2.65×10^{-5}		7.38×10^{-5}
8	1.25	5.61×10^{-9}	1.5	3.27×10^{-9}	1.5	7.36×10^{-9}
12		4.83×10^{-13}		8.31×10^{-13}		1.37×10^{-13}

Table 8.2 Maximum absolute errors of Example 8.1 for the case $\alpha = 2$.

	$n = 3$		$n = 6$		$n = 10$	
t	LGMM	[51]	LGMM	[51]	LGMM	[51]
0.6	9.27×10^{-5}	9.62×10^{-3}	3.67×10^{-7}	4.43×10^{-3}	2.22×10^{-12}	2.19×10^{-3}
0.7	3.84×10^{-5}	1.998×10^{-2}	8.51×10^{-7}	2.67×10^{-3}	1.27×10^{-12}	1.45×10^{-3}
0.8	4.87×10^{-5}	1.29×10^{-2}	5.22×10^{-7}	2.14×10^{-3}	2.17×10^{-12}	1.62×10^{-3}
1	0	0	0	0	0	0

Table 8.3 Maximum pointwise error of Example 8.2.

n	α	E	α	E	α	E	α	E
5		1.47×10^{-5}		2.35×10^{-6}		3.49×10^{-6}		5.13×10^{-7}
10	1.2	5.61×10^{-10}	1.6	3.27×10^{-10}	1.8	1.57×10^{-10}	2	2.14×10^{-11}
15		7.35×10^{-15}		6.28×10^{-15}		5.91×10^{-15}		8.65×10^{-16}

Table 8.4 Maximum pointwise error of Example 8.3.

n	α	E	α	E	α	E	α	E
6		2.4×10^{-7}		3.5×10^{-7}		4.3×10^{-7}		8.1×10^{-9}
9	1.25	5.4×10^{-11}	1.5	7.2×10^{-11}	1.75	5.2×10^{-11}	2	4.7×10^{-12}
12		3.7×10^{-14}		2.8×10^{-14}		4.5×10^{-14}		2.22×10^{-16}

Example 8.3 Consider the following telegraph problem:

$$(D_t^{\alpha} + D_t^{\alpha-1} - D_x^2 + 1)\, u(x,t) = g(x,t), \quad (x,t) \in (0,1) \times (0,1), \tag{8.39}$$

subject to the initial and boundary conditions

$$u(0,t) = u(1,t) = 0, \quad t \in [0,1],$$
$$u(x,0) = u(x,1) = 0, \quad x \in [0,1].$$

Th function $g(x,t)$ is chosen such that the exact solution of (8.39) is $u(x,t) = (1-x)(1-t)\sinh x\ \sin t$. We apply LGMM and list the maximum pointwise errors in Table 8.4.

8.6 Spectral Algorithms for Treating the Space Fractional Diffusion Problem

This section is interested in implementing and analyzing other two spectral algorithms in detail for solving the space fractional diffusion problem. Let us consider the space fractional diffusion problem

$$\frac{\partial\, y(x,t)}{\partial\, t} = p(x)\, \frac{\partial^{\nu}\, y(x,t)}{\partial\, x^{\nu}} + q(x,t); \quad 1 < \nu \leq 2, \quad (x,t) \in \Omega := (0,\ell) \times (0,\tau), \tag{8.40}$$

governed by the following nonhomogeneous Dirichlet boundary conditions:

$$y(0,t) = y_0(t), \quad y(\ell,t) = y_\ell(t); \quad 0 < t < \tau, \tag{8.41}$$

and the following initial condition:

$$y(x,0) = f(x); \quad 0 < x < \ell. \tag{8.42}$$

8.6.1 Transformation of the Problem

First, with the aid of the substitution:

$$y(x,t) = z(x,t) + \left(1 - \frac{x}{\ell}\right) y_0(t) + \frac{x}{\ell}\, y_\ell(t), \tag{8.43}$$

Eq. (8.40) subject to (8.41) and (8.42) can be transformed into the equation

$$\frac{\partial\, z(x,t)}{\partial\, t} = p(x)\, \frac{\partial^\nu\, z(x,t)}{\partial\, x^\nu} + q_1(x,t); \;\; 1 < \nu \leq 2, \;\; (x,t) \in \Omega := (0,\ell) \times (0,\tau), \tag{8.44}$$

subject to the boundary conditions:

$$z(0,t) = z(\ell,t) = 0; \quad 0 < t < \tau, \tag{8.45}$$

and the initial condition:

$$z(x,0) = f_1(x); \quad 0 < x < \ell, \tag{8.46}$$

where

$$\begin{aligned} q_1(x,t) = q(x,t) &+ \frac{x^{-\nu}\, p(x) y_0(t)}{\Gamma(1-\nu)} + \frac{x^{1-\nu}\, p(x)(y_\ell(t) - y_0(t))}{\ell\, \Gamma(2-\nu)} \\ &- \left(1 - \frac{x}{\ell}\right) y_0'(t) - \frac{x}{\ell}\, y_\ell'(t) \end{aligned}$$

and

$$f_1(x) = f(x) - \left(1 - \frac{x}{\ell}\right) y_0(0) - \frac{x}{\ell}\, y_\ell(0).$$

Now, if Eq. (8.44) is integrated over the interval $[0,t]$, then it can be easily shown that Eq. (8.44) subject to (8.45) and (8.46) is equivalent to the following equation

$$\left.\begin{aligned} & z(x,t) = p(x) \int_0^t \frac{\partial^\nu\, z(x,s)}{\partial\, x^\nu}\, ds + Q(x,t); \quad 1 < \nu \leq 2, \\ & (x,t) \in \Omega := (0,\ell) \times (0,\tau), \\ & z(0,t) = z(\ell,t) = 0; \quad 0 < t < \tau, \end{aligned}\right\} \tag{8.47}$$

where

$$Q(x,t) = f_1(x) + \int_0^t q_1(x,s)\, ds. \tag{8.48}$$

8.6.2 Basis Functions Selection

In the following, we consider the two orthogonal polynomial sets

$$\phi_n(x) = x(\ell - x)\, L_{\ell,n}(x), \quad n = 0, 1, 2, \ldots \tag{8.49}$$

and

$$\psi_n(t) = L_{\tau,n}(t), \quad n = 0, 1, 2, \ldots, \tag{8.50}$$

where $L_{\ell,n}(x)$ are the shifted Legendre polynomials on $[0, \ell]$ which defined in (8.7). As in Section 8.3, it is easy to see that the basis $\{\phi_n(x)\}_{n\geq 0}$ satisfy the boundary conditions (8.45), and they are orthogonal on $[0, \ell]$ with respect to the weight function $w(x) = \dfrac{1}{x^2(\ell - x)^2}$. The orthogonality relation is given by

$$\int_0^{\ell} \frac{\phi_m(x)\, \phi_n(x)\, dx}{x^2(\ell - x)^2} = \begin{cases} \frac{\ell}{2n+1}, & m = n, \\ 0, & m \neq n. \end{cases}$$

Now, define the following two spaces

$$\begin{aligned} Z &= \{z \in H^2_{w(x,t)}(\Omega) : z(0, t) = z(\ell, t) = 0, \ t \in [0, \tau]\}, \\ Z_M &= \text{span}\{\phi_m(x)\, L_{\tau,n}(t) \ \ m, n = 0, 1, \ldots, M\}, \end{aligned} \tag{8.51}$$

and $H^2_{w(x,t)}(\Omega)$, $\Omega = (0, \ell) \times (0, \tau]$ is the well-known Sobolev space defined in [54], and

$$w(x, t) = \frac{1}{x^2(\ell - x)^2}.$$

Now, assume that any function $z(x, t) \in Z$ has the following double expansion

$$z(x, t) = \sum_{m=0}^{\infty} \sum_{n=0}^{\infty} c_{mn}\, \phi_m(x)\, \psi_n(t), \tag{8.52}$$

where

$$c_{mn} = \frac{(2m + 1)(2n + 1)}{\ell\, \tau} \int_0^{\tau} \int_0^{\ell} \frac{z(x, t)\, \phi_m(x)\, \psi_n(t)}{x^2(\ell - x)^2}\, dx\, dt, \tag{8.53}$$

and assume that $z(x, t)$ can be approximated by taking the projection of $z(x, t)$ on Y_M as

$$z(x, t) \approx z_M(x, t) = \sum_{m=0}^{M} \sum_{n=0}^{M} c_{mn}\, \phi_m(x)\, \psi_n(t). \tag{8.54}$$

Now we intend to state and prove two theorems. We derive a new formula for the Riemann–Liouville fractional derivatives of the polynomials $\phi_m(x)$ in the first, while we give an integral formula for the polynomials $\psi_n(t)$ in the second.

Theorem 8.5 *For very $\nu \in (1,2)$, one has*

$$\frac{\partial^{\nu}\, \phi_m(x)}{\partial\, x^{\nu}} = \frac{(-1)^{m+1}\, \sin(\nu\, \pi)}{\pi} \times \left[\ell\, \Gamma(1-\nu)\, x^{1-\nu} + \xi_m\, x^{2+m-\nu} + \sum_{r=1}^{m} \eta_{m,r}\, x^{2+m-r-\nu}\right], \tag{8.55}$$

where

$$\xi_m = \frac{4^m\,(1+m)\,(2+m)\,\Gamma\left(m+\frac{1}{2}\right)\Gamma(\nu-m-2)\,\ell^{-m}}{\sqrt{\pi}},$$

$$\eta_{m,r} = \frac{2\,(2+m-r)\,((1+m)^2-r)\,(2m-r)!\,\Gamma(r+\nu-m-2)\,\ell^{r-m}}{r!\,(1+m-r)!}.$$

Proof: With the aid of the power form representation in (8.2), $\phi_m(x)$ is given by

$$\phi_m(x) = \sum_{s=0}^{m} \frac{(-1)^{m+s}\,(m+s)!}{(m-s)!\,(s!)^2\,\ell^{s-1}} x^{s+1} - \sum_{s=0}^{m} \frac{(-1)^{m+s}\,(m+s)!}{(m-s)!\,(s!)^2\,\ell^{s}} x^{s+2}. \tag{8.56}$$

The application of the operator D^{ν} to both sides of (8.56), and making use of the identity

$$\frac{\partial^{\nu}\, x^{r}}{\partial\, x^{\nu}} = \frac{r!}{\Gamma(1+r-\nu)} x^{r-\nu},$$

yield the following relation

$$\frac{\partial^{\nu}\, \phi_m(x)}{\partial\, x^{\nu}} = \sum_{s=0}^{m} \frac{(-1)^{m+s}\,(m+s)!\,(s+1)!}{(m-s)!\,(s!)^2\,\Gamma(s+2-\nu)\,\ell^{s-1}} x^{s+1-\nu} - \sum_{s=0}^{m} \frac{(-1)^{m+s}\,(m+s)!\,(s+2)!}{(m-s)!\,(s!)^2\,\Gamma(s+3-\nu)\,\ell^{s}} x^{s+2-\nu}. \tag{8.57}$$

Now, relation (8.55) follows from relation (8.57) after performing some lengthy manipulations. □

Theorem 8.6 *Let $\psi_n(t)$ be as defined in (8.50). The following integral formula is valid*

$$\int_0^t \psi_n(s)\, ds = \sum_{i=0}^{n} \frac{(-1)^{n+i}\,(n+i)!}{(i+1)\,(n-i)!\,(i!)^2\,\tau^{i}} t^{i+1}. \tag{8.58}$$

Proof: Simple with the aid of relation (8.2). □

8.6.3 A Collocation Scheme for Solving Eq. 8.44

Our objective in this section is to obtain a spectral solution to (8.44). For this purpose, Theorems 8.5 and 8.6 are employed together with the utilization of the collocation method for the sake of transforming Eq. (8.44) into a system of linear algebraic equations in the unknown expansion coefficients c_{mn}, which can be efficiently solved. Now, assume an approximate solution of Eq. (8.44) in the form

$$z_M(x,t) = \sum_{m=0}^{M}\sum_{n=0}^{M} c_{mn}\ \phi_m(x)\ \psi_n(t); \tag{8.59}$$

therefore, the residual of (8.44) is

$$R(x,t) = \sum_{m=0}^{M}\sum_{n=0}^{M} c_{mn}\ \phi_m(x)\ \psi_n(t) - p(x)\sum_{m=0}^{M}\sum_{n=0}^{M} c_{mn}\ \frac{\partial^\nu\ \phi_m(x)}{\partial\ x^\nu}\int_0^t \psi_n(s)\ ds$$
$$- Q(x,t). \tag{8.60}$$

In virtue of the analytic formulas of $\phi_m(x)$ and $\psi_n(t)$, and the two relations in (8.55) and (8.58), the residual $R(x,t)$ can be written as

$$R(x,t) = \sum_{m=0}^{M}\sum_{n=0}^{M}\sum_{r=0}^{m}\sum_{s=0}^{n}$$
$$\times\, c_{mn}\ \frac{(-1)^{m+r+n+s}\ (m+r)!\ (n+s)!}{(m-r)!\ (r!)^2\ (n-s)!\ (s!)^2\ \tau^s\ \ell^r}\ (\ell-x)x^{r+1}\ t^s$$
$$- p(x)\sum_{m=0}^{M}\sum_{n=0}^{M} c_{mn}\ \frac{(-1)^{m+1}\ \sin(\nu\ \pi)}{\pi}$$
$$\times\left[\ell\ \Gamma(1-\nu)\ x^{1-\nu} + \xi_m\ x^{2+m-\nu} + \sum_{r=1}^{m}\eta_{m,r}\ x^{2+m-r-\nu}\right]$$
$$\times\sum_{m=0}^{n}\frac{(-1)^{n+m}\ (n+m)!}{(m+1)\ (n-m)!\ (m!)^2\ \tau^m}t^{m+1} - Q(x,t). \tag{8.61}$$

We apply collocation method by selecting the nodes: $\left\{\left(\frac{m\ \ell}{M+2}, \frac{n\ \tau}{M+2}\right) : 1 \le m,\ n \le M+1\right\}$, then (8.61) constitutes a linear algebraic system of equations whose dimension is $(M+1)^2$ in the unknown expansion coefficients $\{c_{mn} : 0 \le m, n \le M\}$ can be obtained. Finally with the aid of the Gaussian elimination procedure or via any suitable solver, this linear system can be solved to obtain the desired spectral solution.

8.6.4 An Alternative Spectral Petrov–Galerkin Scheme for Solving Eq. (8.44)

In this section, we give an alternative method for obtaining a spectral solution to Eq. (8.44) based on the application of the Petrov–Galerkin method. The philosophy of Petrov–Galerkin method is to select a suitable set of test functions which differs from the set of trial functions. In this respect, we choose the test functions to be:

$$\rho_{\varsigma\iota}(x,t) = x^{\varsigma}\, t^{\iota},$$

and therefore the variational formulation of Eq. (8.44) leads to

$$\int_0^{\tau}\int_0^{\ell} R(x,t)\ \rho_{\varsigma\iota}(x,t)\, dx\, dt = 0, \quad 1 \le \varsigma, \iota \le M+1, \tag{8.62}$$

where $R(x,t)$ is as defined in (8.61).

Now, Eq. (8.62) is equivalent to

$$\begin{aligned} &\sum_{m=0}^{M}\sum_{n=0}^{M}\sum_{r=0}^{m}\sum_{s=0}^{n} c_{mn} \\ &\times \frac{(-1)^{m+r+n+s}\ (m+r)!\ (n+s)!}{(r+\varsigma+2)(r+\varsigma+3)(s+\iota+1)(m-r)!\ (r!)^2\ (n-s)!\ (s!)^2}\ell^{\varsigma+3}\tau^{\iota+1} \\ &-\sum_{m=0}^{M}\sum_{n=0}^{M} c_{mn}\ \frac{(-1)^{m+1}\ \sin(\nu\,\pi)}{\pi} p_{m,\varsigma} \\ &\times \sum_{m=0}^{n} \frac{(-1)^{n+m}\ (n+m)!}{(m+1)\ (n-m)!\ (m!)^2\ (m+\iota+2)}\tau^{\iota+2} = Q_{\varsigma\iota}, \end{aligned} \tag{8.63}$$

$$\varsigma, \iota = 1, 2, \ldots, M+1,$$

where

$$p_{m,\varsigma} = \int_0^{\ell} p(x)\ \left[\ell\ \Gamma(1-\nu)\, x^{1-\nu+\varsigma} + \xi_i\, x^{2+i-\nu+\varsigma} + \sum_{r=1}^{i} \eta_{i,r}\ x^{2+i-r-\nu+\varsigma}\right] dx$$

and

$$Q_{\varsigma\iota} = \int_0^{\tau}\int_0^{\ell} Q(x,t)\ \rho_{\varsigma\iota}(x,t) dx\ dt.$$

Equation (8.64) generates a linear system of dimension $(M+1)^2$ in the unknown expansion coefficients c_{in}, which can be efficiently solved. Gauss elimination technique or any suitable technique may be used for this propose. Hence, the desired spectral solution can be obtained.

8.7 Investigation of Convergence and Error Analysis

This section is confined to introducing a detailed study for the convergence and error analysis of double Legendre expansion, which is suggested in Section 8.6.

For this purpose, two important theorems are stated and proved. In the first, we show that the double Legendre expansion of a function $z(x)$ converges uniformly to $z(x,t)$, while in the second theorem, an upper bound for the global error of the truncated expansion is given.

Theorem 8.7 *Let a function* $z(x,t) = x(\ell - x)\, f(x)\, g(t) \in L^2_{w(x,t)}(\Omega)$, $w(x,t) = \frac{1}{x^2(\ell - x)^2}$, *provided with the conditions:* $|f^{(3)}(x)| \leq M_1$, $|g^{(3)}(x)| \leq M_2$. *This function can be expanded as an infinite sum of the basis* $\{\phi_m(x)\, \psi_n(t)\}_{0\leq m,n\leq M}$. *Furthermore, and the series is uniformly convergent to* $z(x,t)$, *the expansion coefficients in* (8.59) *satisfy the following inequality:*

$$|c_{mn}| < \frac{M_1\, M_2\, \ell^2\, \tau^2}{64\, m^2\, n^2}, \qquad \forall m,n > 1. \tag{8.64}$$

Proof: From relation (8.53), it follows that

$$c_{mn} = \frac{(2m+1)(2n+1)}{\ell\, \tau} \int_0^{\tau} \int_0^{\ell} \frac{z(x,t)\, \phi_m(x)\, \psi_n(t)}{x^2(\ell - x)^2}\, dx\, dt. \tag{8.65}$$

With the aid of Eqs. (8.49) and (8.50), c_{mn} can be written alternatively as

$$c_{mn} = \frac{(2m+1)(2n+1)}{\ell\, \tau} \int_0^{\ell} f(x)\, L_{\ell,m}(x)\, dx \int_0^{\tau} g(t)\, \psi_n(t)\, dt. \tag{8.66}$$

If the right-hand side of Eq. (8.66) is integrated by parts three times, then we get

$$c_{mn} = \ell^2\, \tau^2\, (2m+1)\, (2n+1) \int_0^{\ell} f^{(3)}(x)\, I_{3,\ell}(x)\, dx \int_0^{\tau} g^{(3)}(t)\, I_{3,\tau}(t)\, dt, \tag{8.67}$$

where

$$I_{3,b}(z) = \frac{1}{8}\left[\frac{-L_{m-3,b}(z)}{(2m-3)(2m-1)(2m+1)} + \frac{3\, L_{m-1,b}(z)}{(2m-3)(2m+1)(2m+3)} - \frac{3\, L_{m+1,b}(z)}{(2m-1)(2m+1)(2m+5)} + \frac{L_{m+3,b}(z)}{(2m+1)(2m+3)(2m+5)}\right].$$

Now with the aid of the two assumptions $|f^{(3)}(x)| \leq M_1$, $|g^{(3)}(t)| \leq M_2$, and the well-known inequality $|L_{m,b}(z)| \leq 1$, we have

$$\begin{aligned}
|c_{mn}| &= \left|\ell^2\, \tau^2\, (2m+1)\, (2n+1) \int_0^{\ell} f^{(3)}(x)\, I_{3,\ell}(x)\, dx \int_0^{\tau} g^{(3)}(t)\, I_{3,\tau}(t)\, dt\right| \\
&\leq \ell^2\, \tau^2\, (2m+1)\, (2n+1) \int_0^{\ell} |f^{(3)}(x)\, I_{3,\ell}(x)|\, dx \int_0^{\tau} |g^{(3)}(t)\, I_{3,\tau}(t)|\, dt \\
&< \frac{M_1\, M_2\, \ell^2\, \tau^2}{64(2m-3)(2m+5)(2n-3)(2n+5)}
\end{aligned}$$

$$= \frac{M_1\, M_2\, \ell^2\, \tau^2}{64\, m^2\, n^2},$$

which completes the proof of Theorem 8.7. □

Theorem 8.8 *If $z(x, t)$ satisfies the hypothesis of Theorem 8.7, and if we consider the expansion (8.54), then the truncation error is of $\mathcal{O}(M^{-2})$. Explicitly the following error estimate is obtained*

$$|z(x,t) - z_M(x,t)| < \frac{M_1\, M_2\, \ell^4\, \tau^2}{256\, M^2}. \tag{8.68}$$

Proof: First, we have

$$|z(x,t) - z_M(x,t)| = \left| \sum_{m=M}^{\infty} \sum_{j=M}^{\infty} c_{mn}\, x(\ell - x)\, L_{m,\ell}(x)\, L_{m,\tau}(t) \right|.$$

With the aid of the two identities:

$$x(\ell - x) \leq \frac{\ell^2}{4}, \quad |L_{k,b}(z)| \leq 1,$$

we get

$$|z(x,t) - z_M(x,t)| < \frac{\ell^2}{4} \sum_{m=M}^{\infty} \sum_{n=M}^{\infty} |c_{mn}|.$$

In virtue of Theorem 8.7, and by integral test estimation, we get,

$$|z(x,t) - z_M(x,t)| < \frac{M_1\, M_2\, \ell^4\, \tau^2}{2^8} \left(\int_M^{\infty} \frac{1}{x^2}\, dx \right)^2$$

$$< \frac{M_1\, M_2\, \ell^4\, \tau^2}{2^8\, M^2}.$$

The proof is now complete. □

8.8 Numerical Results and Comparisons

In this section, we present some numerical examples and comparisons with the numerical methods discussed in Refs. [40, 43, 55–58].

Example 8.4 Consider the space fractional diffusion problem [40]:

$$\frac{\partial y}{\partial t} = \frac{\partial^{\nu} y}{\partial x^{\nu}} - e^{-t}(x^{1+\nu} + \Gamma(2+\nu)\, x), \quad (x,t) \in (0,1) \times (0,1), \tag{8.69}$$

subject to the boundary conditions

$$y(0,t) = 0, \quad y(1,t) = e^{-t}, \quad t \in [0,1],$$

Table 8.5 Maximum pointwise error of Example 8.4.

M	ν	E_C	E_{PG}	ν	E_C	E_{PG}	ν	E_C	E_{PG}
5		2.54×10^{-4}	1.02×10^{-4}		2.68×10^{-4}	2.12×10^{-4}		6.38×10^{-4}	3.25×10^{-4}
10	1.1	5.28×10^{-9}	3.24×10^{-9}	1.5	6.33×10^{-9}	4.55×10^{-9}	1.9	5.92×10^{-9}	5.22×10^{-9}
15		3.27×10^{-14}	7.28×10^{-15}		2.22×10^{-14}	4.44×10^{-15}		4.88×10^{-14}	2.22×10^{-15}

Table 8.6 Comparison between best errors of Example 8.4.

ν	M	1.1	1.5	1.9
Method in [40]	512	3.43×10^{-8}	9.60×10^{-9}	2.99×10^{-9}
SLCM	15	3.27×10^{-14}	2.22×10^{-14}	4.88×10^{-14}
SLPGM	15	7.28×10^{-15}	4.44×10^{-15}	2.22×10^{-15}

and the initial condition

$$y(x,0) = x^{1+\nu}, \quad x \in [0,1],$$

with the exact solution $y(x,t) = e^{-t}\, x^{1+\nu}$. We apply the shifted Legendre collocation method (SLCM), and the shifted Legendre Petrov–Galerkin method (SLPGM). Let E_C and E_{PG} represent, respectively, the resulting maximum pointwise errors if the two methods, namely SLCM and SLPGM, are applied. Table 8.5 displays the maximum pointwise error for different values of M and ν, while Table 8.6 displays a comparison between our results with those obtained by the method developed in [40]. In addition, Figure 8.1 displays the absolute error for $\nu = \frac{3}{2}$, $M = 15$, Figure 8.2 displays the absolute error for the case $\nu = 1.5$, $M = 15$ when $t = 1$.

Example 8.5 Consider the space fractional diffusion problem [43, 55, 56]:

$$\frac{\partial y}{\partial t} = \Gamma\left(\frac{6}{5}\right)\, x^{\frac{9}{5}}\, \frac{\partial^{\frac{9}{5}} y}{\partial x^{\frac{9}{5}}} - 3\, x^2(2x-1)\, e^{-t}, \quad (x,t) \in (0,1)\times(0,1), \tag{8.70}$$

subject to the boundary conditions

$$y(0,t) = y(1,t) = 0, \quad t \in [0,1],$$

and the initial condition

$$y(x,0) = x^2(1-x), \quad x \in [0,1],$$

with the exact solution $y(x,t) = x^2(1-x)\, e^{-t}$. In Table 8.7, we list the maximum pointwise error for different values of M and ν if SLCM is applied, while Table 8.8

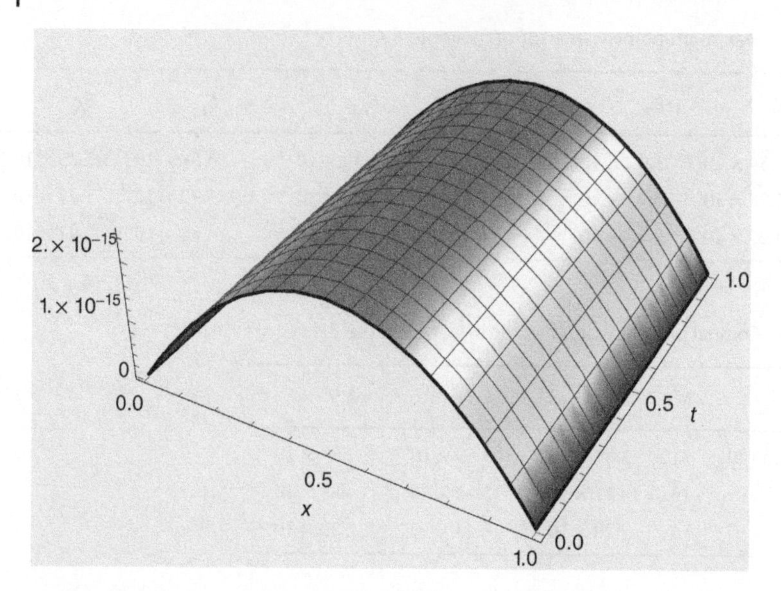

Figure 8.1 Absolute error of Example 8.4 for $\nu = 1.5,\ M = 15$.

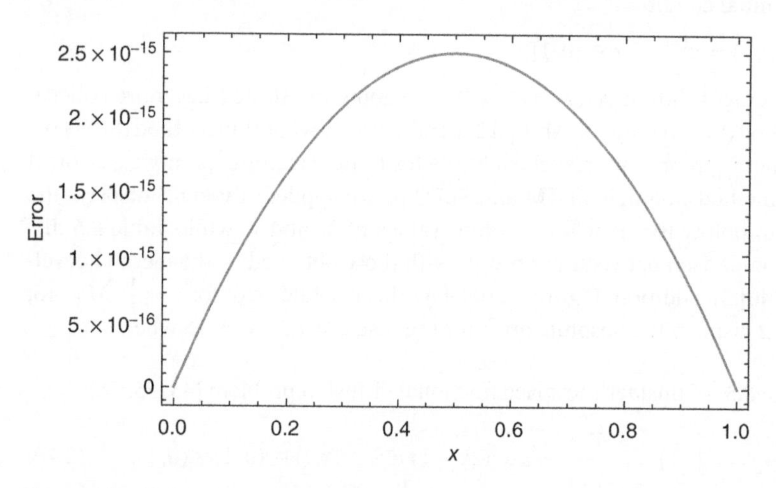

Figure 8.2 Absolute error of Example 8.4 at $t = 1,\ \nu = 1.5,\ M = 15$.

displays a comparison between our results with those obtained by the methods in [43, 55, 56].

Example 8.6 Consider the following fractional diffusion equation:

$$\frac{\partial y}{\partial t} = \frac{\partial^{\frac{3}{2}} y}{\partial x^{\frac{3}{2}}} + q(x,t), \quad (x,t) \in (0,1) \times (0,1), \tag{8.71}$$

Table 8.7 Maximum pointwise error of Example 8.5.

M	E_C	E_{PG}
4	9.19×10^{-7}	2.54×10^{-7}
8	5.88×10^{-10}	3.29×10^{-10}
12	8.67×10^{-15}	8.22×10^{-16}

Table 8.8 Comparison between best errors of Example 8.5.

x	[55] $M = 7$	[56] $M = 7$	[43] $M = 3$	SLCM $M = 3$	SLPGM $M = 3$
0	0	2.81×10^{-5}	0	0	0
0.1	4.26×10^{-5}	4.66×10^{-5}	5.46×10^{-6}	1.03×10^{-7}	2.24×10^{-7}
0.2	5.39×10^{-5}	7.74×10^{-5}	8.51×10^{-6}	2.23×10^{-7}	3.54×10^{-7}
0.3	6.12×10^{-5}	5.00×10^{-5}	9.60×10^{-6}	2.58×10^{-7}	5.22×10^{-7}
0.4	6.48×10^{-5}	2.30×10^{-5}	9.18×10^{-6}	2.41×10^{-7}	1.03×10^{-7}
0.5	6.45×10^{-5}	2.74×10^{-5}	7.69×10^{-6}	2.66×10^{-7}	3.64×10^{-7}
0.6	5.98×10^{-5}	4.38×10^{-5}	5.60×10^{-6}	4.09×10^{-7}	6.24×10^{-7}
0.7	5.23×10^{-5}	3.87×10^{-5}	3.33×10^{-6}	6.69×10^{-7}	2.01×10^{-7}
0.8	4.48×10^{-5}	1.01×10^{-5}	1.34×10^{-6}	9.19×10^{-7}	1.27×10^{-7}
0.9	3.91×10^{-5}	3.35×10^{-5}	8.39×10^{-8}	8.60×10^{-7}	6.47×10^{-7}
1	2.81×10^{-5}	0	0	0	0

subject to the boundary conditions

$$y(0, t) = 1 - t, \quad y(1, t) = 2 - 2t, \quad t \in [0, 1],$$

and the initial condition

$$y(x, 0) = 1 + x^2, \quad x \in [0, 1],$$

where $q(x, t)$ is chosen such that the exact solution of (8.71) is $y(x, t) = (1 - t)(1 + x^2)$. We apply SLCM with $M = 2$, we get the following linear system of equations

$$\begin{aligned}
&(4\sqrt{2} + 2\sqrt{\pi})\, c_{0,0} - 2\sqrt{2}\, c_{0,1} - 4\sqrt{2}\, c_{1,0} + 2\sqrt{2}\, c_{1,1} = -3\sqrt{2} - \sqrt{\pi},\\
&(4\sqrt{2} + \sqrt{\pi})\, c_{0,0} + \sqrt{\pi}\, c_{0,1} - 4\sqrt{2}\, c_{1,0} = -2\sqrt{2},\\
&12\, c_{0,0} - 6\, c_{0,1} + 20\, c_{1,0} - 10\, c_{1,1} = -9,\\
&6\, c_{0,0} - 10\, c_{1,0} = -3.
\end{aligned} \tag{8.72}$$

We solve the system (8.72) to get

$$c_{00} = -c_{01} = -\frac{1}{2}, \quad c_{10} = c_{11} = 0,$$

and consequently

$$z(x,t) = -(1-t)x(1-x),$$

which implies that

$$y(x,t) = z(x,t) + (1-x)y(0,t) + x\,y(1,t) = (1-t)(1+x^2),$$

which is the exact solution.

8.9 Conclusion

This chapter provides some approaches to obtain numerical solutions of some kinds of FDEs. A Galerkin approach is applied for the treatment of time-fractional telegraph equation. Two other spectral methods, namely, Petrov–Galerkin and collocation methods were utilized for handling space fractional linear diffusion problem. The proposed algorithms were built on employing a certain double shifted Legendre basis. The convergence and error analysis for the selected double expansions were discussed carefully. The presented numerical results showed that all the proposed algorithms are efficient, applicable, and easy in implementation.

Bibliography

1 Canuto, C., Hussaini, M.Y., Quarteroni, A., and Zang, T.A. (1988). *Spectral Methods in Fluid Dynamics*. Springer-Verlag.

2 Shizgal, B. (2014). *Spectral Methods in Chemistry and Physics*. Springer.

3 Hesthaven, J., Gottlieb, S., and Gottlieb, D. (2007). *Spectral Methods for Time-Dependent Problems*, vol. **21**. Cambridge University Press.

4 Boyd, J.P. (2001). *Chebyshev and Fourier Spectral Methods*. Courier Corporation.

5 Trefethen, L.N. (2000). *Spectral Methods in MATLAB*, vol. **10**. SIAM.

6 Doha, E.H. and Abd-Elhameed, W.M. (2014). On the coefficients of integrated expansions and integrals of Chebyshev polynomials of third and fourth kinds. *Bull. Malays. Math. Sci. Soc.* **37** (2): 383–398.

7 Abd-Elhameed, W.M., Doha, E.H., and Youssri, Y.H. (2013). Efficient spectral–Petrov–Galerkin methods for third- and fifth-order differential equations using general parameters generalized Jacobi polynomials. *Quaest. Math.* **36**: 15–38.

8 Doha, E.H., Abd-Elhameed, W.M., and Youssri, Y.H. (2013). Second kind Chebyshev operational matrix algorithm for solving differential equations of Lane–Emden type. *New Astron.* **23**: 113–117.

9 Costabile, F. and Napoli, A. (2015). Collocation for high order differential equations with two-points Hermite boundary conditions. *Appl. Numer. Math.* **87**: 157–167.

10 Costabile, F. and Napoli, A. (2015). A method for high-order multipoint boundary value problems with Birkhoff-type conditions. *Int. J. Comput. Math.* **92** (1): 192–200.

11 Abd-Elhameed, W.M. (2014). On solving linear and nonlinear sixth-order two point boundary value problems via an elegant harmonic numbers operational matrix of derivatives. *CMES Comput. Model. Eng. Sci.* **101** (3): 159–185.

12 Esmaeili, S., Shamsi, M., and Luchko, Y. (2011). Numerical solution of fractional differential equations with a collocation method based on Müntz polynomials. *Comput. Math. Appl.* **62** (3): 918–929.

13 Lamnii, A., Mraoui, H., Sbibih, D. et al. (2011). Sextic spline collocation methods for nonlinear fifth-order boundary value problems. *Int. J. Comput. Math.* **88** (10): 2072–2088.

14 Çenesiz, Y., Keskin, Y., and Kurnaz, A. (2010). The solution of the Bagley–Torvik equation with the generalized Taylor collocation method. *J. Franklin Inst.* **347** (2): 452–466.

15 Daftardar-Gejji, V. and Jafari, H. (2007). Solving a multi-order fractional differential equation using Adomian decomposition. *Appl. Math. Comput.* **189** (1): 541–548.

16 Momani, S. and Odibat, Z. (2007). Numerical approach to differential equations of fractional order. *J. Comput. Appl. Math.* **207** (1): 96–110.

17 Meerschaert, M. and Tadjeran, C. (2006). Finite difference approximations for two-sided space-fractional partial differential equations. *Appl. Numer. Math.* **56** (1): 80–90.

18 Das, S. (2009). Analytical solution of a fractional diffusion equation by variational iteration method. *Comput. Math. Appl.* **57** (3): 483–487.

19 Das, D., Ray, P.C., Bera, R.K., and Sarkar, P. (2016). Solution of nonlinear fractional differential equation (NFDE) by homotopy analysis method. *Int. J. Sci. Res. Educ.* **4** (7): 5598–5620.

20 Ghazanfari, B. and Sepahvandzadeh, A. (2015). Homotopy perturbation method for solving fractional Bratu-type equation. *J. Math. Model.* **2** (2): 143–155.

21 Abd-Elhameed, W.M. and Youssri, Y.H. (2015). New spectral solutions of multi-term fractional order initial value problems with error analysis. *CMES-Comput. Model. Eng.* **105**: 375–398.

22 ur Rehman, M. and Khan, R.A. (2011). The Legendre wavelet method for solving fractional differential equations. *Commun. Nonlinear Sci. Numer. Simul.* **16** (11): 4163–4173.

23 Abd-Elhameed, W.M. and Youssri, Y.H. (2016). A novel operational matrix of Caputo fractional derivatives of Fibonacci polynomials: spectral solutions of fractional differential equations. *Entropy* **18** (10): 345.

24 Abd-Elhameed, W.M. and Youssri, Y.H. (2016). Spectral solutions for fractional differential equations via a novel Lucas operational matrix of fractional derivatives. *Rom. J. Phys.* **61** (56): 795–813.

25 Youssri, Y.H. (2017). A new operational matrix of Caputo fractional derivatives of Fermat polynomials: an application for solving the Bagley-Torvik equation. *Adv. Differ. Equ.* **73**: 2017.

26 Hafez, R.M. and Youssri, Y.H. (2018). Jacobi collocation scheme for variable-order fractional reaction-subdiffusion equation. *Comput. Appl. Math.* **37** (4): 5315–5333.

27 Abd-Elhameed, W.M. and Youssri, Y.H. (2018). Fifth-kind orthonormal Chebyshev polynomial solutions for fractional differential equations. *Comput. Appl. Math.* **37**: 2897–2921.

28 Abd-Elhameed, W.M. and Youssri, Y.H. (2019). Sixth-kind Chebyshev spectral approach for solving fractional differential equations. *Int. J. Nonlinear Sci. Numer. Simul.* **20** (2): 191–203.

29 Dehghan, M., Yousefi, S.A., and Lotfi, A. (2011). The use of He's variational iteration method for solving the telegraph and fractional telegraph equations. *Int. J. Numer. Method Biomed. Eng.* **27** (2): 219–231.

30 Momani, S. (2005). Analytic and approximate solutions of the space-and time-fractional telegraph equations. *Appl. Math. Comput.* **170** (2): 1126–1134.

31 Dehghan, M. and Shokri, A. (2008). A numerical method for solving the hyperbolic telegraph equation. *Numer. Methods Part. Differ. Equ.* **24** (4): 1080–1093.

32 Chen, J., Liu, F., and Vo, A. (2008). Analytical solution for the time-fractional telegraph equation by the method of separating variables. *J. Math. Anal. Appl.* **338** (2): 1364–1377.

33 Das, S., Vishal, K., Gupta, P.K., and Yildirim, A. (2011). An approximate analytical solution of time-fractional telegraph equation. *Appl. Math. Comput.* **217** (18): 7405–7411.

34 Saadatmandi, A. and Mohabbati, M. (2015). Numerical solution of fractional telegraph equation via the tau method. *Math. Rep.* **17** (2): 155–166.

35 Atangana, A. (2015). On the stability and convergence of the time-fractional variable order telegraph equation. *J. Comput. Phys.* **293**: 104–114.

36 Sun, Z.Z. and Wu, X.N. (2006). A fully discrete difference scheme for a diffusion-wave system. *Appl. Numer. Math.* **56** (2): 193–209.

37 Yuste, S.B. (2006). Weighted average finite difference methods for fractional diffusion equations. *J. Comput. Phys.* **216** (1): 264–274.
38 Khader, M.M. and Hendy, A.S. (2013). A numerical technique for solving fractional variational problems. *Math. Methods Appl. Sci.* **36** (10): 1281–1289.
39 Dehghan, M. (2005). On the solution of an initial-boundary value problem that combines Neumann and integral condition for the wave equation. *Numer. Methods Part. Differ. Equ.* **21** (1): 24–40.
40 Tian, W., Zhou, H., and Deng, W. (2015). A class of second order difference approximations for solving space fractional diffusion equations. *Math. Comput.* **48**: 1703–1727.
41 Danesh, M. and Safari, M. (2011). Application of Adomian's method for the analytical solution of space fractional diffusion equation. *Adv. Differ. Equ.* **1** (6): 345.
42 Azizi, H. and Loghmani, G.B. (2013). Numerical approximation for space fractional diffusion equations via Chebyshev finite difference method. *J. Fract. Calculus Appl.* **4** (2): 303–311.
43 Sweilam, N.H., Nagy, A.M., and El-Sayed, A.A. (2015). Second kind shifted Chebyshev polynomials for solving space fractional order diffusion equation. *Chaos, Solitons, Fractals* **73**: 141–147.
44 Zhao, Z., Xie, O., You, L., and Meng, Z. (2014). A truncation method based on Hermite expansion for unknown source in space fractional diffusion equation. *Math. Modell. Anal.* **19** (3): 430–442.
45 Bueno-Orovio, A., Kay, D., and Burrage, K. (2014). Fourier spectral methods for fractional-in-space reaction-diffusion equations. *BIT* **54** (4): 937–954.
46 Doha, E.H., Bhrawy, A.H., Baleanu, D., and Ezz-Eldien, S.S. (2014). The operational matrix formulation of the Jacobi tau approximation for space fractional diffusion equation. *Adv. Differ. Equ.* **2014** (1): 231.
47 Doha, E.H., Bhrawy, A.H., and Ezz-Eldien, S.S. (2015). An efficient Legendre spectral tau matrix formulation for solving fractional subdiffusion and reaction subdiffusion equations. *J. Comput. Nonlinear Dyn.* **10** (2): 021019.
48 Rainville, E.D. (1960). *Special Functions*. New York: The Macmillan Company.
49 Napoli, A. and Abd-Elhameed, W.M. (2017). An innovative harmonic numbers operational matrix method for solving initial value problems. *Calcolo* **54** (1): 57–76.
50 Koepf, W. (2014). *Hypergeometric summation*. London: Springer-Verlag.
51 Mollahasani, N., Moghadam, M.M., and Afrooz, K. (2016). A new treatment based on hybrid functions to the solution of telegraph equations of fractional order. *Appl. Math. Modell.* **40** (4: 2804–2814
52 Lanczos, C. (1964). A precision approximation of the gamma function. *SIAM: Series B, Numer. Anal.* **1** 86–96.
53 Boas, R.P. (1978). Estimating remainders. *Math. Mag.* **51** (2): 83–89.

54 Doha, E.H. and Abd-Elhameed, W.M. (2002). Efficient spectral-Galerkin algorithms for direct solution of second-order equations using ultraspherical polynomials. *SIAM J. Sci. Comput.* **24** (2): 548–571.

55 Khader, M.M. (2011). On the numerical solutions for the fractional diffusion equation. *Commun. Nonlinear Sci. Numer. Simul.* **16** (6): 2535–2542.

56 Saadatmandi, A. and Dehghan, M. (2011). A tau approach for solution of the space fractional diffusion equation. *Comput. Math. Appl.* **62** (3): 1135–1142.

57 Tadjeran, C., Meerschaert, M.M., and Scheffler, H.P. (2006). A second-order accurate numerical approximation for the fractional diffusion equation. *J. Comput. Phys.* **213** (1): 205–213.

58 Doha, E.H., Bhrawy, A.H., and Ezz-Eldien, S.S. (2013). Numerical approximations for fractional diffusion equations via a Chebyshev spectral-tau method. *Cent. Eur. J. Phys.* **11** (10): 1494–1503.

9

Mathematical Modeling of an Autonomous Nonlinear Dynamical System for Malaria Transmission Using Caputo Derivative

Abdon Atangana[1,2] *and Sania Qureshi*[3]

[1] *Faculty of Natural and Agricultural Science, Institute for Groundwater Studies, University of Free State, Bloemfontein, South Africa*
[2] *Department of Medical Research, China Medical University Hospital, China Medical University, Taichung, Taiwan*
[3] *Department of Basic Sciences and Related Studies, Mehran University of Engineering and Technology, Jamshoro, Pakistan*

9.1 Introduction

Malaria is a serious global health problem that affects millions of people, particularly kids, pregnant women, patients with other health conditions like HIV and AIDS, and travelers who have had no prior exposure to malaria. Tropical and subtropical regions are hit the hardest, including Latin America, Sub-Saharan Africa, South Asia and Southeast Asia [1, 2]. It is an infection that can be caused by a few different types of plasmodium species with *Plasmodium falciparum* being the dominant species for the infection to be spread around by an infected female Anopheles mosquito. Once the plasmodium gets into the human's bloodstream via mosquito bite, it starts to infect and destroy mainly liver cells and red blood cells, which causes a variety of symptoms (moderate to severe shaking chills, high fever, sweating, headache, vomiting, diarrhea) and sometimes even death.

In 2015 alone, hundreds of millions were infected and almost half a million people died [3, 4]. In addition to being a major cause of death and discomfort in developing countries, malaria also weakens the workforce thereby impacting economic growth negatively.

Despite having substantial amount of available vaccines, there are none that could effectively and completely eradicate the parasitic infection because the epidemiology of malarial virus varies enormously worldwide. Vector control through the use of insecticide treated bed nets, indoor residual spray, and elimination of potential mosquito breeding sites close to humans remain one of the most effective malaria control strategies.

Fractional Order Analysis: Theory, Methods and Applications, First Edition.
Edited by Hemen Dutta, Ahmet Ocak Akdemir, and Abdon Atangana.

Furthermore, getting rid of old tires, broken buckets, or other containers that collect and hold water, checking around taps and air conditioner units to repair leaks, irrigating your gardens carefully to avoid water collection, trimming overgrown plants, not letting water to be accumulated in the saucers of flowerpots, and reporting to the respective government agency in case of any rain gutters or puddles in the public areas could be the most essential steps in preventing the disease from being spread. However, implementation of these measures particularly in developing countries seem to be not only costly but also impossible in most cases.

This is a place where the mathematical models come into play. Mathematical models are tools for thinking about complex problems and they are useful in providing better insights into the transmission dynamics of such vector-borne diseases. In this connection, various mathematical models based upon first order nonlinear system of ordinary differential equations have been proposed [5, 6]. In [7], the authors have proposed model to asses anti-malarial virus strategies whereas the authors in [8] have analyzed a model for the co-interaction of malaria and HIV, which reports decreasing number of HIV and the dual HIV–malaria cases with reduced sexual activity of people carrying malaria disease.

Recently, the authors [9] have presented a model to explore the dynamics of malaria transmission with saturated incidence rates. However, all of the above models are analyzed in the standard integer-order ordinary differential equations whereas the present chapter is aimed at modeling and analyzing the dynamical behavior of such a vector-borne disease in the domain of fractional-order derivative settings using the classical Caputo operator.

Being the generalization of the ordinary derivative from classical calculus, the fractional derivative does not retain all of its properties. In particular, the fractional derivatives do not have clear physical and geometrical interpretations, neither the product nor the chain rules are easy to employ and the index law is applicable only in certain specific functional spaces. However, the "non-local" nature of the fractional operators is what make them superior to the existing ordinary derivative operators which are unable to deal with dynamics of the processes having memory effects.

The reason of increasing popularity of fractional calculus is the natural appearance of its applications in diverse areas of applied sciences and engineering. Fractional differential equations (FDEs) involving real or complex order derivatives have proven to be a useful tool in modeling anomalous dynamics of various physical and biological processes. Using real data, Diethelm in [10] has proposed a Caputo fractional-order model for better understanding of the dynamics of a dengue fever outbreak in the Cape Verde islands. The authors in [11] have fractionalized a model in Caputo sense to get better dynamics of TB virus using real data from 2002 to 2017 in Khyber Pakhtunkhwa, Pakistan whereas the dynamics of Ebola epidemic was better described in [12] using the Caputo fractional derivatives. Various studies in connection with fractional modeling have most recently been carried out in [13–23] to emphasize the preference of fractional-order systems over classical (integer-order) systems.

The basic objective of this chapter is to propose and analyze a Caputo type fractional-order model describing dynamics for the transmission of malaria virus among both human and mosquito population. The paper also aims at to test existence, uniqueness, boundedness and the positivity of the proposed model. Furthermore, with the help of reproductive number; disease-free and endemic equilibrium points have been tested for their stability and instability, respectively. Finally, numerical simulations are carried out for various possible values of the parameters involved in the model while keeping the aim in mind that the proposed fractional-order model performs far better than the original model based upon first order nonlinear ordinary differential equations.

Rest of the chapter is organized as follows: Section 9.2 presents some fundamental concepts from fractional calculus for Caputo type operator followed by formulation of the classical and fractional-order model in Section 9.3 which is further followed by basic properties (reproductive number and equilibrium points) important to be satisfied by the proposed fractional-order model in Section 9.4. Section 9.5 gives proof for the existence, uniqueness, positivity and boundedness for the solutions of the fractional model. Numerical simulations have been carried out in Section 9.6 to illustrate the dynamics of the fractional model. In light of the obtained results, Section 9.7 summarizes all the major findings.

9.2 Mathematical Preliminaries

This section is devoted to some basic but important concepts to carry out further analysis for rest of the chapter. Included are few definitions and theorems from fractional calculus under Caputo operator, in particular.

Definition 9.1 [24] The RL integral of arbitrary real order $\epsilon > 0$ of a function $f(t)$ is defined by the following integral:

$$J^{\epsilon}_{0,t} f(t) = \frac{1}{\Gamma(\epsilon)} \int_0^t (t-\tau)^{\epsilon-1} f(\tau) d\tau, \quad t, \epsilon > 0.$$

Definition 9.2 [24] For a given well-defined absolutely continuous function $f(t) \in C^n[0, T]$ with $\epsilon > 0$, the Caputo fractional derivative of $f(t)$ is defined by the following integral:

$$^{C}D^{\epsilon}_{0,t} f(t) = \frac{1}{\Gamma(n-\epsilon)} \int_0^t (t-\tau)^{n-\epsilon-1} f^{(n)}(\tau) d\tau, \tag{9.1}$$

where $n-1 < \epsilon \leq n, \quad n \in \mathbb{N}$. Note that if $\epsilon \to 1$, then $^{C}D^{\epsilon}_{0,t} f(t)$ approaches to $f'(t)$.

Theorem 9.1 [25] *Let an autonomous fractional-order system be*

$$ {}^{C}D^{\epsilon}_{0,t}\, y(t) = g(y(t)), \quad y(0) = y_0 \in \mathbb{R}^N, \ \epsilon \in (0,1], \tag{9.2} $$

where $y(t) = (y_1(t), y_2(t), \dots, y_N(t)) \in \mathbb{R}^N$ *and* $g : [g_1, g_2, \dots, g_N]^t : \mathbb{R}^N \to \mathbb{R}^N$. *The solutions to* $g(y(t)) = 0$ *are said to be equilibrium points of the system. An equilibrium point E is called the globally asymptotically stable if all the eigenvalues* λ_i *of the Jacobian matrix* $J = \frac{\partial g}{\partial y} = \frac{\partial(g_1, g_2, \dots, g_N)}{\partial(y_1, y_2, \dots, y_N)}$ *evaluated at E satisfy* $|\arg(\lambda_i)| > \frac{\epsilon\pi}{2}$.

Theorem 9.2 [26] ***(Generalized mean value theorem)*** *Let* $g(y) \in C[0,T]$ *and* ${}^{C}D^{\epsilon}_{0,t}\, g(t) \in (0,T]$, *then*

$$ g(t) = g(0) + \frac{1}{\Gamma(\epsilon)} [{}^{C}D^{\epsilon}_{0,t} g](\tau) t^{\epsilon}, \tag{9.3} $$

with $0 \le \tau \le t, \forall t \in (0,T]$.

Corollary 9.1 *Consider that* $g(y) \in C[0,T]$ ${}^{C}D^{\epsilon}_{0,t}\, g(t) \in (0,T]$, *where* $\epsilon \in (0,1]$. *Then, if*

i) ${}^{C}D^{\epsilon}_{0,t}\, g(y) \ge 0, \forall y \in (0,T)$, *then* $g(y)$ *is nondecreasing.*
ii) ${}^{C}D^{\epsilon}_{0,t}\, g(y) \le 0, \forall y \in (0,T)$, *then* $g(y)$ *is nonincreasing.*

9.3 Model Formulation

Recently, the authors in [9] have explored the following seven dimensional malaria transmission dynamics model with nonlinear forces of infection.

$$
\begin{aligned}
\dot{S}_H(t) &= \Lambda_H - \mu_H S_H(t) + \omega R_H(t) - \frac{b\beta_H S_H(t) I_M(t)}{1 + v_H I_M(t)},\\
\dot{E}_H(t) &= -(\alpha_H + \mu_H) E_H(t) + \frac{b\beta_H S_H(t) I_M(t)}{1 + v_H I_M(t)},\\
\dot{I}_H(t) &= \alpha_H E_H(t) - (r + \mu_H + \delta_H) I_H(t),\\
\dot{R}_H(t) &= r I_H(t) - (\mu_H + \omega) R_H(t),\\
\dot{S}_M(t) &= \Lambda_M - \mu_M S_M(t) - \frac{b\beta_M S_M(t) I_H(t)}{1 + v_M I_H(t)},\\
\dot{E}_M(t) &= -(\alpha_M + \mu_M) E_M(t) + \frac{b\beta_M S_M(t) I_H(t)}{1 + v_M I_H(t)},\\
\dot{I}_M(t) &= \alpha_M E_M(t) - (\mu_M + \delta_M) I_M(t).
\end{aligned} \tag{9.4}
$$

In the above model, the unknown quantities $S_H(t), E_H(t), I_H(t), R_H(t), S_M(t), E_M(t), I_M(t)$ at any time t represent number of host humans susceptible (nonvaccinated) to malaria infection, number of host humans exposed to malaria infection, number of infectious host humans, number of recovered host humans, number of susceptible mosquitoes, number of exposed mosquitoes, and the number of infectious mosquitoes, respectively. In this way, the human population has been divided into four categories whereas the mosquito population into three. Description of the parameters with their values is given in Table 9.1.

Having realized importance of the mathematical model (9.4) with all available facts and figures, we develop its modified form in the Caputo fractional-order settings. The strategy is to replace all integer-order differential operators by the Caputo fractional differential operator. Furthermore, there is no room to justify that the human population behavior is exactly the same as that of the mosquitoes'.

It is therefore, various combinations for the fractional-order derivative are possible for each of the groups (humans and mosquitoes). Moreover, the present study employs the use of Caputo derivative operators based upon the fact that they make use of initial conditions in traditional way, that is, the given conditions for integer-order derivative are applicable in real situations unlike Riemann–Liouville operators. Another important reason is the disappearance of constant functions under the Caputo differentiation.

$$
\begin{aligned}
{}^{C}D^{\epsilon}_{0,t}\, S_H(t) &= \Lambda_H - \mu_H S_H(t) + \omega R_H(t) - \frac{b\beta_H S_H(t) I_M(t)}{1+\nu_H I_M(t)},\\
{}^{C}D^{\epsilon}_{0,t}\, E_H(t) &= -(\alpha_H + \mu_H)E_H(t) + \frac{b\beta_H S_H(t) I_M(t)}{1+\nu_H I_M(t)},\\
{}^{C}D^{\epsilon}_{0,t}\, I_H(t) &= \alpha_H E_H(t) - (r + \mu_H + \delta_H) I_H(t),\\
{}^{C}D^{\epsilon}_{0,t}\, R_H(t) &= r I_H(t) - (\mu_H + \omega) R_H(t),\\
{}^{C}D^{\kappa}_{0,t} S_M(t) &= \Lambda_M - \mu_M S_M(t) - \frac{b\beta_M S_M(t) I_H(t)}{1+\nu_M I_H(t)},\\
{}^{C}D^{\kappa}_{0,t} E_M(t) &= -(\alpha_M + \mu_M)E_M(t) + \frac{b\beta_M S_M(t) I_H(t)}{1+\nu_M I_H(t)},\\
{}^{C}D^{\kappa}_{0,t} I_M(t) &= \alpha_M E_M(t) - (\mu_M + \delta_M) I_M(t),
\end{aligned} \tag{9.5}
$$

subject to the initial conditions:

$$
\begin{aligned}
&S_H(0) = c_1, E_H(0) = c_2, I_H(0) = c_3, R_H(0) = c_4, S_M(0) = c_5, E_M(0)\\
&\quad = c_6, I_M(0) = c_7,
\end{aligned} \tag{9.6}
$$

where $\epsilon, \kappa \in (0, 1]$ as exactly one condition is available for each of the differential equations above. The integer-order derivative model is recovered for $\epsilon = \kappa = 1$.

Table 9.1 Description of the parameters and their values [9].

Parameter	Description	Value	Dimension
Λ_H	Recruitment term for the humans' nonvaccinated population	$0.215e-3$	Time^{-1}
Λ_M	Recruitment term for the mosquitoes' nonvaccinated population	$0.7e-1$	Time^{-1}
b	Biting rate per mosquito	0.12	Time^{-1}
μ_H	Human per capita death rate	$0.548e-4$	Time^{-1}
μ_M	Mosquito per capita death rate	1/15	Time^{-1}
δ_H	Disease-induced death rate of human	$0.1e-2$	Time^{-1}
δ_M	Disease-induced death rate of mosquito	$0.1e-1$	Time^{-1}
α_H	Per capita rate of human's transition from exposed to infectious state	1/17	Time^{-1}
α_M	Per capita rate of mosquito's transition from exposed to infectious state	1/18	Time^{-1}
r	Per capita human's recovery rate	$0.5e-1$	Time^{-1}
ω	Per capita loss in human's immunity	1/730	Time^{-1}
ν_H	Antibody proportion by human	0.5	Time^{-1}
ν_M	Antibody proportion by mosquito	0.5	Time^{-1}
β_H	Probability rate of disease transmission from mosquito to human	0.1	—
β_M	Probability rate of disease transmission of parasite to nonvaccinated mosquito	$0.9e-1$	—

9.4 Basic Properties of the Fractional Model

9.4.1 Reproductive Number

An important terminology is the concept of the reproductive number (R_0) that tells us about the number of new cases that an existing case generates on average over the infectious period in a susceptible (nonvaccinated) population [27]. In order to compute R_0, the next generation matrix technique as described in [28] has been employed, which yields the following

$$R_0 = \sqrt{\frac{\alpha_H \beta_H \Lambda_H \alpha_M \beta_M \Lambda_M}{\mu_H(\alpha_H + \mu_H)(r + \delta_H + \mu_H)\mu_M(\delta_M + \mu_M)(\alpha_M + \mu_M)}}\, b. \tag{9.7}$$

Substituting all the given values from Table 9.1, one obtains

$$R_0 \approx 2.4888e - 01 < 1.$$

9.4.2 Existence and Stability of Disease-free Equilibrium Points

Equilibrium points (steady-state solutions) are solutions where the dynamics of the model gets static and no more change is observed. Thus, for the fractional dynamical system (9.5), the steady-state solutions physically refer the complete absence of the malaria infection ($I_H = I_M = 0$) and is found by equating the right-hand side of (9.5) to zero. Hence, the required steady-state solution is found to be

$$E_{\text{steady-state}} = \left(\frac{\Lambda_H}{\mu_H}, 0, 0, 0, \frac{\Lambda_M}{\mu_M}, 0, 0 \right). \tag{9.8}$$

In order to test the stability of this steady-state solution, the following Jacobian matrix is computed at it

$$J(E_{\text{steady-state}}) = \begin{pmatrix} -\mu_H & 0 & 0 & \omega & 0 & 0 & -\frac{b\beta_H \Lambda_H}{\mu_H} \\ 0 & -(\alpha_H + \mu_H) & 0 & 0 & 0 & 0 & \frac{b\beta_H \Lambda_H}{\mu_H} \\ 0 & \alpha_H & -(r + \mu_H + \delta_H) & 0 & 0 & 0 & 0 \\ 0 & 0 & r & -(\mu_H + \omega) & 0 & 0 & 0 \\ 0 & 0 & -\frac{b\beta_M \Lambda_M}{\mu_M} & 0 & -\mu_M & 0 & 0 \\ 0 & 0 & \frac{b\beta_M \Lambda_M}{\mu_M} & 0 & 0 & -(\alpha_M + \mu_M) & 0 \\ 0 & 0 & 0 & 0 & 0 & \alpha_M & -(\mu_M + \delta_M) \end{pmatrix}.$$

Substituting all the values given in Table 9.1, one obtains

$$\begin{pmatrix} -5.48E-05 & 0.00E+00 & 0.00E+00 & 1.37E-03 & 0.00E+00 & 0.00E+00 & -4.71E-02 \\ 0.00E+00 & -5.89E-02 & 0.00E+00 & 0.00E+00 & 0.00E+00 & 0.00E+00 & 4.71E-02 \\ 0.00E+00 & 5.88E-02 & -5.11E-02 & 0.00E+00 & 0.00E+00 & 0.00E+00 & 0.00E+00 \\ 0.00E+00 & 0.00E+00 & 5.00E-02 & -1.43E-03 & 0.00E+00 & 0.00E+00 & 0.00E+00 \\ 0.00E+00 & 0.00E+00 & -1.13E-02 & 0.00E+00 & -6.67E-02 & 0.00E+00 & 0.00E+00 \\ 0.00E+00 & 0.00E+00 & 1.13E-02 & 0.00E+00 & 0.00E+00 & -1.22E-01 & 0.00E+00 \\ 0.00E+00 & 0.00E+00 & 0.00E+00 & 0.00E+00 & 0.00E+00 & 5.56E-02 & -7.67E-02 \end{pmatrix}.$$

The set of the eigenvalues of the above matrix are as follows:

$$\begin{pmatrix} -5.4800e-05 + 0.0000e+00i \\ -1.4247e-03 + 0.0000e+00i \\ -6.6667e-02 + 0.0000e+00i \\ -1.2847e-01 + 0.0000e+00i \\ -3.3329e-02 + 0.0000e+00i \\ -7.3510e-02 + 2.7697e-02i \\ -7.3510e-02 - 2.7697e-02i \end{pmatrix}.$$

Real part of every computed eigenvalue is negative showing that the disease-free steady-state solution (equilibrium point) is locally asymptotically stable if $R_0 < 1$, which is actually the case as computed above. In other words, the condition $|\arg(\lambda_i)| > \left\{\frac{\epsilon\pi}{2}, \frac{\kappa\pi}{2}\right\}$ from above Theorem 9.1 is satisfied for all $\epsilon, \kappa \in (0, 1]$, where $i = 1, 2, \ldots, 7$.

9.4.3 Existence and Stability of Endemic Equilibrium Point

The endemic equilibrium point E_e for the model (9.5) is a point where the disease persists ($I_H \neq 0$, $I_M \neq 0$), and this can be computed by equating the right side of (9.5) to zero. This leads us to

$$E_e = (3.6969e+01, -5.1947e-01, -5.9852e-01, -2.1006e+01, 1.2186e+00, -9.1973e-02, -6.6647e-02). \tag{9.9}$$

The Jacobian matrix of the entire model (9.5) is given as follows:

$$\begin{pmatrix} -\left(\frac{b\beta_H I_M}{1+\nu_H I_M} + \mu_H\right) & 0 & 0 & \omega & 0 & 0 & -\frac{b\beta_H S_H}{(1+\mu_H I_M)^2} \\ \frac{b\beta_H I_M}{1+\nu_H I_M} & -(\alpha_H+\mu_H) & 0 & 0 & 0 & 0 & \frac{b\beta_H S_H}{(1+\mu_H I_M)^2} \\ 0 & \alpha_H & -(r+\mu_H+\delta_H) & 0 & 0 & 0 & 0 \\ 0 & 0 & r & -(\mu_H+\omega) & 0 & 0 & 0 \\ 0 & 0 & -\frac{b\beta_M S_M}{(1+\mu_M I_H)^2} & 0 & -\left(\frac{b\beta_M I_H}{(1+\mu_M I_H)^2} + \mu_M\right) & 0 & 0 \\ 0 & 0 & \frac{b\beta_M S_M}{(1+\mu_M I_H)^2} & 0 & \frac{b\beta_M I_H}{(1+\mu_M I_H)^2} & -(\alpha_M+\mu_M) & 0 \\ 0 & 0 & 0 & 0 & 0 & \alpha_M & -(\mu_M+\delta_M) \end{pmatrix}.$$

The Jacobian computed at this endemic equilibrium point becomes

$$\begin{pmatrix} 7.73E-04 & 0.00E+00 & 0.00E+00 & 1.37E-03 & 0.00E+00 & 0.00E+00 & -4.75E-01 \\ -8.27E-04 & -5.89E-02 & 0.00E+00 & 0.00E+00 & 0.00E+00 & 0.00E+00 & 4.75E-01 \\ 0.00E+00 & 5.88E-02 & -5.11E-02 & 0.00E+00 & 0.00E+00 & 0.00E+00 & 0.00E+00 \\ 0.00E+00 & 0.00E+00 & 5.00E-02 & -1.42E-03 & 0.00E+00 & 0.00E+00 & 0.00E+00 \\ 0.00E+00 & 0.00E+00 & -2.68E-02 & 0.00E+00 & -5.74E-02 & 0.00E+00 & 0.00E+00 \\ 0.00E+00 & 0.00E+00 & 2.68E-02 & 0.00E+00 & -9.22E-03 & -1.22E-01 & 0.00E+00 \\ 0.00E+00 & 0.00E+00 & 0.00E+00 & 0.00E+00 & 0.00E+00 & 5.56E-02 & -7.67E-02 \end{pmatrix}.$$

The set of the eigenvalues of the above matrix are as follows:

$$\begin{pmatrix} -1.6147e-01+0.0000e+00i \\ -7.3704e-02+7.5229e-02i \\ -7.3704e-02-7.5229e-02i \\ 1.1018e-02+0.0000e+00i \\ -6.6838e-05+0.0000e+00i \\ -2.3084e-03+0.0000e+00i \\ -6.6682e-02+0.0000e+00i \end{pmatrix}.$$

Since one of the eigenvalues is positive showing that the endemic equilibrium point E_e is unstable.

9.5 Existence and Uniqueness of the Solutions

The main objective here is to investigate the existence and uniqueness of solutions of the model (9.5) for $\epsilon \in (0,1)$ in a Banach space X with $t \in [0,T]$, $S_H, E_H, I_H, R_H, S_M, E_M, I_M \in C(R,X) \cap L^1_{loc}(R,X)$ and $R = \{(t,x) : t \in [0,T], x \in B(0,r)\}$ for some $T, r > 0$.

Definition 9.3 An ϵ-differentiable function $x \in BC([0,T],X)$ is called a solution of (9.5) if it holds for (9.5).

Theorem 9.3 [29] *Let B be a convex, bounded, and closed subset of a Banach space Y and $\phi : B \to B$ be a condensing map. Then, ϕ has a fixed point in B.*

Consider the initial value problem in (9.5) and (9.6) on the cylinder $R = \{(t,x) \in \mathbb{R} \times X : t \in [0,T], x \in B(0,r)\}$ for some fixed $T > 0, r > 0$, and assume that $\exists\, P \in (0,\epsilon)$, $m_1, m_2, L, L_1 \in L_{1/P}([0,T], \mathbb{R}^+)$ and functions $S_{H1}, E_{H1}, I_{H1}, R_{H1}, S_{M1}, E_{M1}, I_{M1}, S_{H2}, E_{H2}, I_{H2}, R_{H2}, S_{M2}, E_{M2}, I_{M2} \in C(R,X) \cap L^1_{loc}(R,X)$ such that $S_H = S_{H1} + S_{H2}$, $E_H = E_{H1} + E_{H2}$, $I_H = I_{H1} + I_{H2}$, $R_H = R_{H1} + R_{H2}$, $S_M = S_{M1} + S_{M2}$, $E_M = E_{M1} + E_{M2}$, $I_M = I_{M1} + I_{M2}$ and the following assumptions hold

(1) $S_{H1}, E_{H1}, I_{H1}, R_{H1}, S_{M1}, E_{M1}$, and I_{M1} are bounded and Lipschitz.
(2) $S_{H2}, E_{H2}, I_{H2}, R_{H2}, S_{M2}, E_{M2}$, and I_{M2} are compact and bounded.
(3) $|S_H(t,x) - S_H(t,y)| \leq L_1(t)||x - y||$ for all $(t,x), (t,y) \in R$.

Using the Riemann–Liouville integral operator to both sides of (9.5), the following integral representation can be reached.

Lemma 9.1 [30] *The IVP (9.5)–(9.6) is equivalent to the following:*

$$
\begin{aligned}
S_H(t) &= S_H(0) + \frac{1}{\Gamma(\epsilon)} \int_0^t (t - \vartheta)^{\epsilon-1} S_{H1}(\vartheta, S_H(\vartheta)) d\vartheta \\
&\quad + \frac{1}{\Gamma(\epsilon)} \int_0^t (t - \vartheta)^{\epsilon-1} S_{H2}(\vartheta, S_H(\vartheta)) d\vartheta, \\
E_H(t) &= E_H(0) + \frac{1}{\Gamma(\epsilon)} \int_0^t (t - \vartheta)^{\epsilon-1} E_{H1}(\vartheta, E_H(\vartheta)) d\vartheta \\
&\quad + \frac{1}{\Gamma(\epsilon)} \int_0^t (t - \vartheta)^{\epsilon-1} E_{H2}(\vartheta, E_H(\vartheta)) d\vartheta, \\
I_H(t) &= I_H(0) + \frac{1}{\Gamma(\epsilon)} \int_0^t (t - \vartheta)^{\epsilon-1} I_{H1}(\vartheta, I_H(\vartheta)) d\vartheta \\
&\quad + \frac{1}{\Gamma(\epsilon)} \int_0^t (t - \vartheta)^{\epsilon-1} I_{H2}(\vartheta, I_H(\vartheta)) d\vartheta,
\end{aligned}
$$

$$R_H(t) = R_H(0) + \frac{1}{\Gamma(\epsilon)} \int_0^t (t-\vartheta)^{\epsilon-1} R_{H1}(\vartheta, R_H(\vartheta))d\vartheta$$
$$+ \frac{1}{\Gamma(\epsilon)} \int_0^t (t-\vartheta)^{\epsilon-1} I_{H2}(\vartheta, I_H(\vartheta))d\vartheta,$$
$$S_M(t) = S_M(0) + \frac{1}{\Gamma(\kappa)} \int_0^t (t-\vartheta)^{\kappa-1} S_{M1}(\vartheta, S_M(\vartheta))d\vartheta$$
$$+ \frac{1}{\Gamma(\kappa)} \int_0^t (t-\vartheta)^{\kappa-1} S_{M2}(\vartheta, S_M(\vartheta))d\vartheta,$$
$$E_M(t) = E_M(0) + \frac{1}{\Gamma(\kappa)} \int_0^t (t-\vartheta)^{\kappa-1} E_{M1}(\vartheta, E_M(\vartheta))d\vartheta$$
$$+ \frac{1}{\Gamma(\kappa)} \int_0^t (t-\vartheta)^{\kappa-1} E_{M2}(\vartheta, E_M(\vartheta))d\vartheta,$$
$$I_M(t) = I_M(0) + \frac{1}{\Gamma(\kappa)} \int_0^t (t-\vartheta)^{\kappa-1} I_{M1}(\vartheta, I_M(\vartheta))d\vartheta$$
$$+ \frac{1}{\Gamma(\kappa)} \int_0^t (t-\vartheta)^{\kappa-1} I_{M2}(\vartheta, I_M(\vartheta))d\vartheta.$$

This will give us the liberty to prove existence of solutions for (9.5) as follows.

Theorem 9.4 ***(Existence of solution)*** *Based on the assumptions (1) and (2), the IVP (9.5) and (9.6) has at least one solution in [0,T], provided that*

$$\gamma = \frac{c\|L\|_{\frac{1}{P}} T^{\beta}}{\Gamma(\epsilon)} < 1, \quad \text{where } \beta = (\epsilon - p), c = \left(\frac{1-P}{\epsilon - P}\right)^{1-P}. \tag{9.10}$$

Proof: Select r such that $|S_H(0)| + \Gamma(\epsilon)^{-1}c(\|M_1\|_{\frac{1}{P}} + \|M_2\|_{\frac{1}{P}})T^{\beta} \le r$ and let $B_r = \{x : \|x\| \le r\}$ be the closed ball in $BC([0,T], X)$ with sup $\|.\|$. From Lemma 9.1, we have that $S_H : B_r \to BC([0,T], X), x \to S_{H1}x + S_{H2}x$ with

$$S_{H1}(t) = S_H(0) + \frac{1}{\Gamma(\epsilon)} \int_0^t (t-\vartheta)^{\epsilon-1} S_{H1}(\vartheta, x(\vartheta))d\vartheta \quad \text{and}$$
$$S_{H2}(t) = \frac{1}{\Gamma(\epsilon)} \int_0^t (t-\vartheta)^{\epsilon-1} S_{H2}(\vartheta, x(\vartheta))d\vartheta,$$

as a solution of the (9.5) and (9.6). We prove in the following steps that S_H is condensing, the existence of a fixed point of S_H holds from Theorem 9.3. □

First step: We want to show that $S_H(B_r) \subset B_r$. For $x \in B_r$, we attain

$$\|S_H(t)\| \le |S_H(0)| + \frac{1}{\Gamma(\epsilon)} \int_0^t (t-\vartheta)^{\epsilon-1} S_H(\vartheta, x(\vartheta))d\vartheta$$

$$\leq |S_H(0)| + \frac{1}{\Gamma(\epsilon)} \int_0^t (t-\vartheta)^{\epsilon-1} S_{H1}(\vartheta, x(\vartheta))d\vartheta$$
$$+ \frac{1}{\Gamma(\epsilon)} \int_0^t (t-\vartheta)^{\epsilon-1} S_{H2}(\vartheta, x(\vartheta))d\vartheta$$
$$\leq |S_H(0)| + \frac{1}{\Gamma(\epsilon)} \int_0^t (t-\vartheta)^{\epsilon-1} M_1(\vartheta)d\vartheta + \frac{1}{\Gamma(\epsilon)} \int_0^t (t-\vartheta)^{\epsilon-1} M_2(\vartheta)d\vartheta$$
$$\leq |S_H(0)| + \frac{1}{\Gamma(\epsilon)} \left(\int_0^t (t-\vartheta)^{\frac{\epsilon-1}{1-P}} d\vartheta \right)^{1-P} \left(\int_0^t M_1^{\frac{1}{P}}(\vartheta)d\vartheta \right)^P$$
$$+ \frac{1}{\Gamma(\epsilon)} \left(\int_0^t (t-\vartheta)^{\frac{\epsilon-1}{1-P}} d\vartheta \right)^{1-P} \left(\int_0^t M_2^{\frac{1}{P}}(\vartheta)d\vartheta \right)^P$$
$$\leq |S_H(0)| + \frac{c\left(\|M_1\|_{\frac{1}{P}} + \|M_2\|_{\frac{1}{P}} \right)}{\Gamma(\epsilon)} T^\beta \leq r,$$

and therefore, $S_H(B_r) \subset B_r$.

Second step: We want to show that S_{H1} is a contraction. For $x, y \in B_r$ we reach

$$|S_{H1}x(t) - S_{H1}y(t)| \leq \frac{1}{\Gamma(\epsilon)} \int_0^t (t-\vartheta)^{\epsilon-1} L(\vartheta)|x(\vartheta) - y(\vartheta)|d\vartheta$$
$$\leq \frac{1}{\Gamma(\epsilon)} \left(\int_0^t (t-\vartheta)^{\frac{\epsilon-1}{1-P}} d\vartheta \right)^{1-P} \left(L^{\frac{1}{P}}(\vartheta)d\vartheta \right)^P \|x-y\|$$
$$\leq \frac{c\|L\|_{\frac{1}{P}} T^\beta}{\Gamma(\epsilon)} \|x-y\|, \tag{9.11}$$

and therefore, S_{H1} is a contraction with $\|S_{H1}(x) - S_{H2}(y)\| \leq \gamma \|x-y\|$.

Third step: We want to show that S_{H2} is compact. For $0 \leq x_1 \leq x_2 \leq T$, we attain

$$|S_{H2}x(x_1) - S_{H2}y(x_2)| \leq \frac{1}{\Gamma(\epsilon)} \left| \int_0^{x_2} (x_2-\vartheta)^{\epsilon-1} S_{H2}(\vartheta, x(\vartheta))d\vartheta \right.$$
$$\left. - \int_0^{x_1} (x_1-\vartheta)^{\epsilon-1} S_{H2}(\vartheta, x(\vartheta))d\vartheta \right| \leq \frac{1}{\Gamma(\epsilon)} \left| \int_0^{x_1} (x_2-\vartheta)^{\epsilon-1} S_{H2}(\vartheta, x(\vartheta))d\vartheta \right.$$
$$\left. + \int_{x_1}^{x_2} (x_2-\vartheta)^{\epsilon-1} S_{H2}(\vartheta, x(\vartheta))d\vartheta - \int_0^{x_1} (x_1-\vartheta)^{\epsilon-1} S_{H2}(\vartheta, x(\vartheta))d\vartheta \right|$$
$$\leq \frac{1}{\Gamma(\epsilon)} \int_0^{x_1} ((x_1-\vartheta)^{\epsilon-1} - (x_2-\vartheta)^{\epsilon-1})|S_{H2}(\vartheta, x(\vartheta))|d\vartheta$$
$$+ \int_{x_1}^{x_2} (x_2-\vartheta)^{\epsilon-1} |S_{H2}(\vartheta, x(\vartheta))|d\vartheta$$
$$\leq \frac{1}{\Gamma(\epsilon)} \int_0^{x_1} ((x_1-\vartheta)^{\epsilon-1} - (x_2-\vartheta)^{\epsilon-1}) M_2(\vartheta)d\vartheta$$
$$+ \int_{x_1}^{x_2} (x_2-\vartheta)^{\epsilon-1} M_2(\vartheta)d\vartheta$$

$$\leq \frac{1}{\Gamma(\epsilon)}\left(\int_0^{x_1}((x_1-\vartheta)^{\epsilon-1}-(x_2-\vartheta)^{\epsilon-1})^{\frac{1}{1-P}}d\vartheta\right)^{1-P}\left(M_2^{\frac{1}{P}}(\vartheta)d\vartheta\right)^P$$
$$+\frac{1}{\Gamma(\epsilon)}\left(\int_{x_1}^{x_2}(x_2-\vartheta)^{\frac{\epsilon-1}{1-P}}d\vartheta\right)^{1-P}\times\left(M_2^{\frac{1}{P}}(\vartheta)d\vartheta\right)^P$$
$$\leq \frac{c}{\Gamma(\epsilon)}\left(x_1^{\frac{\epsilon-P}{1-P}}-x_2^{\frac{\epsilon-P}{1-P}}+(x_2-x_1)^{\frac{\epsilon-P}{1-P}}\right)^{1-P}\|M_2\|_{\frac{1}{P}}$$
$$+\frac{c}{\Gamma(\epsilon)}(x_2-x_1)^{\epsilon-P}\|M_2\|_{\frac{1}{P}}\leq\frac{c}{\Gamma(\epsilon)}\left((x_2-x_1)^{\frac{\epsilon-P}{1-P}}\right)^{1-P}\|M_2\|_{\frac{1}{P}}$$
$$+\frac{c}{\Gamma(\epsilon)}(x_2-x_1)^{\epsilon-P}\|M_2\|_{\frac{1}{P}}\leq\frac{2c\|M_2\|_{\frac{1}{P}}}{\Gamma(\epsilon)}(x_2-x_1)^{\epsilon-P}.$$

One can observe that the right-hand side of the above inequality is free from x. Therefore, by Arzela–Ascoli principle for equicontinuous functions [31], we conclude that $S_{H2}(B_r)$ is relatively compact implies S_{H2} is compact. Since S_{H1} is a contraction and S_{H2} is compact and hence completely continuous, as presented in [32], the map $S_H = S_{H1} + S_{H2}$ is condensing on B_r and Lemma 9.1 implies the existence of a fixed point of S_H. It is clearly followed for $E_H, I_H, R_H, S_M, E_M, I_M$.

Theorem 9.5 ***(Uniqueness of solution)*** *Suppose that assumption (3) and $\gamma = \frac{c\|L\|_{\frac{1}{P}}T^{\beta}}{\Gamma(\alpha)} < 1$. Then, the IVP (9.5) and (9.6) possesses solution on [0,T].*

Proof: Consider the map F given by

$$FS_H(t) = S_H(0) + \frac{1}{\Gamma(\epsilon)}\int_0^t (t-\vartheta)^{\epsilon-1}S_H(\vartheta, S_H(\vartheta))d\vartheta. \tag{9.12}$$

For $S_H(t), S_{H1}(t) \in B_r$, we attain

$$|FS_H(t) - FS_{H1}(t)| \leq \frac{1}{\Gamma(\epsilon)}\int_0^t (t-\vartheta)^{\epsilon-1}L_1(\vartheta)|S_H(\vartheta) - S_{H1}(\vartheta)|d\vartheta$$
$$\leq \frac{1}{\Gamma(\epsilon)}\left(\int_0^t (t-\vartheta)^{\frac{\epsilon-1}{1-P}}d\vartheta\right)^{1-P}\left(\int_0^t L_1^{\frac{1}{P}}(\vartheta)d\vartheta\right)^P \leq \frac{c\|L_1\|_{\frac{1}{P}}T^{\beta}}{\Gamma(\epsilon)}\|S_H - S_{H1}\|. \tag{9.13}$$

Hence, $\gamma = \frac{c\|L\|_{\frac{1}{P}}T^{\beta}}{\Gamma(\epsilon)} < 1$ is guaranteed and verifies the existence of a unique solution for the proposed fractional-order model (9.5). □

9.5.1 Positivity of the Solutions

In order to prove the positivity of the solutions, i.e. the non-negative hyperoctant $\mathbb{R}^7_+$ is a positively invariant region for the model (9.5), let us consider that

$$\mathbb{R}^7_+ = \{\mathbf{y} \in \mathbb{R}^7 | \mathbf{y} \geq 0\}. \tag{9.14}$$

Furthermore,

$$\boldsymbol{y} = (S_H(t), E_H(t), I_H(t), R_H(t), S_M(t), E_M(t), I_M(t))^T. \tag{9.15}$$

It has to be shown that on each hyperplane bounding the non-negative hyperoctant, the vector field points into $\mathbb{R}^7_+$. From the fractional model, we obtain

$$\begin{aligned}
&{}^{C}D^{\epsilon}_{0,t}\, S_H(t)|_{S_H=0} = \Lambda_H + \omega R_H(t) \geq 0, \quad {}^{C}D^{\epsilon}_{0,t}\, E_H(t)|_{E_H=0} = \frac{b\beta_H S_H(t) I_M(t)}{1+\nu_H I_M(t)} \geq 0,\\
&{}^{C}D^{\epsilon}_{0,t}\, I_H(t)|_{I_H=0} = \alpha_H E_H(t) \geq 0, \qquad {}^{C}D^{\epsilon}_{0,t}\, R_H(t)|_{R_H=0} = r I_H(t) \geq 0,\\
&{}^{C}D^{\kappa}_{0,t} S_M(t)|_{S_M=0} = \Lambda_M \geq 0, \qquad {}^{C}D^{\kappa}_{0,t} E_M(t)|_{E_M=0} = \frac{b\beta_M S_M(t) I_H(t)}{1+\nu_M I_H(t)} \geq 0,\\
&{}^{C}D^{\kappa}_{0,t} I_M(t)|_{I_M=0} = \alpha_M E_M(t) \geq 0.
\end{aligned} \tag{9.16}$$

Hence, using Corollary 9.1, it has been ascertained that the solutions will remain in $\mathbb{R}^7_+$. This completes the proof of the positivity of the solutions of the model (9.5).

9.6 Numerical Simulations

This section presents the numerical treatment of the proposed fractional-order model (9.5) using the Adams-type predictor–corrector algorithm under Caputo fractional operator as proposed in [33] and well explored in [34]. The values of parameters used in the present study are taken from Table 9.1 maintaining $R_0 < 1$ whereas the initial conditions are chosen to be as follows:

$$\begin{aligned}
&S_H(0) = 100, E_H(0) = 20, I_H(0) = 10, R_H(0) = 0, S_M(0) = 1000,\\
&E_M(0) = 20, I_M(0) = 30.
\end{aligned}$$

The step size h for the simulation has been chosen to be 10^{-2} days. For the sake of convenience, we write

$$\begin{aligned}
{}^{C}D^{\epsilon}_{0,t}\, S_H(t) &= G_1(t, S_H, E_H, I_H, R_H, S_M, E_M, I_M),\\
{}^{C}D^{\epsilon}_{0,t}\, E_H(t) &= G_2(t, S_H, E_H, I_H, R_H, S_M, E_M, I_M),\\
{}^{C}D^{\epsilon}_{0,t}\, I_H(t) &= G_3(t, S_H, E_H, I_H, R_H, S_M, E_M, I_M),\\
{}^{C}D^{\epsilon}_{0,t}\, R_H(t) &= G_4(t, S_H, E_H, I_H, R_H, S_M, E_M, I_M),\\
{}^{C}D^{\kappa}_{0,t} S_M(t) &= G_5(t, S_H, E_H, I_H, R_H, S_M, E_M, I_M),\\
{}^{C}D^{\kappa}_{0,t} E_M(t) &= G_6(t, S_H, E_H, I_H, R_H, S_M, E_M, I_M),\\
{}^{C}D^{\kappa}_{0,t} I_M(t) &= G_7(t, S_H, E_H, I_H, R_H, S_M, E_M, I_M).
\end{aligned}$$

Thus, the biological system in the present study with Caputo operator takes the following form for its predictor part

$$
\begin{aligned}
S^P_{H,n+1} &= S_H(0) + \sum_{j=0}^{n} b_{\epsilon j,n+1} G_1(t_j, S_{Hj}, E_{Hj}, I_{Hj}, R_{Hj}, S_{Mj}, E_{Mj}, I_{Mj}),\\
E^P_{H,n+1} &= E_H(0) + \sum_{j=0}^{n} b_{\epsilon j,n+1} G_2(t_j, S_{Hj}, E_{Hj}, I_{Hj}, R_{Hj}, S_{Mj}, E_{Mj}, I_{Mj}),\\
I^P_{H,n+1} &= I_H(0) + \sum_{j=0}^{n} b_{\epsilon j,n+1} G_3(t_j, S_{Hj}, E_{Hj}, I_{Hj}, R_{Hj}, S_{Mj}, E_{Mj}, I_{Mj}),\\
R^P_{H,n+1} &= R_H(0) + \sum_{j=0}^{n} b_{\epsilon j,n+1} G_4(t_j, S_{Hj}, E_{Hj}, I_{Hj}, R_{Hj}, S_{Mj}, E_{Mj}, I_{Mj}),\\
S^P_{M,n+1} &= S_M(0) + \sum_{j=0}^{n} b_{\kappa j,n+1} G_5(t_j, S_{Hj}, E_{Hj}, I_{Hj}, R_{Hj}, S_{Mj}, E_{Mj}, I_{Mj}),\\
E^P_{M,n+1} &= E_M(0) + \sum_{j=0}^{n} b_{\kappa j,n+1} G_6(t_j, S_{Hj}, E_{Hj}, I_{Hj}, R_{Hj}, S_{Mj}, E_{Mj}, I_{Mj}),\\
I^P_{M,n+1} &= I_M(0) + \sum_{j=0}^{n} b_{\kappa j,n+1} G_7(t_j, S_{Hj}, E_{Hj}, I_{Hj}, R_{Hj}, S_{Mj}, E_{Mj}, I_{Mj}).
\end{aligned}
$$

Similarly, the corrector part will be

$$
\begin{aligned}
S^C_{H,n+1} &= S_H(0) + a_{\epsilon,n+1,n+1} G_1(t_j, S^P_{Hj}, E^P_{Hj}, I^P_{Hj}, R^P_{Hj}, S^P_{Mj}, E^P_{Mj}, I^P_{Mj})\\
&\quad + \sum_{j=0}^{n} a_{\epsilon j,n+1} G_1(t_j, S_{Hj}, E_{Hj}, I_{Hj}, R_{Hj}, S_{Mj}, E_{Mj}, I_{Mj}),\\
E^C_{H,n+1} &= E_H(0) + a_{\epsilon,n+1,n+1} G_2(t_j, S^P_{Hj}, E^P_{Hj}, I^P_{Hj}, R^P_{Hj}, S^P_{Mj}, E^P_{Mj}, I^P_{Mj})\\
&\quad + \sum_{j=0}^{n} a_{\epsilon j,n+1} G_2(t_j, S_{Hj}, E_{Hj}, I_{Hj}, R_{Hj}, S_{Mj}, E_{Mj}, I_{Mj}),\\
I^C_{H,n+1} &= I_H(0) + a_{\epsilon,n+1,n+1} G_3(t_j, S^P_{Hj}, E^P_{Hj}, I^P_{Hj}, R^P_{Hj}, S^P_{Mj}, E^P_{Mj}, I^P_{Mj})\\
&\quad + \sum_{j=0}^{n} a_{\epsilon j,n+1} G_3(t_j, S_{Hj}, E_{Hj}, I_{Hj}, R_{Hj}, S_{Mj}, E_{Mj}, I_{Mj}),\\
R^C_{H,n+1} &= R_H(0) + a_{\epsilon,n+1,n+1} G_4(t_j, S^P_{Hj}, E^P_{Hj}, I^P_{Hj}, R^P_{Hj}, S^P_{Mj}, E^P_{Mj}, I^P_{Mj})\\
&\quad + \sum_{j=0}^{n} a_{\epsilon j,n+1} G_4(t_j, S_{Hj}, E_{Hj}, I_{Hj}, R_{Hj}, S_{Mj}, E_{Mj}, I_{Mj}),
\end{aligned}
$$

$$
\begin{aligned}
S^C_{M,n+1} &= S_M(0) + a_{\kappa,n+1,n+1} G_5(t_j, S^P_{H,j}, E^P_{H,j}, I^P_{H,j}, R^P_{H,j}, S^P_{M,j}, E^P_{M,j}, I^P_{M,j}) \\
&\quad + \sum_{j=0}^{n} a_{\kappa,j,n+1} G_5(t_j, S_{H,j}, E_{H,j}, I_{H,j}, R_{H,j}, S_{M,j}, E_{M,j}, I_{M,j}), \\
E^C_{M,n+1} &= E_M(0) + a_{\kappa,n+1,n+1} G_6(t_j, S^P_{H,j}, E^P_{H,j}, I^P_{H,j}, R^P_{H,j}, S^P_{M,j}, E^P_{M,j}, I^P_{M,j}) \\
&\quad + \sum_{j=0}^{n} a_{\kappa,j,n+1} G_6(t_j, S_{H,j}, E_{H,j}, I_{H,j}, R_{H,j}, S_{M,j}, E_{M,j}, I_{M,j}), \\
I^C_{M,n+1} &= I_M(0) + a_{\kappa,n+1,n+1} G_7(t_j, S^P_{H,j}, E^P_{H,j}, I^P_{H,j}, R^P_{H,j}, S^P_{M,j}, E^P_{M,j}, I^P_{M,j}) \\
&\quad + \sum_{j=0}^{n} a_{\kappa,j,n+1} G_7(t_j, S_{H,j}, E_{H,j}, I_{H,j}, R_{H,j}, S_{M,j}, E_{M,j}, I_{M,j}),
\end{aligned}
$$

where

$$
\begin{aligned}
a_{\epsilon,j,n+1} &= \frac{h^\epsilon}{\Gamma(\epsilon+2)} \begin{cases} n^{\epsilon+1} - (n-\epsilon)(n+1)^\epsilon, \ j = 0, \\ (n-j+2)^{\epsilon+1} - 2(n-j+1)^{\epsilon+1} + (n-j)^{\epsilon+1}, \ 1 \le j \le n, \\ 1, j = n+1, \end{cases} \\
a_{\kappa,j,n+1} &= \frac{h^\kappa}{\Gamma(\kappa+2)} \begin{cases} n^{\kappa+1} - (n-\kappa)(n+1)^\kappa, \ j = 0, \\ (n-j+2)^{\kappa+1} - 2(n-j+1)^{\kappa+1} + (n-j)^{\kappa+1}, \ 1 \le j \le n, \\ 1, j = n+1, \end{cases} \\
b_{\epsilon,j,n+1} &= \frac{1}{\Gamma(\epsilon+1)}[(n-j+1)^\epsilon - (n-j)^\epsilon], \\
b_{\kappa,j,n+1} &= \frac{1}{\Gamma(\kappa+1)}[(n-j+1)^\kappa - (n-j)^\kappa].
\end{aligned}
$$

Table 9.2 shows the combination of fractional-order values selected for simulation to understand the dynamics of the model (9.5).

Figures 9.1–9.4 and 9.5–9.7 show, respectively, the simulation of the model (9.5) for the human and mosquito population with varying values of fractional order (ϵ, κ) and increasing amount of antibody in humans. In particular, Figure 9.1 shows that an increase in the antibody proportion in humans affects the behavior of susceptible individuals, which is clear from the plot (a) in the figure with

Table 9.2 Values for the fractional orders.

ϵ	1	0.99	0.95	0.90	0.88
κ	1	0.95	0.90	0.85	0.80

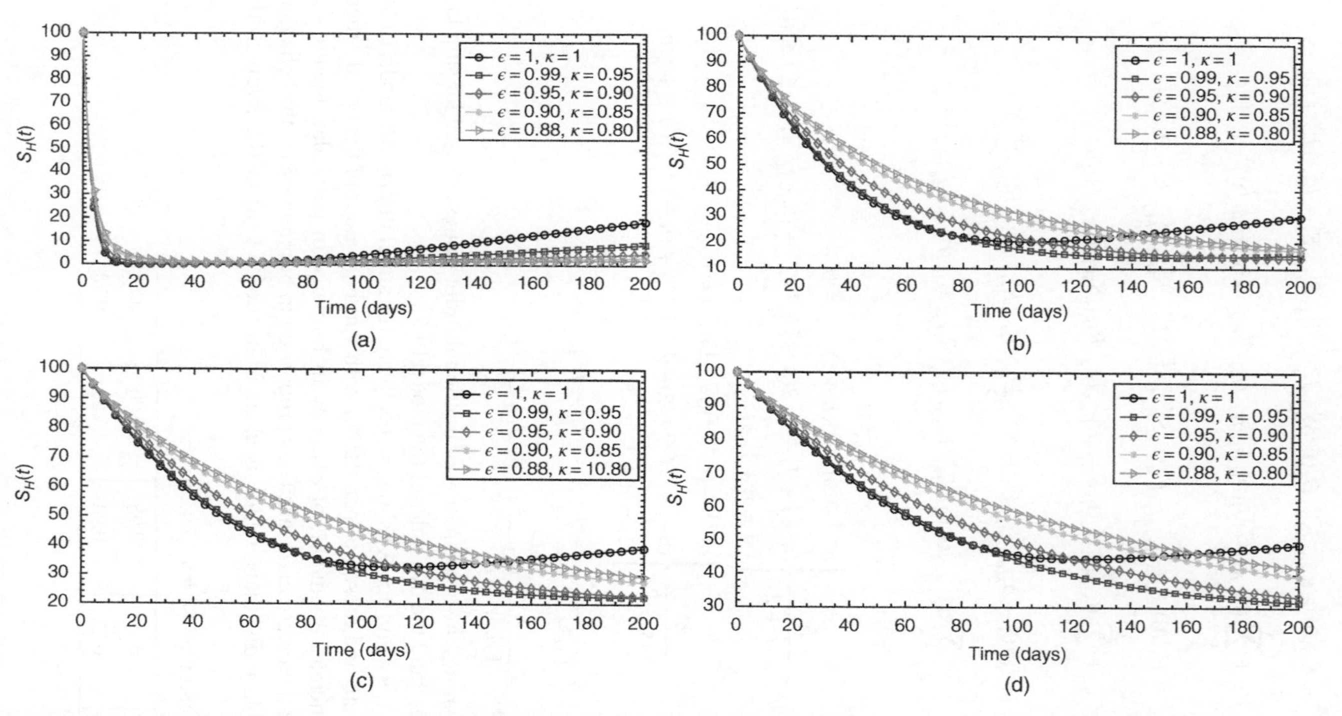

Figure 9.1 The behavior of nonvaccinated human population for varying fractional orders. (a) $\nu_H = 0.0$. (b) $\nu_H = 0.5$. (c) $\nu_H = 0.8$. (d) $\nu_H = 1.2$.

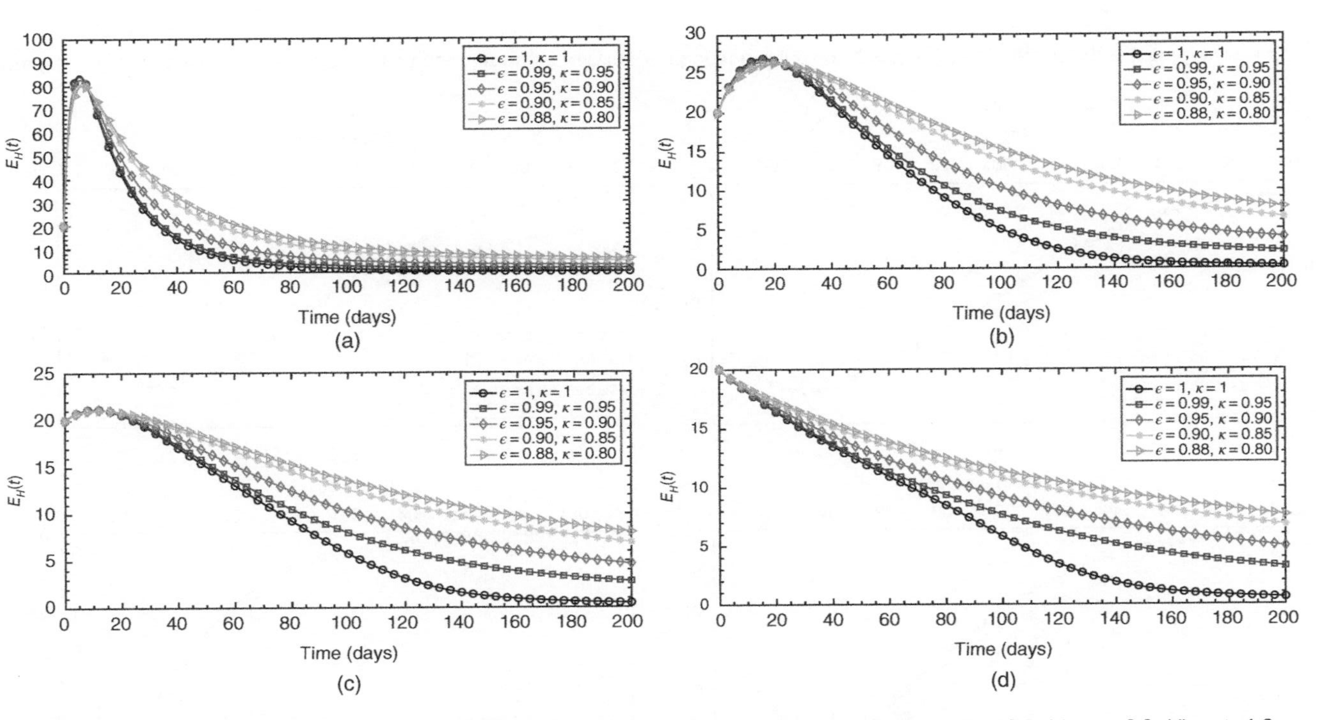

Figure 9.2 The behavior of exposed human population for varying fractional orders. (a) $\nu_H = 0.0$. (b) $\nu_H = 0.5$. (c) $\nu_H = 0.8$. (d) $\nu_H = 1.2$.

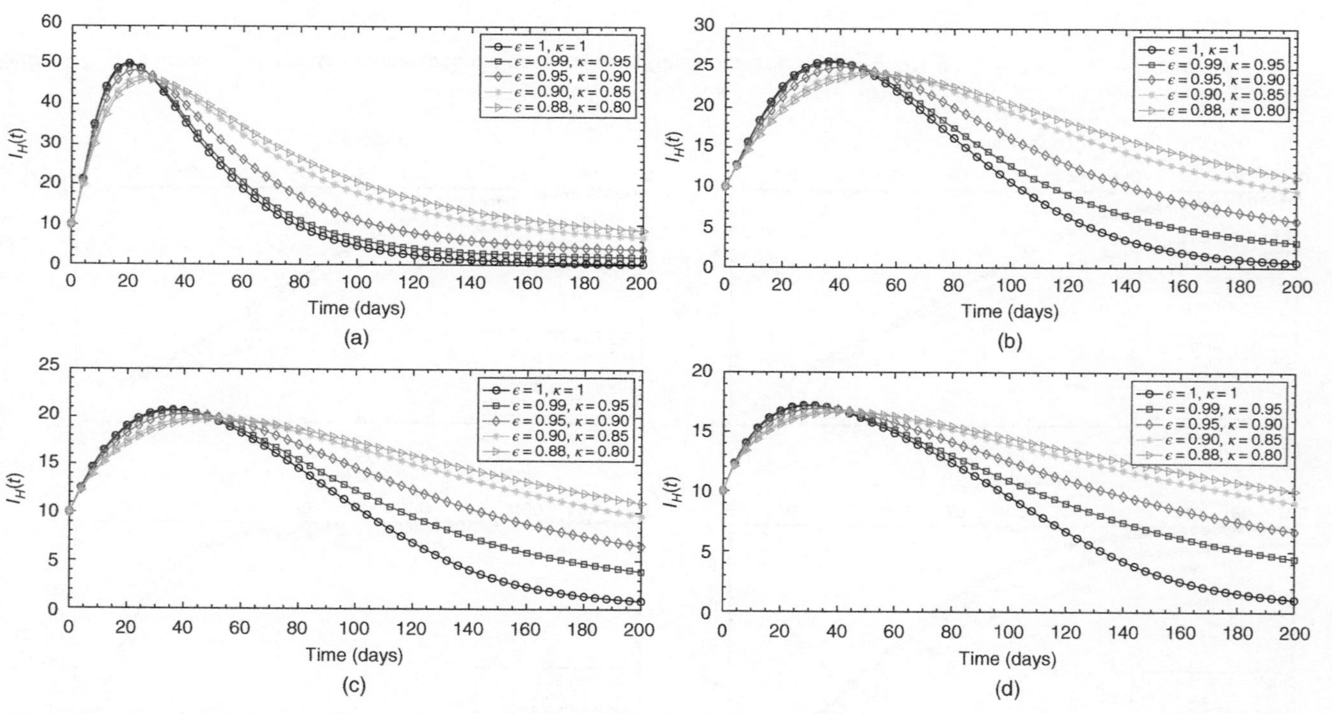

Figure 9.3 The behavior of infectious human population for varying fractional orders. (a) $\nu_H = 0.0$. (b) $\nu_H = 0.5$. (c) $\nu_H = 0.8$. (d) $\nu_H = 1.2$.

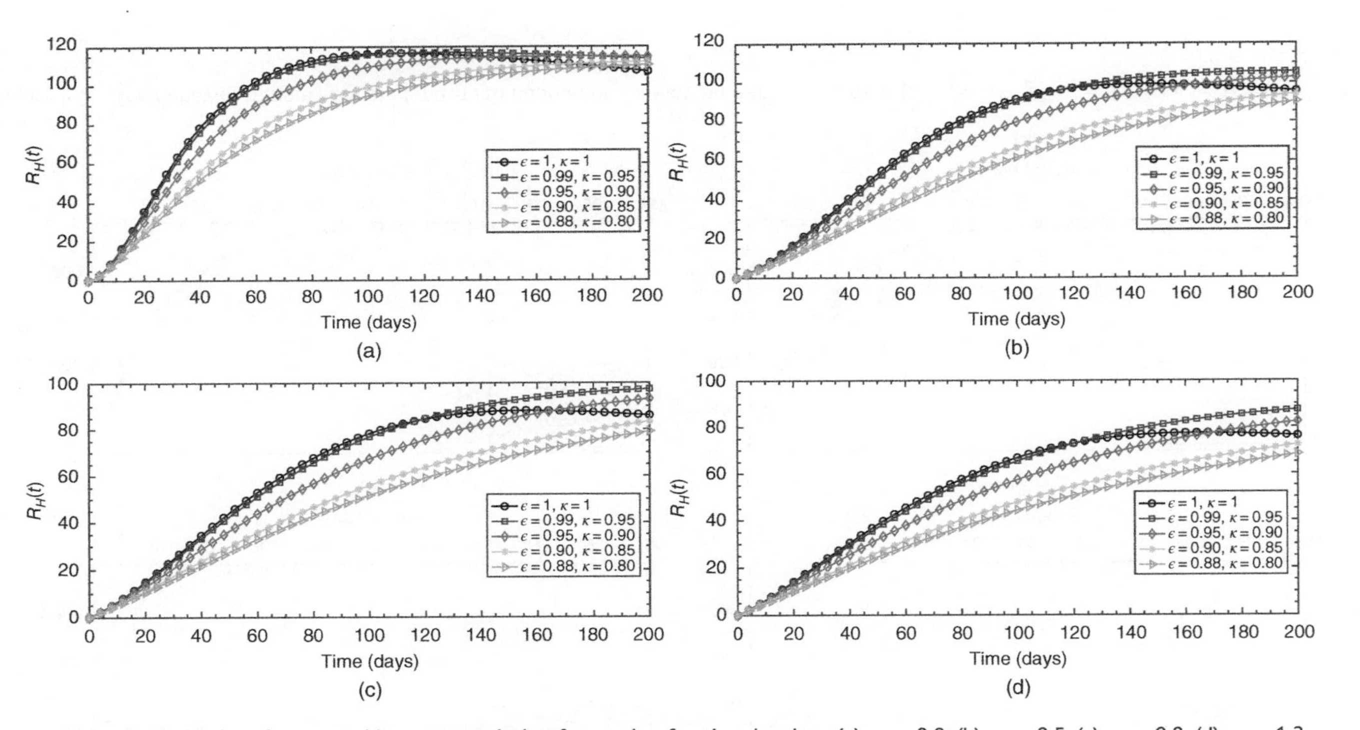

Figure 9.4 The behavior of recovered human population for varying fractional orders. (a) $\nu_H = 0.0$. (b) $\nu_H = 0.5$. (c) $\nu_H = 0.8$. (d) $\nu_H = 1.2$.

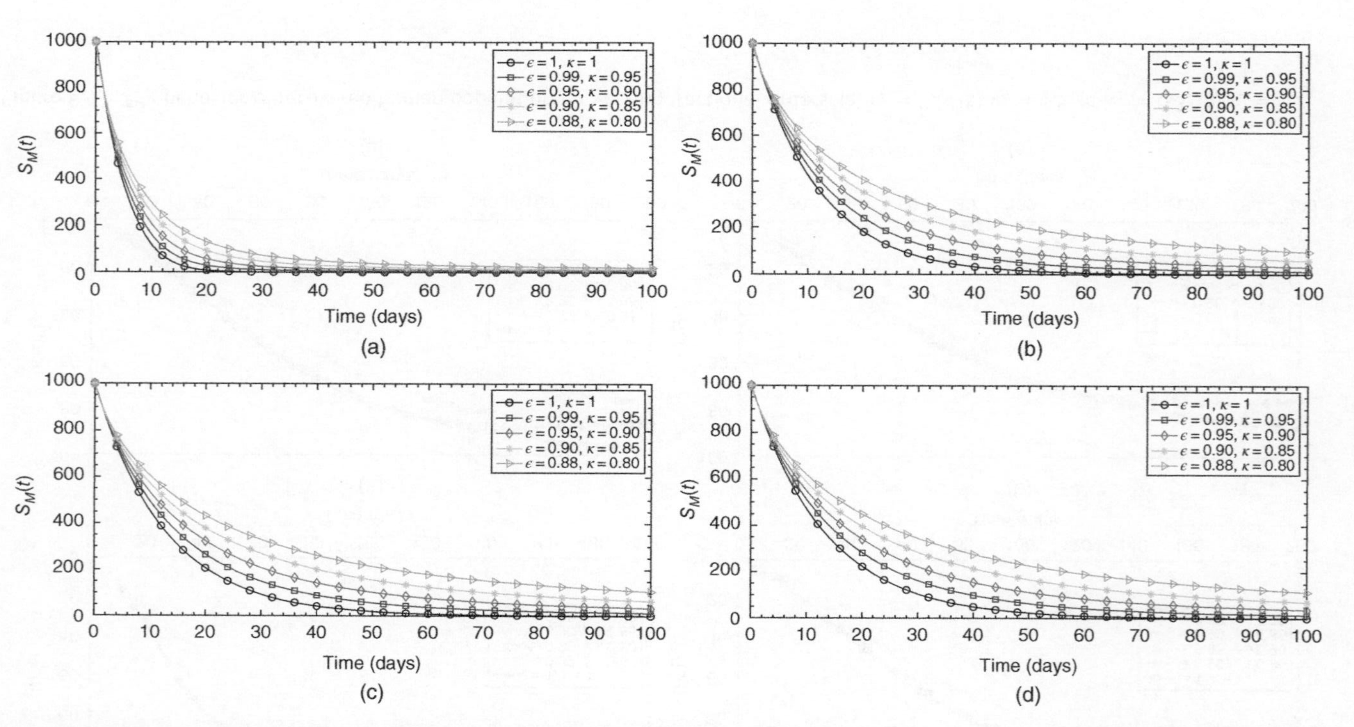

Figure 9.5 The behavior of nonvaccinated mosquito population for varying fractional orders. (a) $\nu_M = 0.0$. (b) $\nu_M = 0.5$. (c) $\nu_M = 0.8$. (d) $\nu_M = 1.2$.

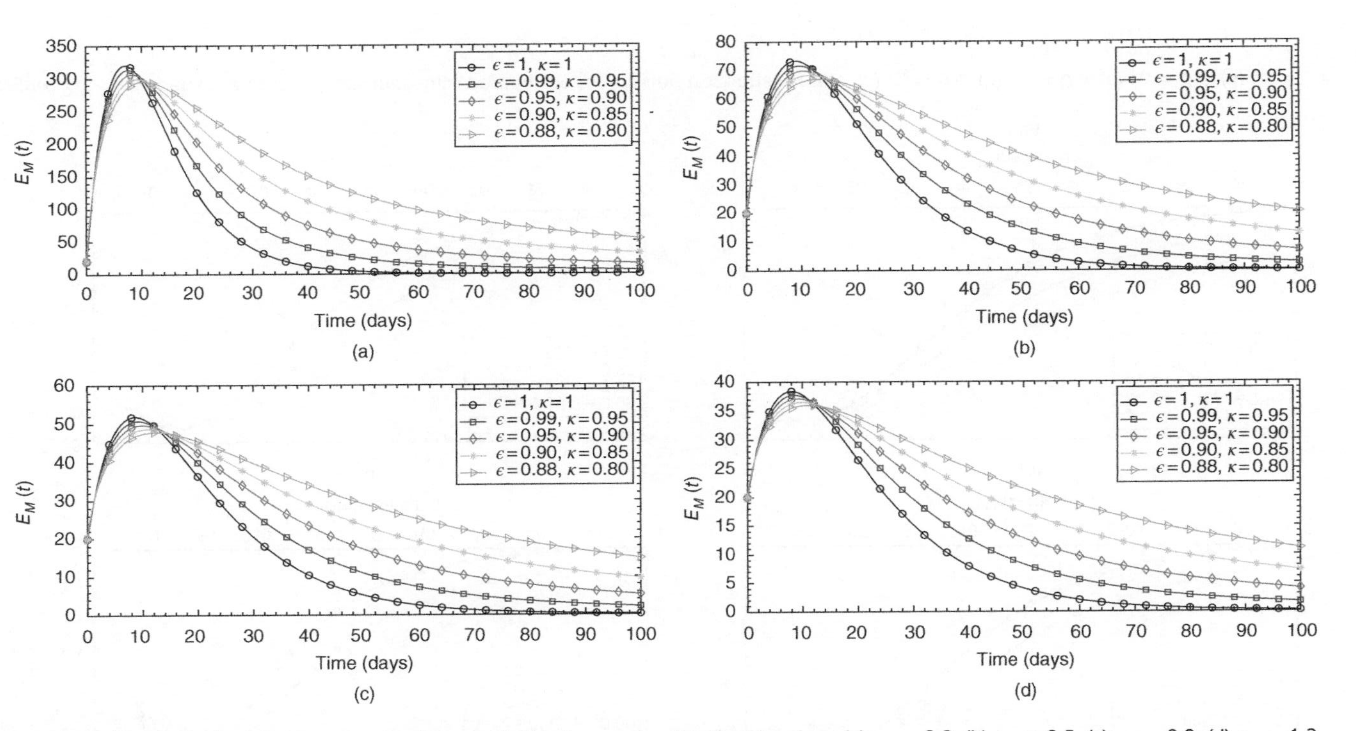

Figure 9.6 The behavior of exposed mosquito population for varying fractional orders. (a) $\nu_M = 0.0$. (b) $\nu_M = 0.5$. (c) $\nu_M = 0.8$. (d) $\nu_M = 1.2$.

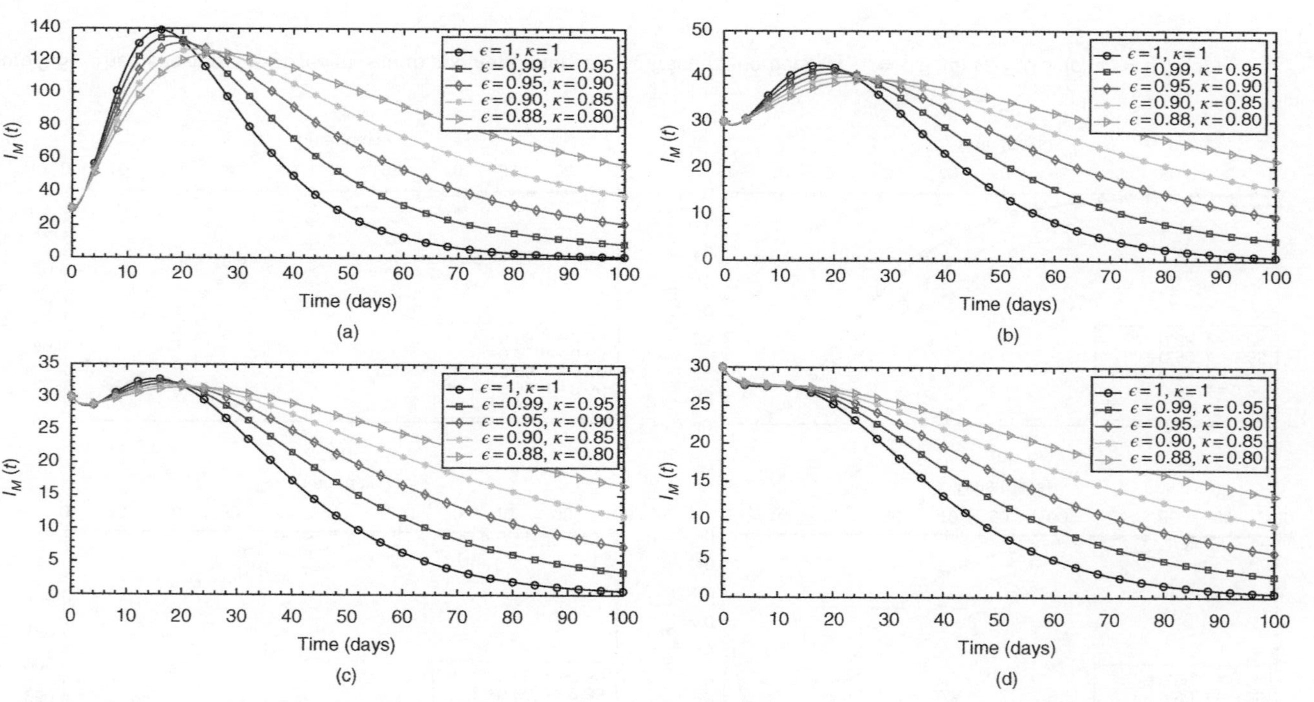

Figure 9.7 The behavior of infectious mosquito population for varying fractional orders. (a) $\nu_M = 0.0$. (b) $\nu_M = 0.5$. (c) $\nu_M = 0.8$. (d) $\nu_M = 1.2$.

decreasing values of ϵ and κ depicting that an the absence of antibody proportion affects susceptibility negatively in humans with time.

It should be noted here that integer-order model was not much suitable to capture this behavior for longer period of time as evident in Figure 9.1. Similarly, the number of exposed humans will start decreasing with an increasing dose of antibody and this behavior is more clearly captured with fractional-orders in Figure 9.2. In line to this, the infection – although decreases – does not completely die out with a mere increase of antibody amount in humans as the recruitment rate of mosquitoes is more than humans ($\Lambda_M > \Lambda_H$). This behavior is depicted well by fractional orders in Figure 9.3. The reduction of infection and increased antibody amount have positive effects on recovery of human population as shown by fractional orders in Figure 9.4.

Similar analysis could be carried out in case of mosquito population as shown in Figures 9.5–9.7. In Figure 9.8, the plots (a) and (b) reveal the decreasing behavior of the infection for $\epsilon = 0.99$ and $\kappa = 0.95$ if the recovery rate of humans increase whereas the plots (c) and (d) show an increase in infectious human population due to increasing biting rate for about 20 initial days and then it gets stabilized toward its initial position.

Finally, it is clearly observed in (a) and (d) plots of Figure 9.9 that the infection increases if either the biting rate or the mosquito recruitment term (Λ_M) starts to increase, and it doesn't spread if obviously mosquito death rate increases as shown in (b) and (c) plots. Thus, from these graphical illustrations, it can be concluded that the dynamics of the model significantly depends upon the fractional-order derivatives, which are in turn proved to be more useful and realistic in the biological systems.

The real-world phenomenon reveals in some instance that the spread of malaria virus follows a process similar to that of non-Gaussian distribution. To capture this process, one needs to make use of a differential operator with non-Gaussian settings. There exists, in the present literature, several mathematical operators called differential operators, only that of Caputo and Riemann-Liouville, are able to capture the power law processes very accurately. We, therefore, made use of fractional differential operator, that is the Caputo fractional derivative operator to model the spread of malaria virus in the present research.

9.7 Conclusion

Fractional version of the model (9.4) in the Caputo sense has extensively been analyzed in the present study. Computation of the basic reproductive number of the fractional-order model (9.5) shows that the infectious disease can be controlled under values of the parameters given in Table 9.1.

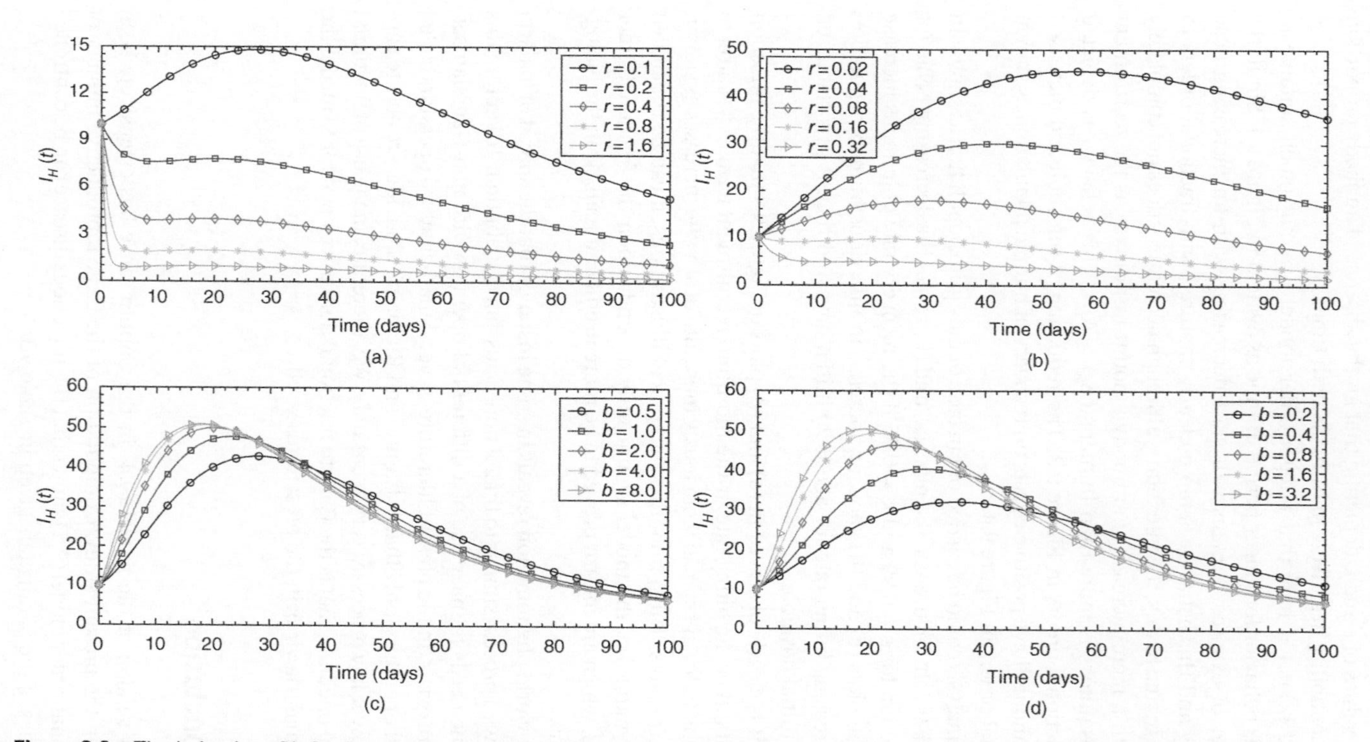

Figure 9.8 The behavior of infectious human population for varying human recovery and mosquito biting rates where $R_0 > 1$ for $b > 0.4$. (a) $\epsilon = 0.99, \kappa = 0.95$. (b) $\epsilon = 0.99, \kappa = 0.95$. (c) $\epsilon = 0.99, \kappa = 0.95$. (d) $\epsilon = 0.99, \kappa = 0.95$.

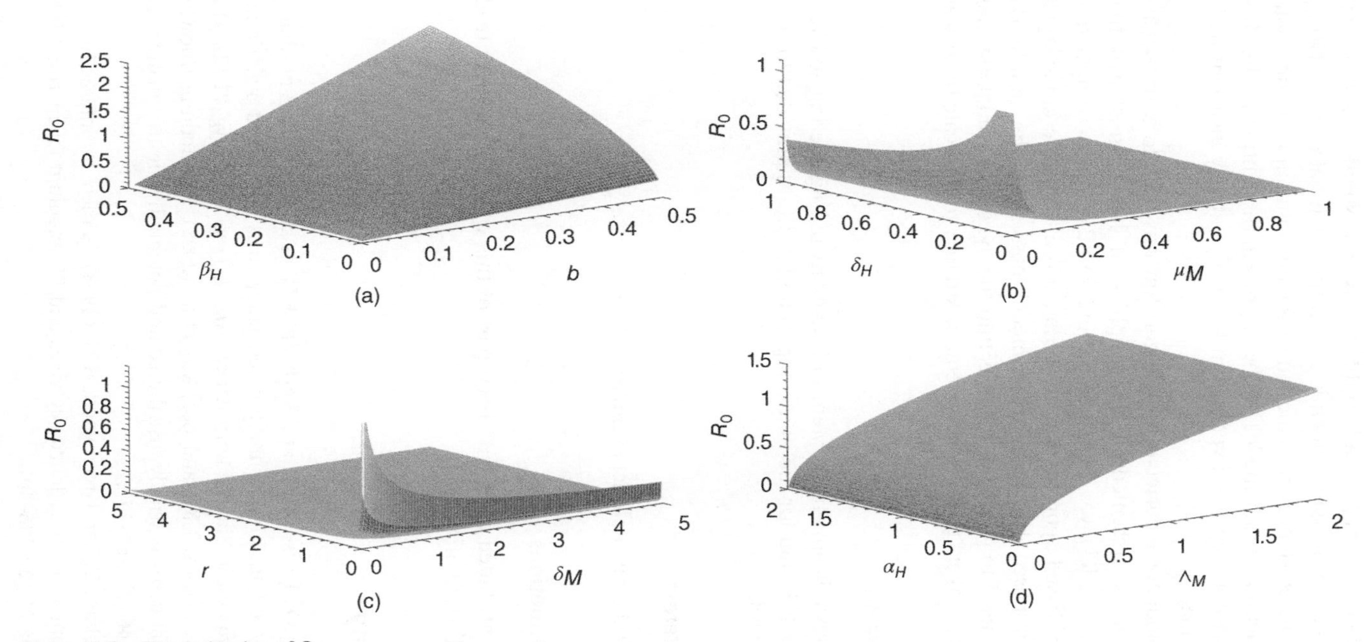

Figure 9.9 The behavior of R_0.

A unique globally asymptotically stable steady-state solution exists for the proposed model and an unstable endemic equilibrium point also exists. Using fixed point theory, the existence and uniqueness for the solution including boundedness of the fractional-order model has been proved. In addition to this, it is also shown that the non-negative hyperoctant $\mathbb{R}_+^7$ is a positively invariant region for the fractional model.

Finally, a predictor – corrector type algorithm was used to carry out numerical simulation as presented in the graphics. It has been shown that the fractional-order model gives more realistic and sound results in comparison to its classical model based upon first-order nonlinear ordinary differential equations. Hence, the proposed fractional-order model possessing the non-Markovian nature can be used to get better insights into the dynamics of malaria infection and help policy makers to devise a systematic strategy to eradicate the disease.

Acknowledgments

Ms. Sania Qureshi is grateful to Mehran University of Engineering and Technology, Jamshoro, Pakistan for the kind support and facilities provided to carry out this research work.

Conflict of Interest

The authors declare no conflict of interest.

Authors' Contributions

All authors contributed equally to each part of this work. All authors read and approved the final manuscript.

Bibliography

1 Ter Kuile, F.O., Parise, M.E., and Verhoeff, F.H. et al. (2004). The burden of co-infection with human immunodeficiency virus type 1 and malaria in pregnant women in Sub-Saharan Africa. *Am. J. Trop. Med. Hyg.* **71** (2): 414–5.

2 Cox-Singh, J., Davis, T.M., and Lee, K.S. et al. (2008). Plasmodium Knowlesi malaria in humans is widely distributed and potentially life threatening. *Clin. Infect. Dis.* **46** (2): 165–171.

3 Chandramohan, D. and Greenwood, B.M. (1998). Is there an interaction between human immunodeficiency virus and Plasmodium falciparum? *Int. J. Epidemiol.* **27** (2): 296–301.

4 Sullivan, A.D., Nyirenda, T., and Cullinan, T. et al. (1999). Malaria infection during pregnancy: intrauterine growth retardation and preterm delivery in Malawi. *J. Infect. Dis.* **179** (6): 1580–1583.

5 Beretta, E., Capasso, V., and Garao, D.G. (2018). A mathematical model for malaria transmission with asymptomatic carriers and two age groups in the human population. *Math. Biosci.* **300**: 87–101.

6 Shcherbacheva, A., Haario, H., and Killeen, G.F. (2018). Modeling host-seeking behavior of African malaria vector mosquitoes in the presence of long-lasting insecticidal nets. *Math. Biosci.* **295**: 36–47.

7 Bowman, C., Gumel, A.B., and Van den Driessche, P. et al. (2005). A mathematical model for assessing control strategies against West Nile virus. *Bull. Math. Biol.* **67** (5): 1107–1133.

8 Mukandavire, Z., Gumel A.B., and Garira W. et al. (2009). Mathematical analysis of a model for HIV-malaria co-infection. *Math. Biosci. Eng.* **6** (2): 333–362.

9 Olaniyi, S. and Obabiyi, O.S. (2013). Mathematical model for malaria transmission dynamics in human and mosquito populations with nonlinear forces of infection. *Int. J. Pure Appl. Math.* **88** (1): 125–156.

10 Diethelm, K. (2013). A fractional calculus based model for the simulation of an outbreak of dengue fever. *Nonlinear Dyn.* **71** (4): 613–619.

11 Ullah, S., Khan, M.A., and Farooq, M. (2018). A fractional model for the dynamics of TB virus. *Chaos, Solitons Fractals* **116**: 63–71.

12 Area, I., Batarfi, H., and Losada, J. et al. (2015). On a fractional order Ebola epidemic model. *Adv. Differ. Equ.* **2015** (1): 278.

13 Yusuf, A., Qureshi, S., Inc, M. et al. (2018). Two-strain epidemic model involving fractional derivative with Mittag–Leffler kernel. *Chaos: Interdisciplin. J. Nonlinear Sci.* **28** (12): 123–121.

14 Abro, K.A., Memon, A.A., and Uqaili, M.A. (2018). A comparative mathematical analysis of RL and RC electrical circuits via Atangana–Baleanu and Caputo–Fabrizio fractional derivatives. *Eur. Phys. J. Plus* **133** (3): 113.

15 Abro, K.A., Memon, A.A., and Memon, A.A. (2019). Functionality of circuit via modern fractional differentiations. *Analog Integr. Circuits Signal Process.* **99**: 111. https://doi.org/10.1007/s10470-018-1371-6.

16 Owolabi, K.M. and Atangana, A. (2018). Chaotic behaviour in system of noninteger-order ordinary differential equations. *Chaos, Solitons Fractals* **115**: 362–370.

17 Singh, J., Kumar, D., Hammouch, Z., and Atangana, A. (2018). A fractional epidemiological model for computer viruses pertaining to a new fractional derivative. *Appl. Math. Comput.* **316**: 504–515.

18 Owolabi, K.M. (2018). Modelling and simulation of a dynamical system with the Atangana–Baleanu fractional derivative. *Eur. Phys. J. Plus* **133** (1): 15.

19 Inc, M., Yusuf, A., Aliyu, A.I., and Baleanu, D. (2018). Investigation of the logarithmic-KdV equation involving Mittag–Leffler type kernel with Atangana–Baleanu derivative. *Physica A* **506**: 520–531.

20 Atangana, A. and Jain, S. (2018). The role of power decay, exponential decay and Mittag–Leffler function's waiting time distribution: application of cancer spread. *Physica A* **512**: 330–351.

21 Atangana, A. and Alqahtani, R.T. (2018). New numerical method and application to Keller–Segel model with fractional order derivative. *Chaos, Solitons Fractals* **116**: 14–21.

22 Owolabi, K.M. and Atangana, A. (2018). Modelling and formation of spatiotemporal patterns of fractional predation system in subdiffusion and superdiffusion scenarios. *Eur. Phys. J. Plus* **133** (2): 43.

23 Qureshi, S., Yusuf, A., Shaikh, A.A. et al. (2019). Fractional modeling of blood ethanol concentration system with real data application. *Chaos: Interdiscipl. J. Nonlinear Sci.* **29** (1): 013143.

24 Podlubny, I. (1999). Fractional differential equations. In: *Mathematics in Science and Engineering*, **198**. New York and London: Academic Press.

25 Petráš, I. (2011). *Fractional-Order Nonlinear Systems: Modeling, Analysis and Simulation*. Springer Science & Business Media.

26 Odibat, Z.M. and Shawagfeh, N.T. (2007). Generalized Taylor's formula. *Appl. Math. Comput.* **186** (1): 286–293.

27 Heffernan, J.M., Smith, R.J., and Wahl, L.M. (2005). Perspectives on the basic reproductive ratio. *J. R. Soc. Interface* **2** (4): 281–293.

28 Diekmann, O., Heesterbeek, J.A.P., and Metz, J.A. (1990). On the definition and the computation of the basic reproduction ratio R_0 in models for infectious diseases in heterogeneous populations. *J. Math. Biol.* **28** (4): 365–382.

29 Sadovskii, B.N. (1967). On a fixed point principle. *Funct. Anal. Appl.* **1**: 74–76.

30 Kilbas, A.A., Srivastava, H.M., and Trujillo, J.J. (2006). *Theory and Applications of Fractional Differential Equations, North-Holland Mathematics Studies*, vol. **204**. Elsevier.

31 Diethelm, K. (2010). *The Analysis of Fractional Differential Equation*. Springer-Verlag.

32 Zeidler, E. (1986). *Non-Linear Functional Analysis and its Application*. New York: Springer-Verlag.

33 Diethelm, K., Ford, N.J., and Freed, A.D. (2002). A predictor-corrector approach for the numerical solution of fractional differential equations. *Nonlinear Dyn.* **29** (14): 322.

34 Diethelm, K., Ford, N.J., and Freed, A.D. (2004). Detailed error analysis for a fractional Adams method. *Numer. Algorithms* **36** (1): 31–52.

10

MHD-free Convection Flow Over a Vertical Plate with Ramped Wall Temperature and Chemical Reaction in View of Nonsingular Kernel

Muhammad B. Riaz[1,2], *Abdon Atangana*[2,3], *and Syed T. Saeed*[4]

[1] *Department of Mathematics, University of Management and Technology, Lahore, Pakistan*
[2] *Institute for Groundwater Studies (IGS), University of the Free State, Bloemfontein, South Africa*
[3] *Department of Medical Research, China Medical University Hospital, China Medical University, Taichung, Taiwan*
[4] *Department of Mathematics, National University of Computer and Emerging Sciences, Lahore, Pakistan*

10.1 Introduction

The Maxwell fluid model is one of the most important model among researchers as compared to other fluid models because it is simplest rate-type fluid model [1–5]. Maxwell fluids play important role in polymeric industry. However, they are some restrictions. In a simple shear flow, this model does not explain the link between shear stress and shear rate [6–8]. Maxwell fluids have been considered in many research articles due to its simplicity and have great importance in momentum transfer [9–15]. Fetecau and Fetecau [11] investigated exact solution for Maxwell fluid over a moving plate. Numerical solution of Maxwell fluid over a shrinking sheet was explored by Motsa et al. [16]. Khan et al. [17] proposed exact solution of MHD Maxwell fluid in porous medium. Exact solutions of Maxwell fluid on an infinite plate and oscillating plane were considered by Abro et al. [18–20]. Maxwell nanofluid with ramped temperature was observed by Aman et al. [21]. Karra et al. [22] investigated solution of generalized Maxwell fluid with pressure-dependent material. Riaz et al. [23] discovered exact solution of generalized Maxwell fluid. Generalized Maxwell fluid over a moving plate having slip effect was considered by Liu and Guo [24]. Imran et al. [25] investigated analytical solution of Maxwell fluid in the presence of Newtonian heating and slip effect. Comparative analysis for Maxwell fluid with Newtonian heating was investigated by Raza and Ullah [26].

The phenomena of heat and mass transfer in MHD boundary layer flow have gained attention in chemical engineering, geophysical environments, etc. Different models of fluid with heat and mass transfer have been studies by taking various

Fractional Order Analysis: Theory, Methods and Applications, First Edition.
Edited by Hemen Dutta, Ahmet Ocak Akdemir, and Abdon Atangana.

conditions on temperature and concentration. Singh and Kumar [27] explored the heat and mass transfer in viscous fluid with slip condition. Furthermore, chemical reaction and thermal radiation have been considered. Tahir et al. [28] observed heat transfer in Maxwell fluid with wall slip on oscillating plate. Exact solution of transient-free convective mass transfer flow under ramped wall temperature and plate velocity has investigated by Ahmed and Dutta [29].

Ghara et al. [30] considered Laplace transformation in order to investigate exact solution of MHD-free convection flow under the influence of ramped wall temperature on a moving plate. Seth et al. [31] observed heat and mass transfer in viscous fluid on a moving plate. Soret and Hall effects are also considered. Heat and mass transfer in incompressible fluid along with oscillatory suction velocity was observed by Reddy [32]. Second-grade fluid having ramped wall temperature under magnetic field was considered by Ahmad et al. [33]. Narahari and Debnath [34] explored free convection flow with heat generation/absorption on a vertical plate. Furthermore, for some significant results, see [35–39]. Fractional differential operators with new trend of fractional kernel have been introduced very recently and have been proven to be very efficient due to their ability to capture heterogeneity that cannot be captured by the power law, for instance, the heterogeneity associated to statistic, diffusion with crossover behavior, and diffusion with two types of densities of probability. In particular, the kernel generalized Mittag–Leffler function has many other properties that will not be listed here as it is not the subject of investigation; however, one will realize that this kernel is very efficient in modeling movement of fluid.

All the above discussion is done in the integer-order models solutions and their effects. In the present chapter, we show the strength of fractional-order models and their solutions. Fractional-order Caputo, Caputo–Fabrizio, and Atangana–Baleanu time derivatives are used to study the effect of fractional parameters (memory effect) on the dynamics of fluid. We found that fractional-order model is best to explain the memory effect and flow behavior of the fluid with reference to ordinary order model. Furthermore, it is noted that ABC is the best to highlight the dynamics of fluid. The influence of transverse magnetic fields studied. Moreover, the effects of system parameters on the field velocity are analyzed through numerical simulation and graphs. Some results present in literature are also recovered as limiting cases from our general results.

10.2 Mathematical Model

Let incompressible magnetohydrodynamic flow of Maxwell fluid along an infinite plate. Motion of the plate is rectilinear. A magnetic field acting on the plate having strength B_0. By considering very small Reynolds number, external electric and induced magnetic field is negligible. Initially, system is at rest. After some time, plate starts moving with velocity $U_0 f(t)$. Mathematical modeling of the problem is given below [38, 39]:

$$\left(1+\lambda\frac{\partial}{\partial t}\right)\frac{\partial u(y,t)}{\partial t} = \upsilon\frac{\partial^2 u(y,t)}{\partial y^2} + g\beta_T(T(y,t)-T_\infty) + g\beta_C(C(y,t)-C_\infty) - \frac{\sigma B_0^2}{\rho}\left(1+\lambda\frac{\partial}{\partial t}\right)u(y,t), \tag{10.1}$$

$$\rho C_p\frac{\partial T(y,t)}{\partial t} = k\frac{\partial^2 T(y,t)}{\partial y^2} - Q(T(y,t)-T_\infty), \tag{10.2}$$

$$\frac{\partial C(y,t)}{\partial t} = D_m\frac{\partial^2 C(y,t)}{\partial y^2} - R(C(y,t)-C_\infty). \tag{10.3}$$

The appropriate initial and boundary conditions are

$$u(y,0)=0,\ T(y,0)=T_\infty,\ C(y,0)=C_\infty,\quad y\geq 0, \tag{10.4}$$

$$u(0,t)=U_0 f(t),\ T(0,t)=\begin{cases} T_\infty+(T_w-T_\infty)\frac{t}{t_0}, & 0<t\leq t_0; \\ T(0,t)=T_w, & t>t_0 \end{cases},\ C(0,t)=C_w, \tag{10.5}$$

$$u(y,t)<\infty,\ T(y,t)\to\infty,\ C(y,t)\to\infty\ as\ y\to\infty. \tag{10.6}$$

Following dimensionless variables are used to form the problem-free from geometric regime

$$y^*=\frac{y}{\sqrt{\upsilon t_0}},\ t^*=\frac{t}{t_0},\ u^*=\frac{u}{U_0},\ T^*=\frac{T-T_\infty}{T_w-T_\infty},\ C^*=\frac{C-C_\infty}{C_w-C_\infty},$$
$$G_r=\frac{g\beta_T t_0(T-T_\infty)}{U_0},\ G_m=\frac{g\beta_C t_0(C-C_\infty)}{U_0},\ M=\sqrt{\upsilon t_0}B_0\sqrt{\frac{\sigma}{\mu}},$$
$$P_r=\frac{\upsilon C_p}{k},\ Q^*=\frac{Qt_0}{\rho C_p},\ S_c=\frac{\upsilon}{D_m},\ R^*=Rt_0,\ f^*(t^*)=f(t_0t^*), \tag{10.7}$$

and dimensionless set of governing equations are

$$\left(1+\lambda\frac{\partial}{\partial t}\right)\frac{\partial u(y,t)}{\partial t} = \frac{\partial^2 u(y,t)}{\partial y^2} + G_rT(y,t) + G_mC(y,t) - M^2\left(1+\lambda\frac{\partial}{\partial t}\right)u(y,t), \tag{10.8}$$

$$\frac{\partial T(y,t)}{\partial t} = \frac{1}{P_r}\frac{\partial^2 T(y,t)}{\partial y^2} - QT(y,t), \tag{10.9}$$

$$\frac{\partial C(y,t)}{\partial t} = \frac{1}{S_c}\frac{\partial^2 C(y,t)}{\partial y^2} - RC(y,t) \tag{10.10}$$

and corresponding set of initial and boundary conditions are

$$u(y,0) = 0, \ T(y,0) = 0, \ C(y,0) = 0, \tag{10.11}$$

$$u(0,t) = f(t), \ T(0,t) = \begin{cases} t, & 0 < t \leq 1; = tH(t) - (t-1)H(t-1) \\ 1, & t > 1 \end{cases}, \ C(0,t) = 1, \tag{10.12}$$

$$u(y,t) \to 0, \ T(y,t) \to 0, \ C(y,t) \to 0 \ as \ y \to \infty. \tag{10.13}$$

10.2.1 Preliminaries

The Caputo time derivative is defined as:

$$^{C}D_t^{\alpha} f(y,t) = \frac{1}{\Gamma(n-\alpha)} \int_a^t \left(\frac{f^{(n)}(\tau)}{(t-\tau)^{\alpha+1-n}} \right) d\tau, \tag{10.14}$$

where $\Gamma(\cdot)$ is gamma function.

The Laplace transform of Caputo derivative is

$$\mathcal{L}(^{C}D_t^{\alpha} f(y,t)) = s^{\alpha} \mathcal{L}(f(y,t)) - s^{\alpha-1} f(y,0). \tag{10.15}$$

The Caputo–Fabrizio time derivative is defined as:

$$^{CF}D_t^{\alpha} f(y,t) = \frac{1}{1-\alpha} \int_0^t \exp\left(-\frac{\alpha(t-\alpha)}{1-\alpha} \right) \frac{\partial f(y,\tau)}{\partial \tau} d\tau, \quad 0 < \alpha < 1. \tag{10.16}$$

The Laplace transform of Caputo–Fabrizio derivative is

$$\mathcal{L}(^{CF}D_t^{\alpha} f(y,t)) = \frac{s\mathcal{L}(f(y,t)) - f(y,0)}{(1-\alpha)s + \alpha}. \tag{10.17}$$

The Atangana–Baleanu time derivative in Caputo sense (ABC) is defined as:

$$^{abc}D_t^{\alpha} f(y,t) = \frac{1}{1-\alpha} \int_0^t E_{\alpha} \left(-\frac{\alpha(t-\alpha)^{\alpha}}{1-\alpha} \right) \frac{\partial f(y,\tau)}{\partial \tau} d\tau. \tag{10.18}$$

The Laplace transform of Atangana–Baleanu Caputo derivative is

$$\mathcal{L}(^{abc}D_t^{\alpha} f(y,t)) = \frac{s^{\alpha}\mathcal{L}(f(y,t)) - s^{\alpha-1} f(y,0)}{(1-\alpha)s^{\alpha} + \alpha}. \tag{10.19}$$

10.3 Solution

We use integral transformation to get solution of Eqs. (10.8)–(10.10) using the definitions of time-fractional derivatives (Caputo, Caputo–Fabrizio, and Atangana–Baleanu) and their Laplace transforms given by Eqs. (10.14)–(10.19) by applying initial and boundary conditions (10.11)–(10.13). In order to get solution of velocity, we have to find solutions of temperature and concentration.

10.3.1 Concentration Fields

10.3.1.1 Concentration Field with Caputo Time-Fractional Derivative

Solving Eq. (10.10) using (10.14) and (10.15), we have

$$\frac{\partial^2 \overline{C}_c(y,q)}{\partial y^2} - S_c(q^\alpha + R)\overline{C}_c(y,q) = 0, \tag{10.20}$$

the solution of above second-order differential equation is given by

$$\overline{C}_c(y,q) = c_1 e^{y\sqrt{S_c(q^\alpha+R)}} + c_2 e^{-y\sqrt{S_c(q^\alpha+R)}}, \tag{10.21}$$

in order to find the values of constants c_1 and c_2, boundary conditions for concentration given by Eqs. (10.12) and (10.13) are applied, we have

$$\overline{C}_c(y,q) = \frac{1}{q} e^{-y\sqrt{S_c(q^\alpha+R)}}. \tag{10.22}$$

10.3.1.2 Concentration Field with Caputo–Fabrizio Time-Fractional Derivative

Concentration field with Caputo–Fabrizio time-fractional is given as follow after using (10.16) and (10.17) with (10.10)

$$\frac{\partial^2 \overline{C}_{cf}(y,q)}{\partial y^2} - S_c\left[\frac{q}{(1-\alpha)q+\alpha} + R\right]\overline{C}_{cf}(y,q) = 0, \tag{10.23}$$

$$\overline{C}_{cf}(y,q) = c_1 e^{y\sqrt{S_c\left[\frac{q}{(1-\alpha)q+\alpha}+R\right]}} + c_2 e^{-y\sqrt{S_c\left[\frac{q}{(1-\alpha)q+\alpha}+R\right]}} \tag{10.24}$$

and

$$\overline{C}_{cf}(y,q) = \frac{1}{q} e^{-y\sqrt{S_c\left(\frac{q}{(1-\alpha)q+\alpha}+R\right)}}, \tag{10.25}$$

where $c_1 = 0$ and $c_2 = \frac{1}{q}$.

10.3.1.3 Concentration Field with Atangana–Baleanu Time-Fractional Derivative

Concentration field with Atangana–Baleanu time-fractional derivative is given by applying (10.18) and (10.19) on (10.10), we have

$$\frac{\partial^2 \overline{C}_{abc}(y,q)}{\partial y^2} - S_c\left(\frac{q^\alpha}{(1-\alpha)q^\alpha+\alpha} + R\right)\overline{C}_{abc}(y,q) = 0, \tag{10.26}$$

$$\overline{C}_{abc}(y,q) = c_1 e^{y\sqrt{S_c\left(\frac{q^\alpha}{(1-\alpha)q^\alpha+\alpha}+R\right)}} + c_2 e^{-y\sqrt{S_c\left(\frac{q^\alpha}{(1-\alpha)q^\alpha+\alpha}+R\right)}}, \tag{10.27}$$

by using boundary conditions on concentration, we have following simplified form

$$\overline{C}_{abc}(y,q) = \frac{1}{q} e^{-y\sqrt{S_c\left(\frac{q^\alpha}{(1-\alpha)q^\alpha+\alpha}+R\right)}}, \tag{10.28}$$

By making $\alpha \to 1$ in Eqs. (10.22), (10.25), and (10.28), we get the expression/result for integer-order model same as obtained by Iftikhar [39]. This shows the validity of our results with the literature.

10.3.2 Temperature Fields

10.3.2.1 Temperature Field with Caputo Time-Fractional Derivative

Solving Eq. (10.9) using (10.14) and (10.15), we have

$$\frac{\partial^2 \overline{T}_c(y,q)}{\partial y^2} - P_r(q^\alpha + Q)\overline{T}_c(y,q) = 0, \tag{10.29}$$

and its solution is given by

$$\overline{T}_c(y,q) = c_1 e^{y\sqrt{P_r(q^\alpha+Q)}} + c_2 e^{-y\sqrt{P_r(q^\alpha+Q)}} \tag{10.30}$$

in order to find the values of constants c_1 and c_2, boundary conditions for temperature given by Eqs. (10.12) and (10.13) are applied, we have

$$\overline{T}_c(y,q) = \left(\frac{1-e^{-q}}{q^2}\right) e^{-y\sqrt{P_r(q^\alpha+Q)}}. \tag{10.31}$$

10.3.2.2 Temperature Field with Caputo–Fabrizio Time-Fractional Derivative

Temperature field with Caputo–Fabrizio time-fractional is given as follow after using (10.16) and (10.17) with (10.9)

$$\frac{\partial^2 \overline{T}_{cf}(y,q)}{\partial y^2} - P_r\left[\frac{q}{(1-\alpha)q+\alpha} + Q\right]\overline{T}_{cf}(y,q) = 0, \tag{10.32}$$

and

$$\overline{T}_{cf}(y,q) = c_1 e^{y\sqrt{P_r\left[\frac{q}{(1-\alpha)q+\alpha}+Q\right]}} + c_2 e^{-y\sqrt{P_r\left[\frac{q}{(1-\alpha)q+\alpha}+Q\right]}}, \tag{10.33}$$

$$\overline{T}_{cf}(y,q) = \left(\frac{1-e^{-q}}{q^2}\right) e^{-y\sqrt{P_r\left(\frac{q}{(1-\alpha)q+\alpha}+Q\right)}}. \tag{10.34}$$

10.3.2.3 Temperature Field with Atangana–Baleanu Time-Fractional Derivative

Temperature field with Atangana–Baleanu time-fractional derivative is given by applying (10.18) and (10.19) on (10.9), we have

$$\frac{\partial^2 \overline{T}_{abc}(y,q)}{\partial y^2} - P_r\left(\frac{q^\alpha}{(1-\alpha)q^\alpha+\alpha} + Q\right)\overline{T}_{abc}(y,q) = 0, \tag{10.35}$$

or

$$\overline{T}_{abc}(y,q) = c_1 e^{y\sqrt{P_r\left(\frac{q^\alpha}{(1-\alpha)q^\alpha+\alpha}+Q\right)}} + c_2 e^{-y\sqrt{P_r\left(\frac{q^\alpha}{(1-\alpha)q^\alpha+\alpha}+Q\right)}}, \tag{10.36}$$

$$\overline{T}_{abc}(y,q) = \left(\frac{1-e^{-q}}{q^2}\right) e^{-y\sqrt{P_r\left(\frac{q^\alpha}{(1-\alpha)q^\alpha+\alpha}+Q\right)}}, \tag{10.37}$$

by making $\alpha \to 1$ in Eqs. (10.31), (10.34), and (10.37), we get the expression/result for integer-order model same as obtained by Iftikhar et al. [39]. This shows the validity of our results with the literature.

10.3.3 Velocity Fields

10.3.3.1 Velocity Field with Caputo Time-Fractional Derivative

Solution of Eq. (10.8) using (10.14) and (10.15) with the help of initial condition on velocity given in (10.11), we have

$$(1+\lambda q^\alpha)(q+M^2)\overline{u}_c(y,q) = \frac{\partial^2 \overline{u}_c(y,q)}{\partial y^2} + G_r \overline{T}_c(y,q) + G_m \overline{C}_c(y,q), \tag{10.38}$$

by using Eqs. (10.31) and (10.22) for the values $\overline{T}_c(y,q)$ and $\overline{C}_c(y,q)$, solution of Eq. (10.38) can be written as

$$\begin{aligned}\overline{u_c}(y,q) &= c_1 e^{y\sqrt{(q+M^2)(1+\lambda q^\alpha)}} + c_2 e^{-y\sqrt{(q+M^2)(1+\lambda q^\alpha)}} \\ &\quad - \left(\frac{G_r(1-e^{-q})}{P_r(q^\alpha+Q)-(q+M^2)(1+\lambda q^\alpha)}\right) \\ &\quad \times \frac{e^{-y\sqrt{P_r(q^\alpha+Q)}}}{q^2} - \left(\frac{G_m}{S_c(q^\alpha+R)-(q+M^2)(1+\lambda q^\alpha)}\right)\frac{e^{-y\sqrt{S_c(q^\alpha+R)}}}{q}.\end{aligned} \tag{10.39}$$

Apply boundary conditions on velocity, we get

$$\begin{aligned}\overline{u_c}(y,q) &= F(q)e^{-y\sqrt{(q+M^2)(1+\lambda q^\alpha)}} \\ &\quad - \left(\frac{G_r(1-e^{-q})}{q^2(P_r(q^\alpha+Q)-(q+M^2)(1+\lambda q^\alpha))}\right) \\ &\quad \times \left(e^{-y\sqrt{(q+M^2)(1+\lambda q^\alpha)}} - e^{-y\sqrt{P_r(q^\alpha+Q)}}\right) \\ &\quad - \left(\frac{G_m}{q(S_c(q^\alpha+R)-(q+M^2)(1+\lambda q^\alpha))}\right) \\ &\quad \times \left(e^{-y\sqrt{(q+M^2)(1+\lambda q^\alpha)}} - e^{-y\sqrt{S_c(q^\alpha+R)}}\right).\end{aligned} \tag{10.40}$$

10.3.3.2 Velocity Field with Caputo–Fabrizio Time-Fractional Derivative

Solution of Eq. (10.8) using (10.16) and (10.17) with the help of initial condition on velocity given in (10.11), we have

$$\left(1+\frac{\lambda q}{(1-\alpha)q+\alpha}\right)(q+M^2)\overline{u}_{cf}(y,q) = \frac{\partial^2 \overline{u}_{cf}(y,q)}{\partial y^2} + G_r\overline{T}(y,q) + G_m\overline{C}(y,q), \tag{10.41}$$

solution of homogeneous part of Eq. (10.41) is given below

$$\overline{u_h}(y,q) = c_1 e^{y\sqrt{(q+M^2)\frac{(1-\alpha+\lambda)q+\alpha}{(1-\alpha)q+\alpha}}} + c_2 e^{-y\sqrt{(q+M^2)\frac{(1-\alpha+\lambda)q+\alpha}{(1-\alpha)q+\alpha}}}, \tag{10.42}$$

and particular solution can be give as follow after making use of Eqs. (10.34) and (10.25) for the values $\overline{T}_{cf}(y,q)$ and $\overline{C}_{cf}(y,q)$

$$\begin{aligned}\overline{u_p}(y,q) = &-\frac{G_r(1-e^{-q})((1-\alpha)q+\alpha)}{q^2(P_r[(1+Q+\alpha Q)q+\alpha Q]) - (q+M^2)((1-\alpha+\lambda)q+\alpha)}\\ &\times e^{-y\sqrt{P_r\left(\frac{q}{(1-\alpha)q+\alpha}+Q\right)}}\\ &-\frac{G_m((1-\alpha)q+\alpha)}{q(S_c[(1+R+\alpha R)q+\alpha R]) - (q+M^2)((1-\alpha+\lambda)q+\alpha)}\\ &\times e^{-y\sqrt{S_c\left(\frac{q}{(1-\alpha)q+\alpha}+R\right)}},\end{aligned} \tag{10.43}$$

$$\begin{aligned}\overline{u_p}(y,q) = &\frac{G_r(1-e^{-q})(a_4q+\alpha)}{q^2(a_3q^2+(\alpha-a_1P_r+a_3M^2)q-(a_2P_r-\alpha M^2))}e^{-y\sqrt{P_r\left(\frac{a_1q+a_2}{a_4q+\alpha}\right)}}\\ &+\frac{G_m(a_4q+\alpha)}{q(a_3q^2+(\alpha-a_5S_c+a_3M^2)q-(a_6S_c-\alpha M^2))}e^{-y\sqrt{S_c\left(\frac{a_5q+a_6}{a_4q+\alpha}\right)}},\end{aligned} \tag{10.44}$$

and solution of Eq. (10.41) can be given as follow:

$$\overline{u}_{cf}(y,q) = \overline{u_h}(y,q) + \overline{u_p}(y,q) \tag{10.45}$$

or

$$\begin{aligned}\overline{u}_{cf}(y,q) = &c_1 e^{y\sqrt{(q+M^2)\frac{a_3q+\alpha}{a_4q+\alpha}}} + c_2 e^{-y\sqrt{(q+M^2)\frac{a_3q+\alpha}{a_4q+\alpha}}}\\ &+\frac{G_r(1-e^{-q})(a_4q+\alpha)}{q^2(a_3q^2+(\alpha-a_1P_r+a_3M^2)q-(a_2P_r-\alpha M^2))}e^{-y\sqrt{P_r\left(\frac{a_1q+a_2}{a_4q+\alpha}\right)}}\\ &+\frac{G_m(a_4q+\alpha)}{q(a_3q^2+(\alpha-a_5S_c+a_3M^2)q-(a_6S_c-\alpha M^2))}e^{-y\sqrt{S_c\left(\frac{a_5q+a_6}{a_4q+\alpha}\right)}},\end{aligned} \tag{10.46}$$

using boundary conditions for velocity in order to find constants, we have

$$\begin{aligned}\overline{u}_{cf}(y,q) = &F(q)e^{-y\sqrt{(q+M^2)\left(\frac{a_3q+\alpha}{a_4q+\alpha}\right)}} - \frac{G_r(1-e^{-q})(a_4q+\alpha)}{q^2(a_3q^2+A_1q-A_2)}\\ &\times\left(e^{-y\sqrt{\frac{(q+M^2)(a_3q+\alpha)}{a_4q+\alpha}}} - e^{-y\sqrt{P_r\frac{a_1q+a_2}{a_4q+\alpha}}}\right)\\ &-\frac{G_m(a_4q+\alpha)}{q(a_3q^2+A_3q-A_4)}\left(e^{-y\sqrt{\frac{(q+M^2)(a_3q+\alpha)}{a_4q+\alpha}}} - e^{-y\sqrt{S_c\frac{a_5q+a_6}{a_4q+\alpha}}}\right)\end{aligned} \tag{10.47}$$

or

$$\bar{u}_{cf}(y,q) = F(q)e^{-y\sqrt{(q+M^2)\left(\frac{a_3q+\alpha}{a_4q+\alpha}\right)}} - \frac{G_r(1-e^{-q})(a_4q+\alpha)}{q^2(q^2+a_7q-a_8)}$$
$$\times\left(e^{-y\sqrt{\frac{(q+M^2)(a_3q+\alpha)}{a_4q+\alpha}}} - e^{-y\sqrt{P_r\frac{a_1q+a_2}{a_4q+\alpha}}}\right)$$
$$-\frac{G_m(a_4q+\alpha)}{q(q^2+a_9q-a_{10})}\left(e^{-y\sqrt{\frac{(q+M^2)(a_3q+\alpha)}{a_4q+\alpha}}} - e^{-y\sqrt{S_c\frac{a_5q+a_6}{a_4q+\alpha}}}\right). \tag{10.48}$$

In simplified form, suitable for inversion algorithm, we have

$$\bar{u}_{cf}(y,q) = F(q)e^{-y\sqrt{(q+M^2)\left(\frac{a_3q+\alpha}{a_4q+\alpha}\right)}} - (G_r(1-e^{-q}))$$
$$\times\left(\frac{H_1}{q}+\frac{H_2}{q^2}+\frac{H_3}{q+H_5}+\frac{H_4}{q+H_6}\right)$$
$$\times\left(e^{-y\sqrt{\frac{(q+M^2)(a_3q+\alpha)}{a_4q+\alpha}}} - e^{-y\sqrt{P_r\frac{a_1q+a_2}{a_4q+\alpha}}}\right) - G_m\left(\frac{G_1}{q}+\frac{G_2}{q+G_4}+\frac{G_3}{q+G_5}\right)$$
$$\times\left(e^{-y\sqrt{\frac{(q+M^2)(a_3q+\alpha)}{a_4q+\alpha}}} - e^{-y\sqrt{S_c\frac{a_5q+a_6}{a_4q+\alpha}}}\right), \tag{10.49}$$

where $a_1 = 1+Q-\alpha Q$, $a_2 = \alpha Q$, $a_3 = 1-\alpha+\lambda$, $a_4 = 1-\alpha$, $a_5 = 1+R-\alpha R$, $a_6 = \alpha R$,

$$A_1 = \alpha - a_1P_r + a_3M^2,\ A_2 = a_2P_r - \alpha M^2,\ A_3 = \alpha - a_5S_c + a_3M^2,$$
$$A_4 = a_6S_c - a_3M^2,$$
$$a_7 = \frac{A_1}{a_3},\ a_8 = \frac{A_2}{a_3},\ a_9 = \frac{A_3}{a_3},\ a_{10} = \frac{A_4}{a_3},\ H = \sqrt{\frac{a_7^2+4a_8}{4}},$$
$$G = \sqrt{\frac{a_9^2+4a_{10}}{4}},$$
$$H_1 = -\frac{4a_4}{4H^2-a_7^2} + \frac{256a_7\alpha}{(a_7+2H)^2(-a_7+2H)^2},\ H_2 = \frac{4\alpha}{a_7^2-4H^2},$$
$$H_3 = \frac{a_4}{H(a_7+2H)} - \frac{2\alpha}{H(a_7+2H)^2},\ H_4 = \frac{a_4}{H(-a_7+2H)} + \frac{2\alpha}{H(-a_7+2H)^2},$$
$$H_5 = \frac{a_7}{2}+H,\ H_6 = \frac{a_7}{2}-H,\ G_1 = \frac{4\alpha}{a_9^2-4G^2},\ G_2 = \frac{a_4(-a_9-2G)+2\alpha}{2(a_9+2G)G},$$
$$G_3 = \frac{a_4}{2G} + \frac{\alpha}{G(-a_9+2G)},\ G_4 = \frac{a_9}{2}+G,\ G_5 = \frac{a_9}{2}-G. \tag{10.50}$$

10.3.3.3 Velocity Field with Atangana–Baleanu Time-Fractional Derivative

Solution of Eq. (10.8) using (10.18) and (10.19) with the help of initial condition on velocity given in (10.11), we have

$$\left(1+\frac{\lambda q^{\alpha}}{(1-\alpha)q^{\alpha}+\alpha}\right)(q+M^{2})\overline{u}_{abc}(y,q)=\frac{\partial^{2}\overline{u}}{\partial y^{2}}+G_{r}\overline{T}_{abc}(y,q)+G_{m}\overline{C}_{abc}(y,q), \tag{10.51}$$

solution of Eq. (10.51) is given as (after lengthy calculations)

$$\overline{u}_{abc}(y,q)=c_{1}e^{y\sqrt{(q+M^{2})\frac{a_{3}q^{\alpha}+\alpha}{a_{4}q^{\alpha}+\alpha}}}+c_{2}e^{-y\sqrt{(q+M^{2})\frac{a_{3}q^{\alpha}+\alpha}{a_{4}q^{\alpha}+\alpha}}}+\frac{G_{r}(1-e^{-q})(a_{4}q^{\alpha}+\alpha)}{q^{2}(P_{r}(a_{1}q^{\alpha}+a_{2})-(q+M^{2})(a_{3}q^{\alpha}+\alpha))}e^{-y\sqrt{P_{r}\left(\frac{a_{1}q^{\alpha}+a_{2}}{a_{4}q^{\alpha}+\alpha}\right)}}+\frac{G_{m}(a_{4}q^{\alpha}+\alpha)}{q(S_{c}(a_{5}q^{\alpha}+a_{6})-(q+M^{2})(a_{3}q^{\alpha}+\alpha))}e^{-y\sqrt{S_{c}\left(\frac{a_{5}q^{\alpha}+a_{6}}{a_{4}q^{\alpha}+\alpha}\right)}} \tag{10.52}$$

or

$$\overline{u}(y,q)=F(q)e^{-y\sqrt{(q+M^{2})\left(\frac{a_{3}q^{\alpha}+\alpha}{a_{4}q^{\alpha}+\alpha}\right)}}+\frac{G_{r}(1-e^{-q})(a_{4}q^{\alpha}+\alpha)}{q^{2}(P_{r}(a_{1}q^{\alpha}+a_{2})-(q+M^{2})(a_{3}q^{\alpha}+\alpha))}\times\left(e^{-y\sqrt{(q+M^{2})\left(\frac{a_{3}q^{\alpha}+\alpha}{a_{4}q^{\alpha}+\alpha}\right)}}-e^{-y\sqrt{P_{r}\left(\frac{a_{1}q^{\alpha}+a_{2}}{a_{4}q^{\alpha}+\alpha}\right)}}\right)-\frac{G_{m}(a_{4}q^{\alpha}+\alpha)}{q(S_{c}(a_{5}q^{\alpha}+a_{6})-(q+M^{2})(a_{3}q^{\alpha}+\alpha))}\times\left(e^{-y\sqrt{(q+M^{2})\left(\frac{a_{3}q^{\alpha}+\alpha}{a_{4}q^{\alpha}+\alpha}\right)}}-e^{-y\sqrt{S_{c}\left(\frac{a_{5}q^{\alpha}+a_{6}}{a_{4}q^{\alpha}+\alpha}\right)}}\right) \tag{10.53}$$

by making $\alpha \to 1$ in Eqs. (10.31), (10.34), and (10.53), we get the expression/result for integer-order model same as obtained by Iftikhar et al. [39]. It is also noted that as $\lambda \to 0$ and $u_0 = 0$ similar results are obtain by Narahari and Debnath [34] with $a_0 = 0$ and Tokis [40]. This shows that the strength of our obtained general results. In our flow models, we use classical computational technique (Laplace transform) to solve the given models using different definitions fractional derivatives. In order to find their inverses, we use Stehfest's and Tzou's algorithms for semianalytical solutions [25]. In their paper, they showed the accuracy of this Laplace inversion algorithms. Tzou's calculation for approval of our numerical inverse Laplace $v(r,t)=\frac{e^{4.7}}{t}\left[\frac{1}{2}\overline{v}\left(r,\frac{4.7}{t}\right)+Re\left\{\sum_{k=1}^{N_1}(-1)^{k}\overline{v}\left(r,\frac{4.7+k\pi i}{t}\right)\right\}\right]$, where $Re(\cdot)$ is the real part, i is the imaginary unit and N_1 is a natural number [25].

10.4 Results and Discussion

The MHD-free convection flow over vertical plate with chemical reaction in view of nonsingular kernel is studied. The solution of concentration, temperature, and velocity field of the fluid are obtained by Caputo (*C*), Caputo–Fabrizio (*CF*), and Atangana–Baleanu (*ABC*) fractional time derivative. The flow of fluids depends on magnetic forces with influences of physical parameters. The numerical results for velocity are discussed graphically for different parameters such as magnetic field M, Prandtl number P_r, Grashof number G_r, Maxwell parameter λ, and fractional parameter α with variation of time.

The significant impact of fractional parameter α on fluid velocity is compared with Caputo, Caputo–Fabrizio, and Atangana–Baleanu models as shown in Figures 10.1–10.3. The reaction of α is remarkable to vertical plate with temperature. The velocity minimize with increase of α as shown in figures. The nature of velocities for all three fractional models are same in small and large time. The fractional parameters (*C*), (*CF*), and (*ABC*) reduce to integer order by $\alpha \to 1$. It is worth mentioning that fractional-order models are best to explain the history (memory) of the fluids. Furthermore, it is noted that ABC model have more velocity as compared to the others. The velocity near the plate shows significance behavior and goes converges to a point as y tends to infinite. Velocity is also decreasing function of time.

The effect of magnetic field M is depicted in Figures 10.4–10.6. By increase the value of M, the resultant velocity decrease. The flow of fluid velocity is slow down due to impact of magnetic flux. The behavior of *C*, *CF*, and *ABC* in magnetic field are same at different time. Figures 10.7–10.9 illustrate the effect of Prandtl number P_r. The impact of P_r shows a relation between viscous forces and thermal forces. By increasing the value of P_r, the velocity decrease due to thermal forces. The influence of λ shown in Figures 10.9–10.12. The large value of λ, the velocity decreases. The influence of thermal Grashof number G_r is highlighted in Figures 10.13–10.15. Grashof number is defined as the ratio between buoyancy forces to viscid forces to motion of fluid. With increase of G_r, the velocity also increase.

10.5 Conclusion

The comprehensive analysis of MHD-free convection flow with ramped temperature and chemical reaction via Caputo (*C*), Caputo–Fabrizio (*CF*), and Atangana–Baleanu (*ABC*) is investigated. The comparison between (*C*), (*CF*), and

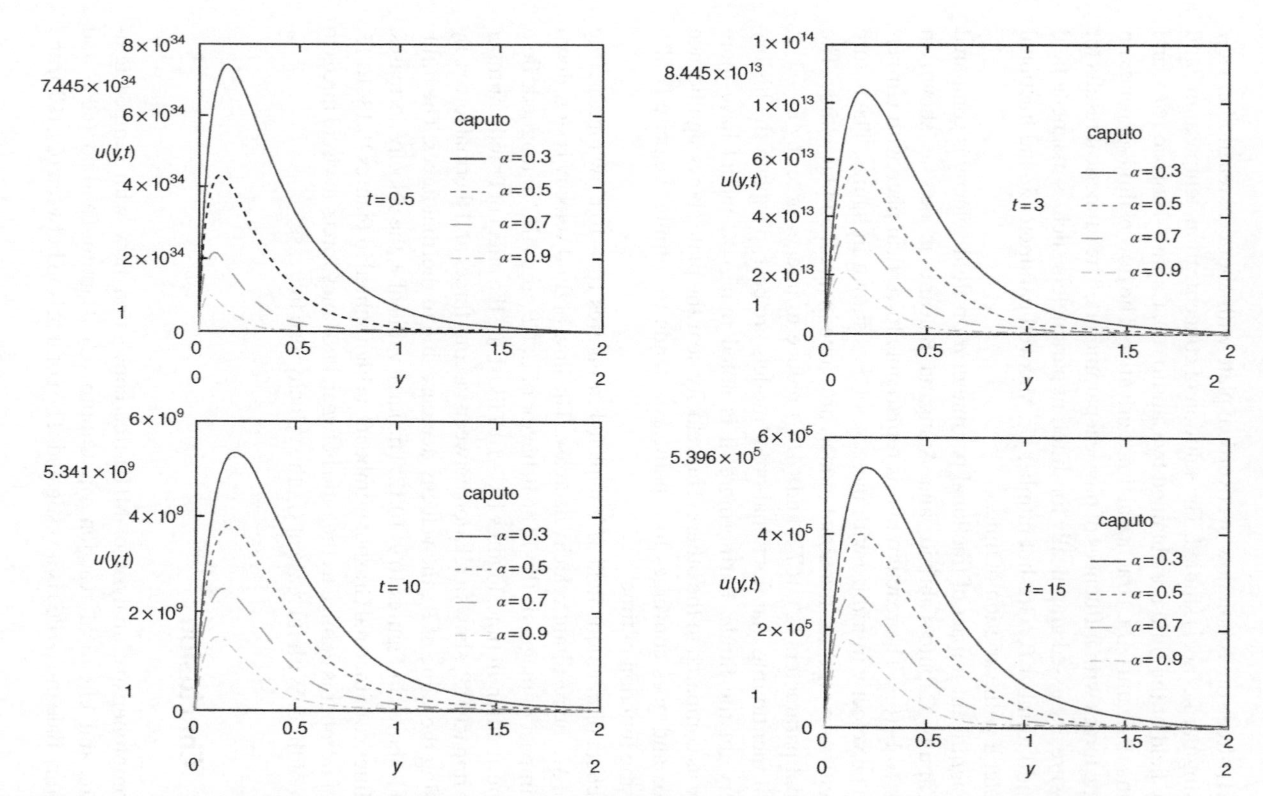

Figure 10.1 Caputo velocity profile for α variation with small and large time and other parameters are as $G_r = 3.5$, $P_r = 1.5$, $M = 2.0$, $\lambda = 0.6$.

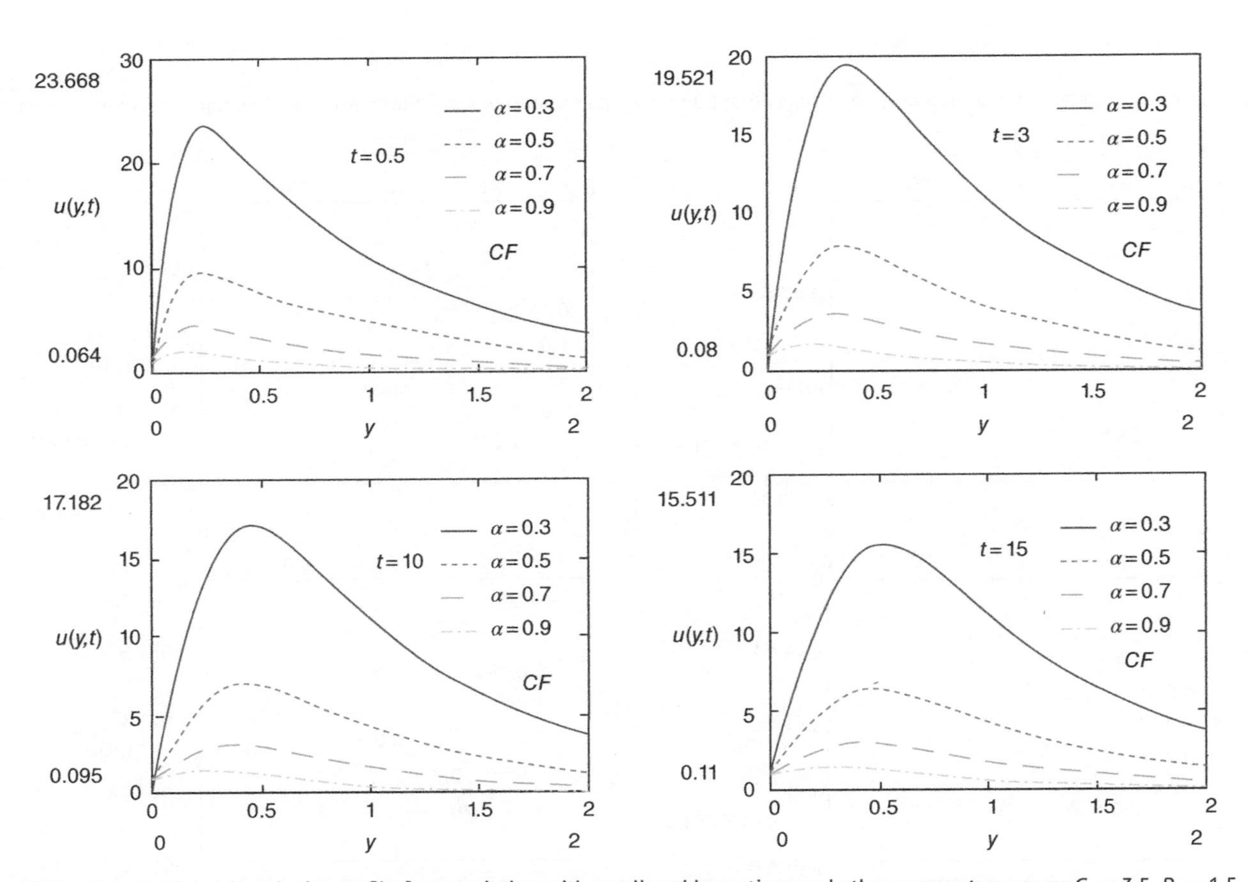

Figure 10.2 Caputo–Fabrizio velocity profile for α variation with small and large time and other parameters are as $G_r = 3.5$, $P_r = 1.5$, $M = 2.0$, $\lambda = 0.6$.

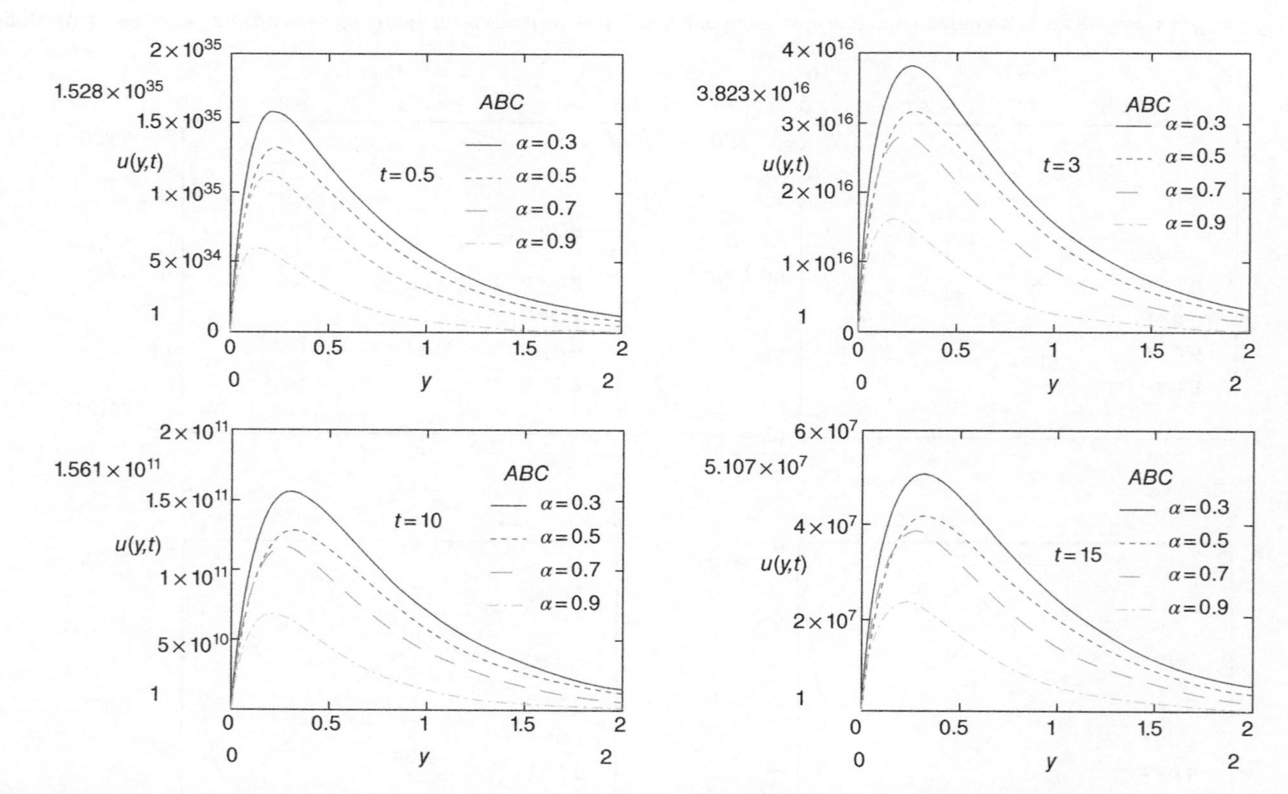

Figure 10.3 Antangana–Baleanu velocity profile for α variation with small and large time and other parameters are as $G_r = 3.5$, $P_r = 1.5$, $M = 2.0$, $\lambda = 0.6$.

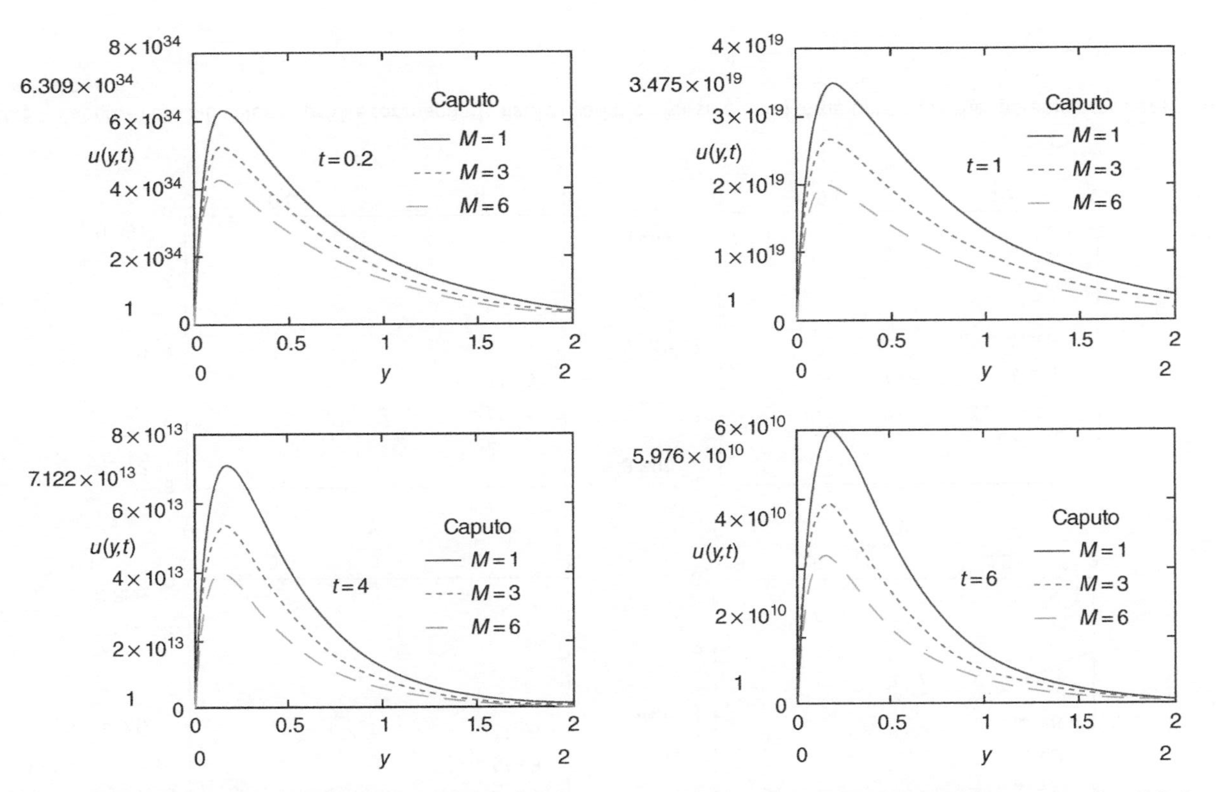

Figure 10.4 Caputo velocity profile for magnetic field M variation for small and large time and other parameters are as $G_r = 3.5$, $P_r = 1.5$, $\alpha = 0.5$, $\lambda = 0.6$.

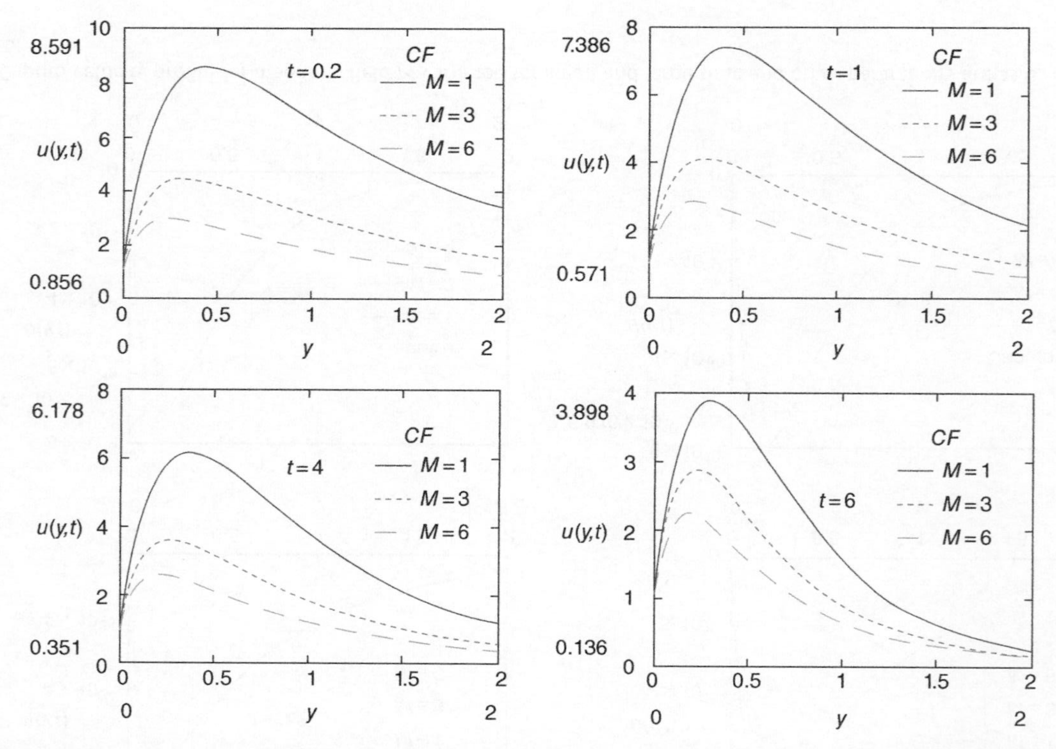

Figure 10.5 Caputo–Fabrizio velocity profile for magnetic field M variation for small and large time and other parameters are as $G_r = 3.5$, $P_r = 1.5$, $\alpha = 0.5$, $\lambda = 0.6$.

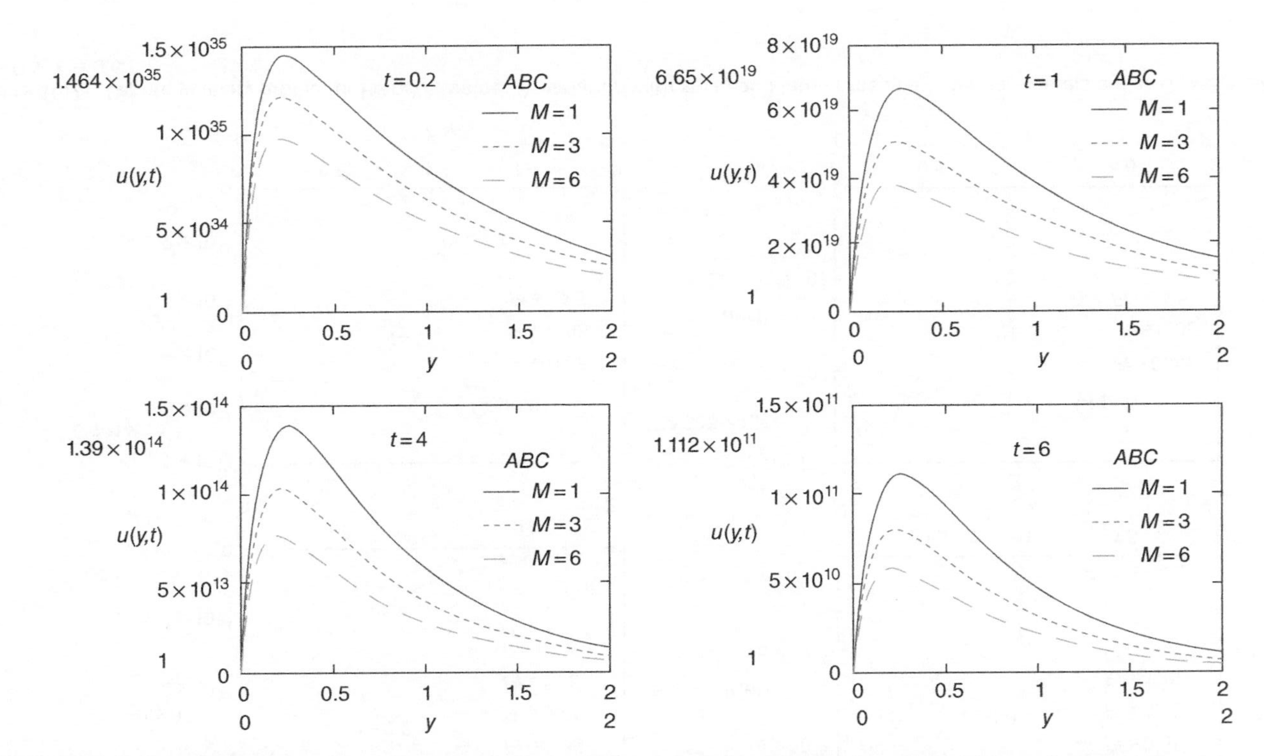

Figure 10.6 Atangana–Baleanu velocity profile for magnetic field M variation with for small and large time and other parameters are as $G_r = 3.5$, $P_r = 1.5$, $\alpha = 0.5$, $\lambda = 0.6$.

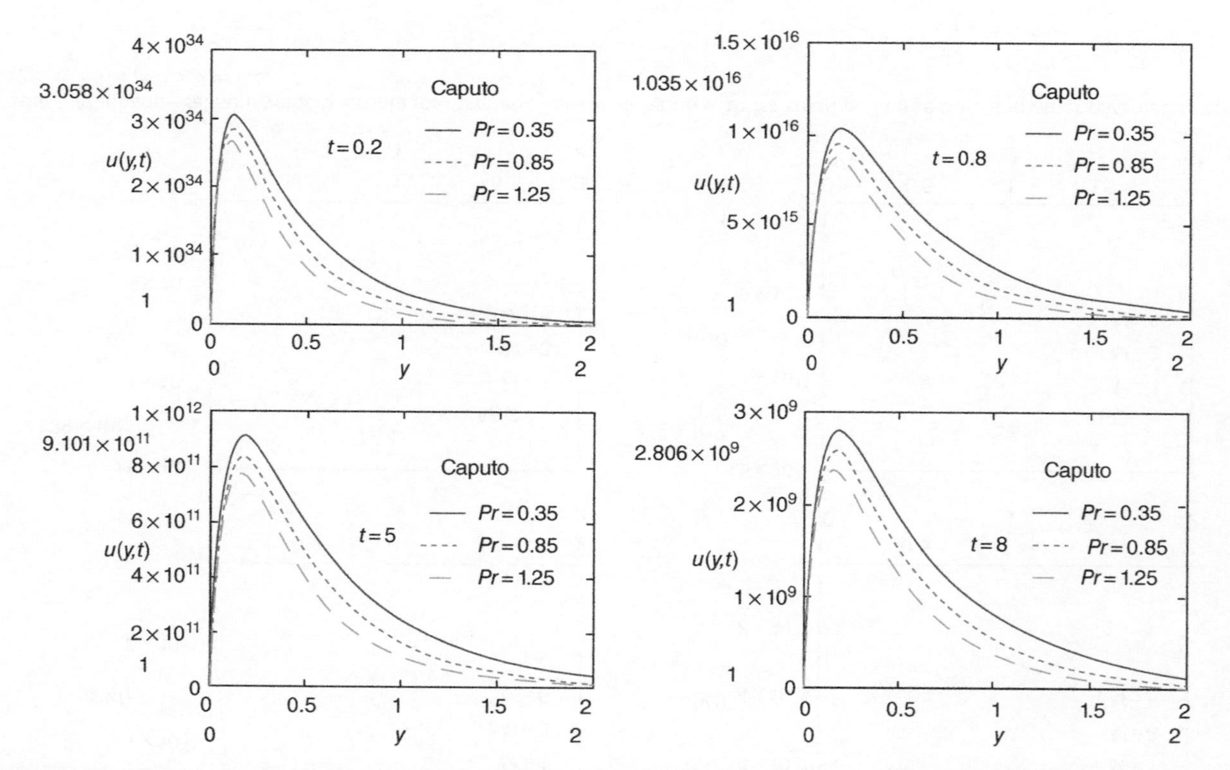

Figure 10.7 Caputo velocity profile for Prandtl number P_r variation with small and large time and other parameters are as $G_r = 3.5$, $M = 2$, $\alpha = 0.5$, $\lambda = 0.6$.

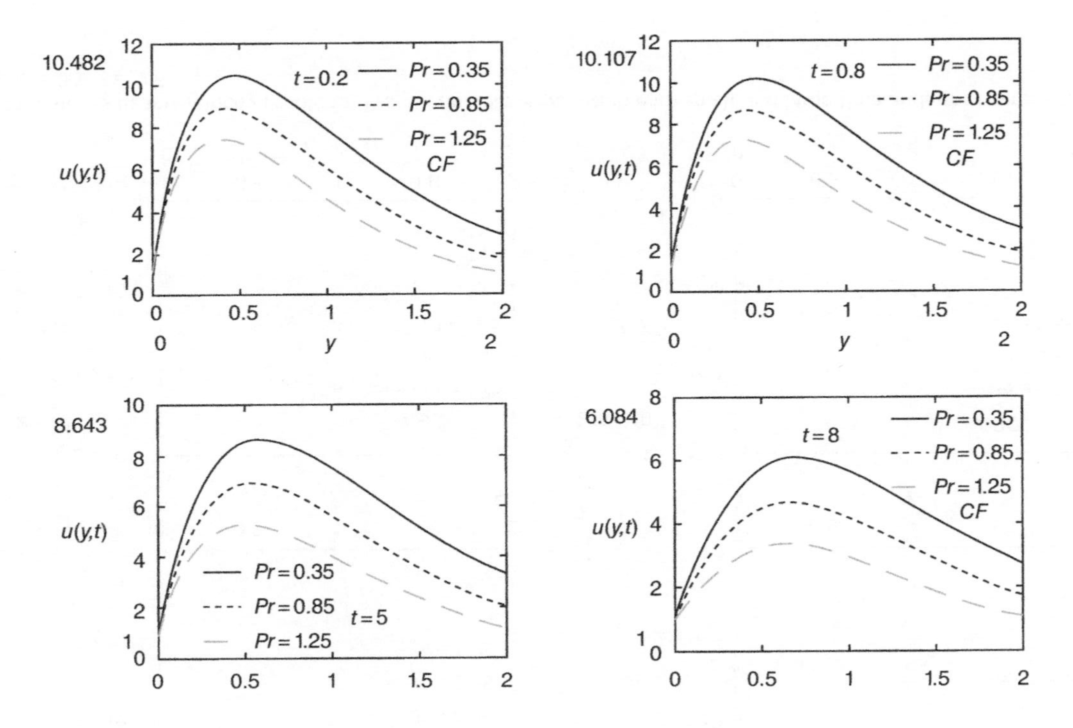

Figure 10.8 Caputo–Fabrizio velocity profile for Prandtl number P_r variation with small and large time and other parameters are as $G_r = 3.5$, $M = 2$, $\alpha = 0.5$, $\lambda = 0.6$.

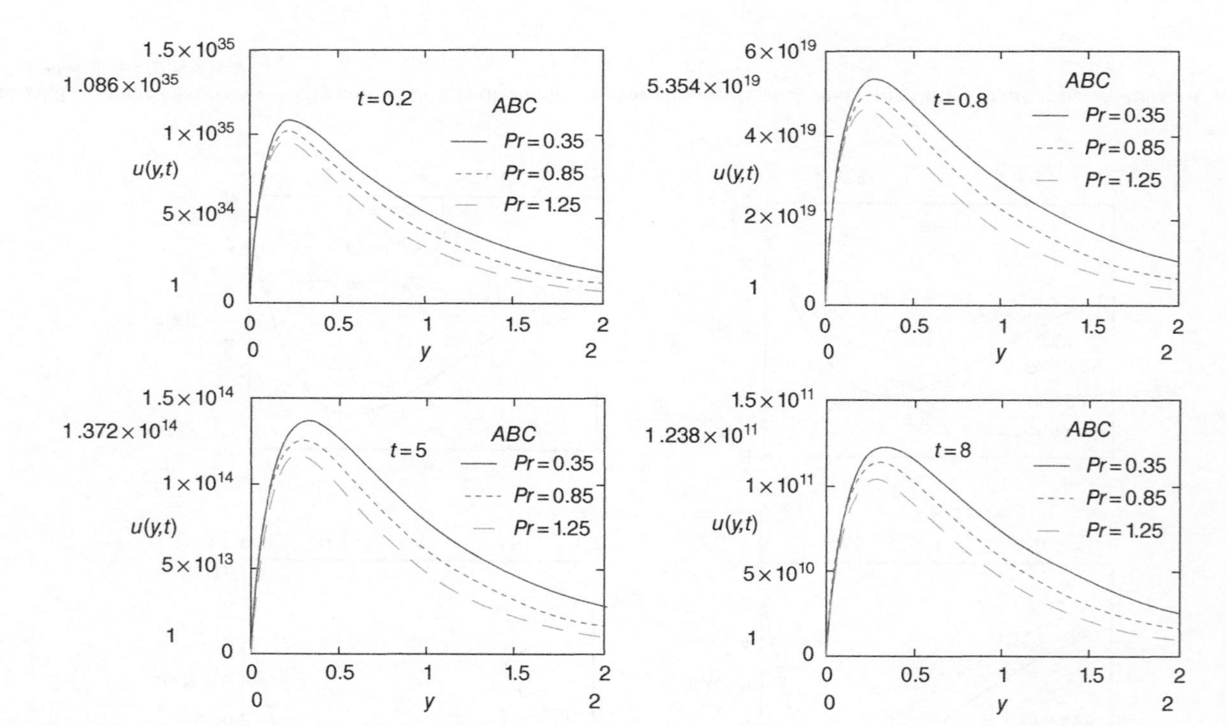

Figure 10.9 Atangana–Baleanu velocity profile for Prandtl number P_r variation with small and large time and other parameters are as $G_r = 3.5$, $M = 2$, $\alpha = 0.5$, $\lambda = 0.6$.

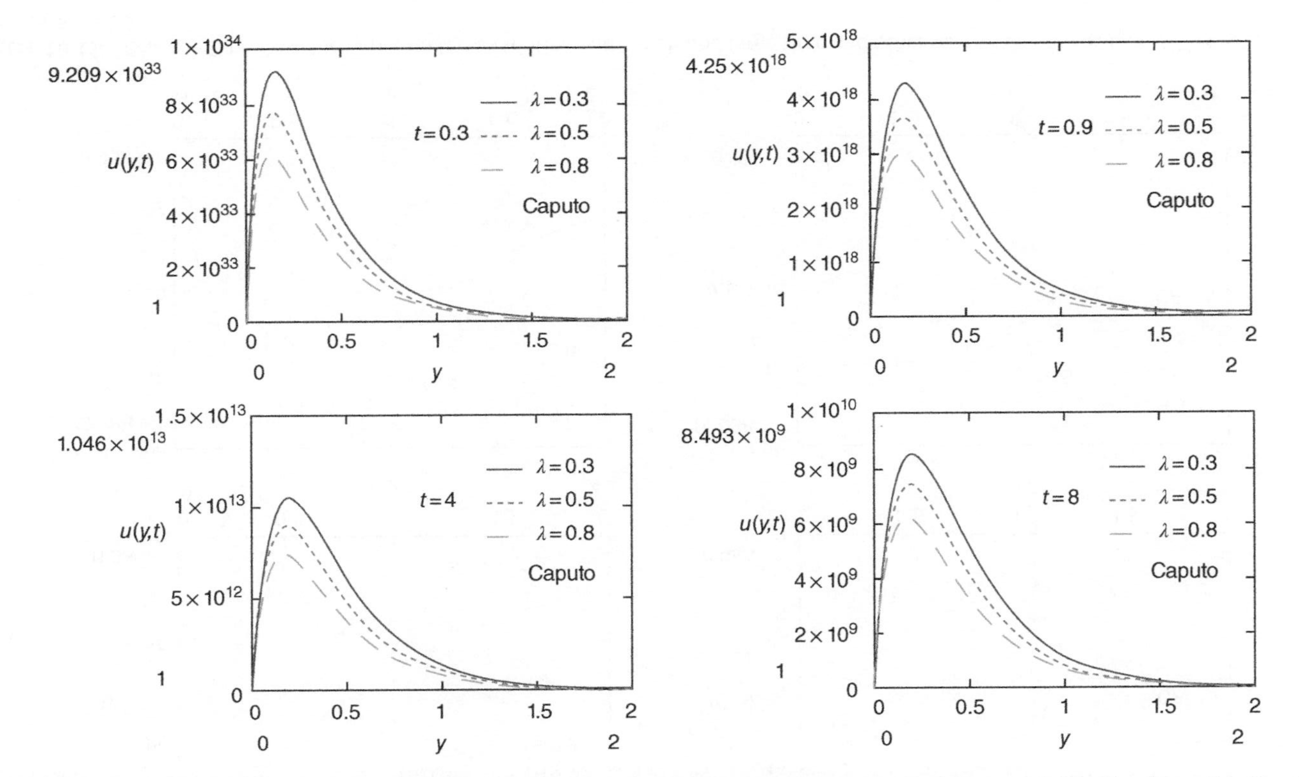

Figure 10.10 Caputo velocity profile for λ variation with small and large time and other parameters are as $G_r = 3.5$, $M = 2$, $\alpha = 0.5$, $P_r = 1.5$.

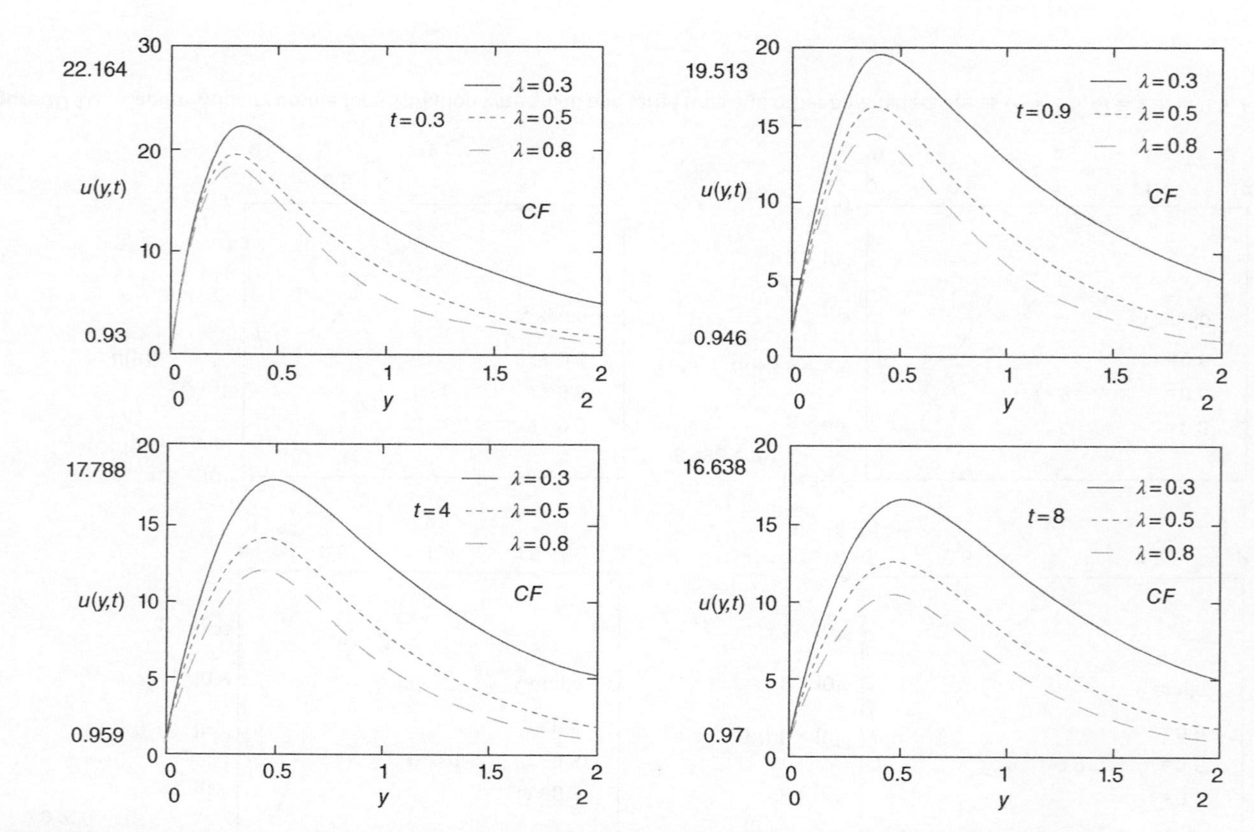

Figure 10.11 Caputo–Fabrizio velocity profile for λ variation with small and large time and other parameters are as $G_r = 3.5$, $M = 2$, $\alpha = 0.5$, $P_r = 1.5$.

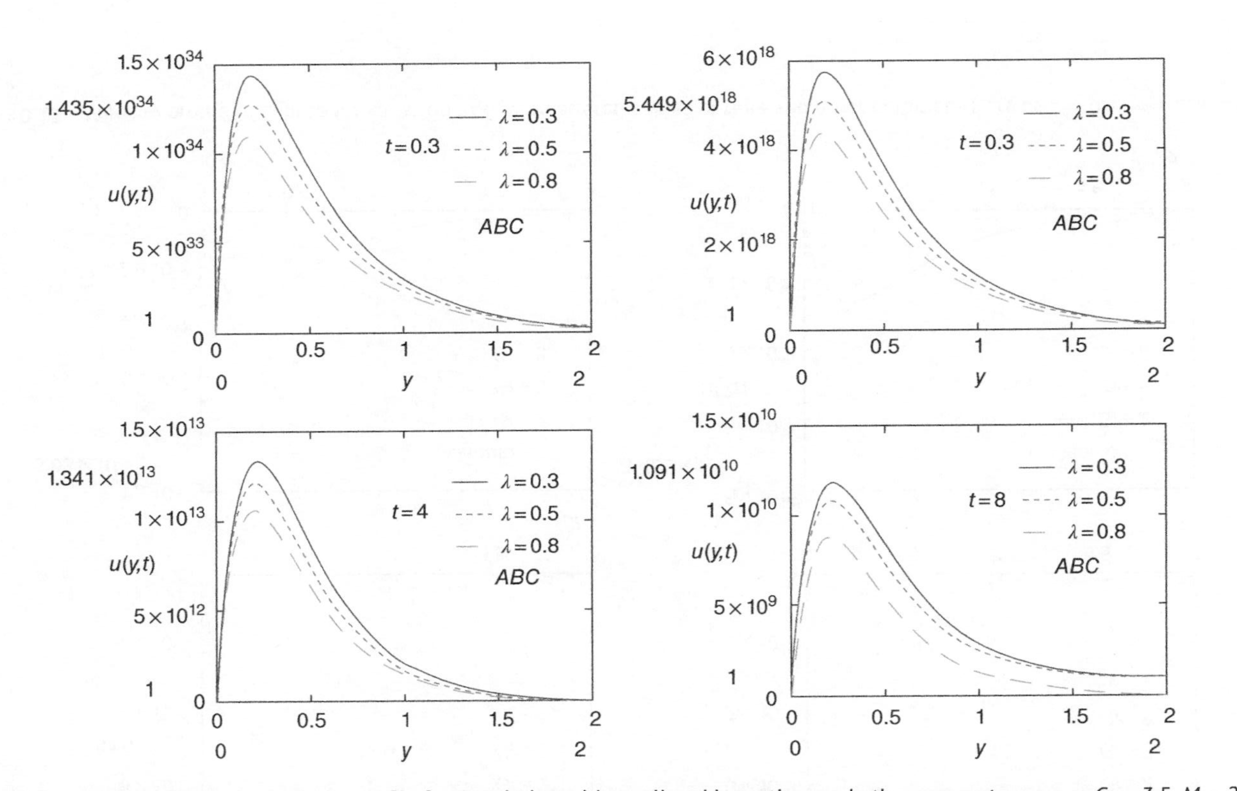

Figure 10.12 Atangana–Baleanu velocity profile for λ variation with small and large time and other parameters are as $G_r = 3.5$, $M = 2$, $\alpha = 0.5$, $P_r = 1.5$.

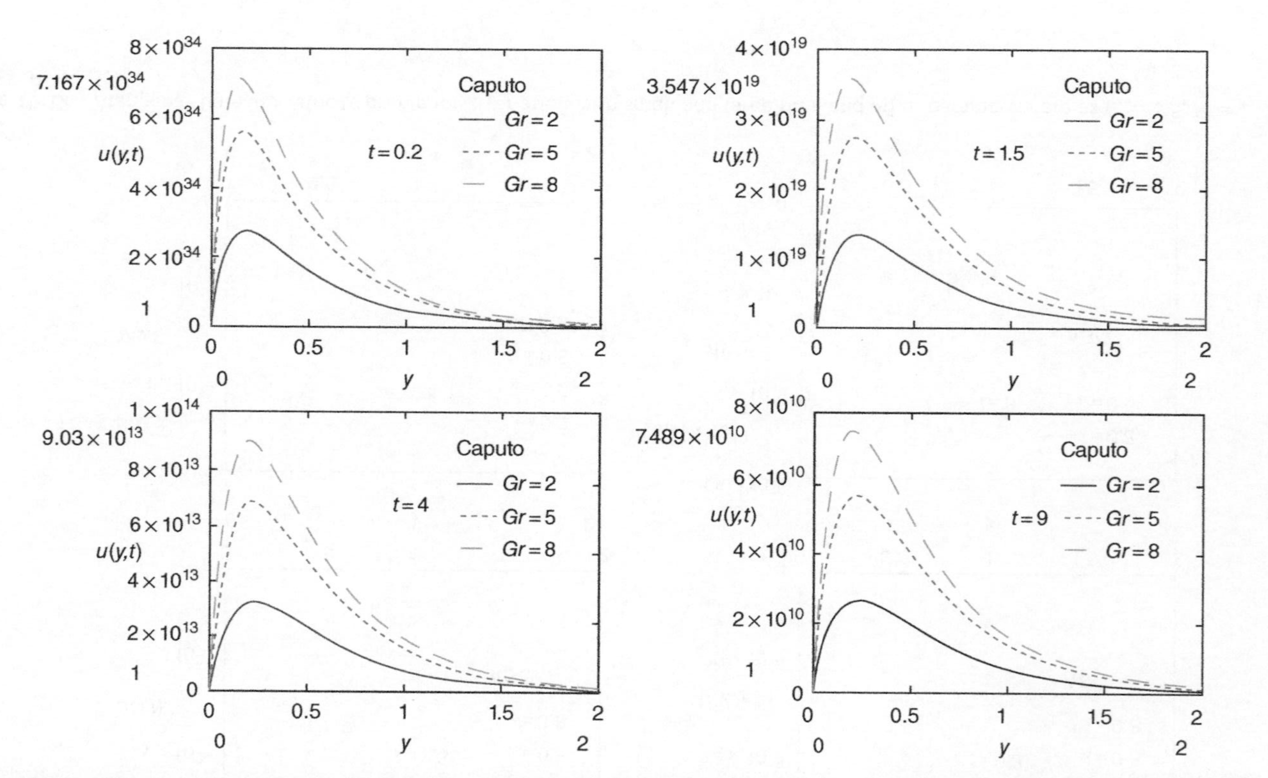

Figure 10.13 Velocity profile of Caputo for variation of G_r with altered values of time and other parameters are as $\lambda = 0.6$, $M = 2$, $\alpha = 0.5$, $P_r = 1.5$.

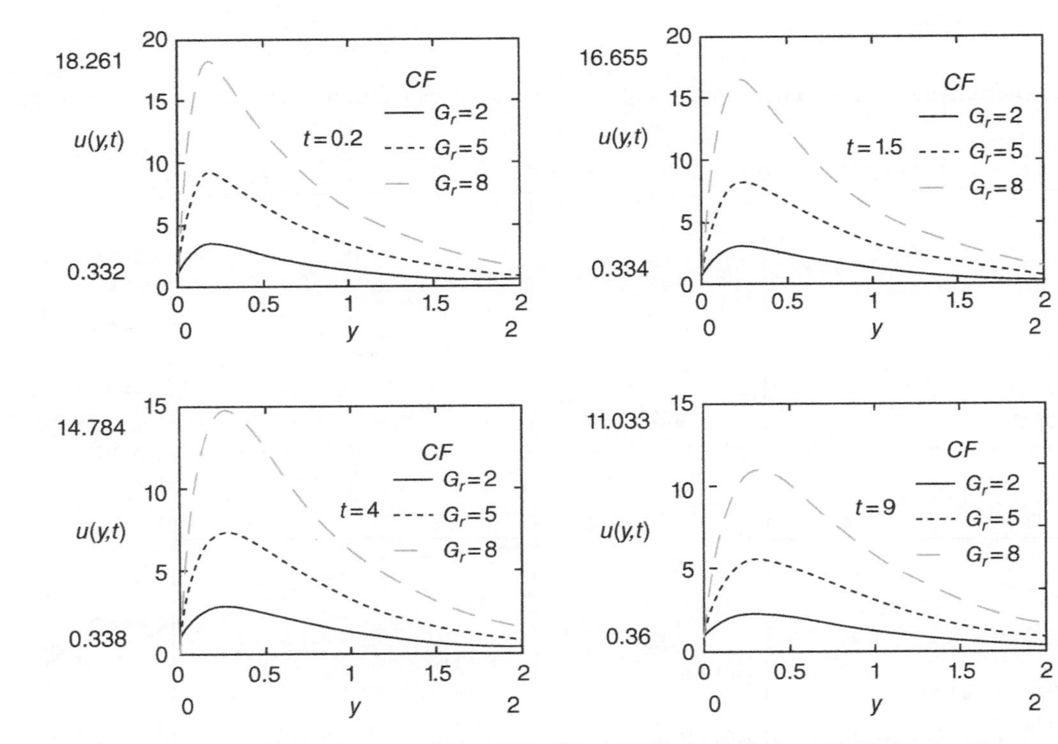

Figure 10.14 Velocity profile of Caputo–Fabrizio for variation of G_r with altered values of time and other parameters are as $\lambda = 0.6$, $M = 2$, $\alpha = 0.5$, $P_r = 1.5$.

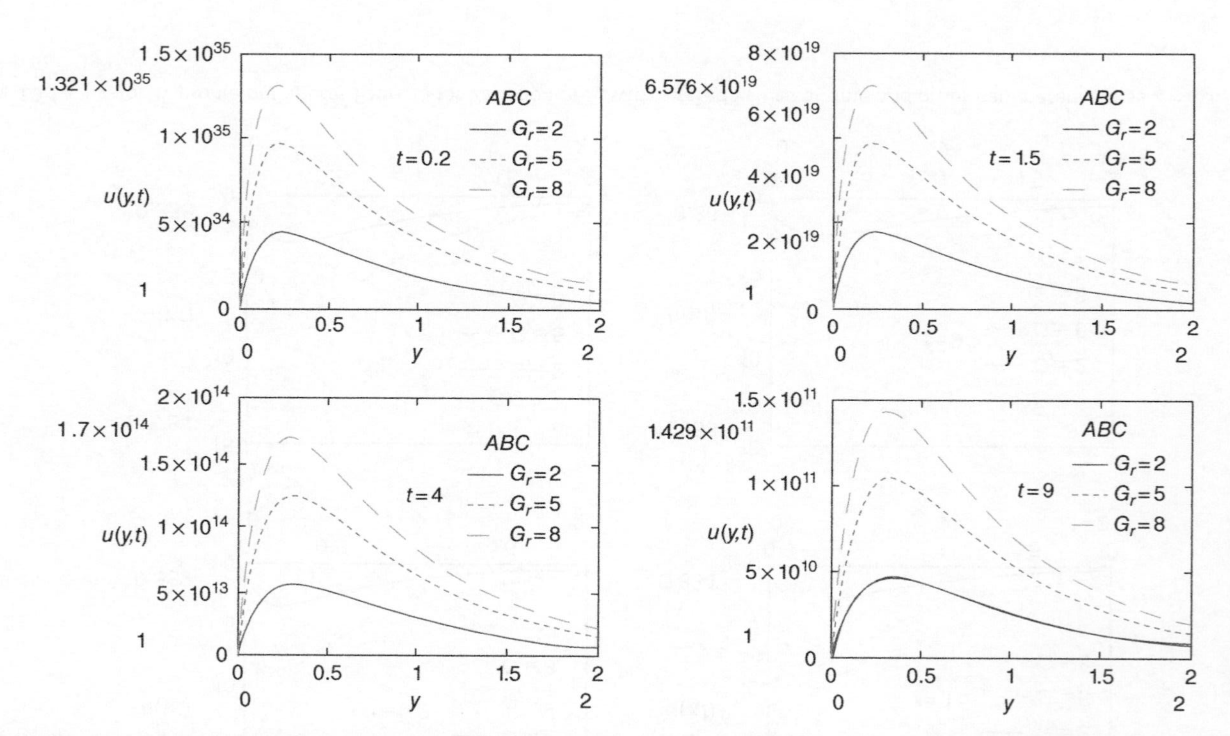

Figure 10.15 Velocity profile of Atangana–Baleanu for variation of G_r with altered values of time and other parameters are as $\lambda = 0.6$, $M = 2$, $\alpha = 0.5$, $P_r = 1.5$.

(*ABC*) for temperature and velocity field are discussed. The following useful key points extracted from this study.

- The increase in fractional parameter α, velocity decreases with variation of time for Caputo, Caputo–Fabrizio, and Atangana–Baleanu models.
- The value of Prandtl number P_r increases, the resultant velocity decreases for Caputo, Caputo–Fabrizio, and Atangana–Baleanu models.
- The velocity decrease by magnify the value of magnetic field.
- Velocity is decreasing function of time.
- The increase in velocity by increase in the value λ.
- Increase the value of Grashof number G_r, the velocity field increases.
- Atangana–Baleanu (*ABC*) is appropriate for all physical parameters then the other fractional derivatives.

Bibliography

1 Sheikholeslami, M., Hayat, T., and Alsaedi, A. (2016). MHD free convection of water nanofluid considering thermal radiation: a numerical study. *Int. J. Heat Mass Transfer* **96**: 513–524.

2 Sheikholeslami, M. and Rashidi, M.M. (2015). Effect of space dependent magnetic field on free convection of water nanofluid. *J. Taiwan Inst. Chem. Eng.* **56**: 615.

3 Sheikholeslami, M., Vajravelu, K., and Rashidi, M.M. (2016). Forced convection heat transfer in a semi annulus under the influence of a variable magnetic field. *Int. J. Heat Mass Transfer* **92**: 339.

4 Sheikholeslami, M., Gorji-Bandpy, M., Ganji, D.D. et al. (2014). Magnetohydrodynamic free convection of water nanofluid considering Thermophoresis and Brownian motion effects. *Comput. Fluids* **94**: 147160.

5 Sheikholeslami, M. and Ellahi, R. (2015). Three dimensional mesoscopic simulation of magnetic field effect on natural convection of nanofluid. *Int. J. Heat Mass Transfer* **89**: 799808.

6 Takashima, M. (1970). The effect of a magnetic field on thermal instability in a layer of Maxwell fluid. *Phys. Lett. A* **33** (6): 371372.

7 Maxwell, J.C. (1867). II. On the dynamical theory of gases. *Proc. R. Soc. London* **15**: 167171.

8 Friedrich, C.H.R. (1991). Relaxation and retardation functions of the Maxwell model with fractional derivatives. *Rheol. Acta* **30** (2): 151158.

9 Olsson, F. and Ystrom, J. (1993). Some properties of the upper convected Maxwell model for viscoelastic fluid flow. *J. Non-Newtonian Fluid Mech.* **48** (12): 125145.

10 Choi, J.J., Rusak, Z., and Tichy, J.A. (1999). Maxwell fluid suction flow in a channel. *J. Non-Newtonian Fluid Mech.* **85** (23): 165187.
11 Fetecau, C. and Fetecau, C. (2003). A new exact solution for the flow of a Maxwell fluid past an infinite plate. *Int. J. Non-Linear Mech.* **38** (3): 423427.
12 Fetecau, C. and Fetecau, C. (2003). The RayleighStokes-Problem for a fluid of Maxwellian type. *Int. J. Non-Linear Mech.* **38** (4): 603607.
13 Jordan, P.M., Puri, A., and Boros, G. (2004). On a new exact solution to Stokes first problem for Maxwell fluids. *Int. J. Non-Linear Mech.* **39** (8): 13711377.
14 Zierep, J. and Fetecau, C. (2007). Energetic balance for the Rayleigh Stokes problem of a Maxwell fluid. *Int. J. Eng. Sci.* **45** (28): 617627.
15 Fetecau, C., Jamil, M., Fetecau, C., and Siddique, I. (2009). A note on the second problem of Stokes for Maxwell fluids. *Int. J. Non-Linear Mech.* **44** (10): 10851090.
16 Motsa, S.S., Khan, Y., and Shateyi, S. (2012). A new numerical solution of Maxwell fluid over a shrinking sheet in the region of a stagnation point. *Math. Prob. Eng.* **2012**: 111.
17 Khan, I., Ali, F., and Shafie, S. (2013). Exact Solutions for Unsteady Magnetohydrodynamic oscillatory flow of a Maxwell fluid in a porous medium. *Z. Naturforsch., A* **68** (1011): 635645.
18 Abro, K.A., Shaikh, A.A., Junejo, I.A., and Chandio, M.S. (2015). Analytical solutions under no slip effects for accelerated flows of Maxwell fluids. *Sindh Univ. Res. J.-SURJ (Science Series)* **3**: 47.
19 Abro, K.A. and Shaikh, A.A. (2015). Exact analytical solutions for Maxwell fluid over an oscillating plane. *Sci. Int.(Lahore) ISSN* **27**: 923929.
20 Abro, K.A., Shaikh, A.A., and Dehraj, S. (2016). Exact solutions on the oscillating plate of Maxwell fluids. *Mehran Univ. Res. J. Eng. Technol.* **35** (1): 155.
21 Aman, S., Salleh, M.Z., Ismail, Z., and Khan, I. (2017). Exact solution for heat transfer free convection flow of Maxwell nanofluids with graphene nanoparticles. *J. Phys. Conf. Ser.* **890** (1): 012004.
22 Karra, S., Prusa, V., and Rajagopal, K.R. (2011). On Maxwell fluids with relaxation time and viscosity depending on the pressure. *Int. J. Non-Linear Mech.* **46** (6): 819827.
23 Riaz, M.B., Imran, M.A., and Shabbir, K. (2016). New exact solutions for the flow of generalized Maxwell fluid. *J. Comput. Theor. Nanosci.* **13** (8): 52545257.
24 Liu, Y. and Guo, B. (2017). Effects of second-order slip on the flow of a fractional Maxwell MHD fluid. *J. Assoc. Arab Univ. Basic Appl. Sci.* **24**: 232241.
25 Imran, M.A., Riaz, M.B., Shah, N.A., and Zafar, A.A. (2018). Boundary layer flow of MHD generalized Maxwell fluid over an exponentially accelerated infinite vertical surface with slip and Newtonian heating at the boundary. *Results Phys.* **8**: 10611067.

26 Raza, N. and Ullah, M.A. (2020). A comparative study of heat transfer analysis of fractional Maxwell fluid by using Caputo and CaputoFabrizio derivatives. *Can. J. Phys.* **98** (1): 89–101. https://doi.org/10.1139/cjp-2018-0602.

27 Singh, K.D. and Kumar, R. (2011). Fluctuating heat and mass transfer on unsteady MHD free convection flow of radiating and reacting fluid past a vertical porous plate in slip-flow regime. *J. Appl. Fluid Mech.* **4** (4): 101106.

28 Tahir, M., Imran, M.A., Raza, N. et al. (2017). Wall slip and non-integer order derivative effects on the heat transfer flow of Maxwell fluid over an oscillating vertical plate with new definition of fractional CaputoFabrizio derivatives. *Results Phys.* **7**: 18871898.

29 Ahmed, N. and Dutta, M. (2013). Transient mass transfer flow past an impulsively started infinite vertical plate with ramped plate velocity and ramped temperature. *Int. J. Phys. Sci.* **8** (7): 254263.

30 Ghara, N., Das, S., Maji, S.L., and Jana, R.N. (2012). Effect of radiation on MHD free convection flow past an impulsively moving vertical plate with ramped wall temperature. *Am. J. Sci. Ind. Res.* **3** (6): 376386.

31 Seth, G.S., Kumbhakar, B., and Sarkar, S. (2015). Soret and Hall effects on unsteady MHD free convection flow of radiating and chemically reactive fluid past a moving vertical plate with ramped temperature in rotating system. *Int. J. Eng. Sci. Technol.* **7** (2): 94108.

32 Reddy, B.P. (2014). Effects of thermal diffusion and viscous dissipation on unsteady MHD free convection flow past a vertical porous plate under oscillatory suction velocity with heat sink. *Int. J. Appl. Mech. Eng.* **19** (2): 303320.

33 Ahmad, S., Vieru, D., Khan, I., and Shafie, S. (2014). Unsteady magnetohydrodynamic free convection flow of a second grade fluid in a porous medium with ramped wall temperature. *PLoS One* **9** (5): e88766.

34 Narahari, M. and Debnath, L. (2013). Unsteady magnetohydrodynamic free convection flow past an accelerated vertical plate with constant heat flux and heat generation or absorption. *ZAMM-J. Appl. Math. Mech./Z. Angew. Math. Mech.* **93** (1): 3849.

35 Khalid, A., Khan, I., Khan, A., and Shafie, S. (2015). Unsteady MHD free convection flow of Casson fluid past over an oscillating vertical plate embedded in a porous medium. *Eng. Sci. Technol. Int. J.* **18** (3): 309317.

36 Sarkar, B.C., Das, S., and Jana, R.N. (2013). Hall effects on unsteady MHD free convective flow past an accelerated moving vertical plate with viscous and Joule dissipations. *Int. J. Comput. Appl.* **24**: 70.

37 Seth, G.S., Ansari, M.S., and Nandkeolyar, R. (2011). MHD natural convection flow with radiative heat transfer past an impulsively moving plate with ramped wall temperature. *Heat Mass Transfer* **47** (5): 551561.

38 Shah, N.A., Zafar, A.A., and Akhtar, S. (2018). General solution for MHD-free convection flow over a vertical plate with ramped wall temperature and chemical reaction. *Arabian J. Math.* **7** (1): 4960.

39 Iftikhar, N., Husnine, S.M., and Riaz, M.B. (2019). Heat and mass transfer in MHD Maxwell Fluid over an infinite vertical plate. *J. Prime Res. Math.* **15**: 63–80.

40 Tokis, J.N. (1985). A class of exact solutions of the unsteady magnetohydrodynamic free-convection flows. *Astrophys. Space Sci.* **112** (2): 413422.

11

Comparison of the Different Fractional Derivatives for the Dynamics of Zika Virus

Muhammad Altaf Khan

Faculty of Natural and Agricultural Sciences, Institute of Ground water studies, University of the Free State, Bloemfontein, South Africa

11.1 Introduction

The understanding of infectious diseases through mathematical modeling gains too much attentions day by day. It is proven that the disease dynamics can be best described through a mathematical model, and some recommendations can be made easily regarding the disease status, whether to control or it can spread. Mathematical modeling related to the Zika virus is very important in order to highlight their complications and the related information. The Zika virus is related to the other diseases such as the dengue, yellow fever, Japanese encephalitis, etc. The Zika virus belongs to the *Flaviviridae* family. It spreads same as the virus of the dengue, etc. that is mainly by the female *Aedes aegypti* mosquito. There also another route of transmission of the Zika virus is the *sexual one*, where the infected pregnant women can transfer the virus to her fetus. Many outbreaks and epidemics were observed for the Zika virus in the world with many infection cases [1–4].

Mathematical models that were published on the Zika virus are many in literature, see [5–8] and the references therein. For example, a Zika virus model with prevention and control is studied in [5]. The dynamics of Zika virus transmission through a mathematical model is discussed in [6]. The dynamics of Zika virus in Brazil through mathematical formulation is studied in [7]. The rapid spread of the Zika virus in the pacific area are studied in [8]. More mathematical models associated to the Zika virus are studied in [9–14]. The above-described mathematical models are limited to investigate the Zika dynamics in integer-order models, which sometimes difficult to capture their dynamics well.

Fractional-order modeling gaining much attention day by day due to their applications and applicability to real-life problems. It is well known that the fractional order models are more better than that of the integer-order models, due to the

Fractional Order Analysis: Theory, Methods and Applications, First Edition.
Edited by Hemen Dutta, Ahmet Ocak Akdemir, and Abdon Atangana.

heredity and memory properties. The dynamics of a mathematical model for any arbitrary derivative and their suitability to some real data is also best described through fractional modeling. In literature, a variety of fractional-order modeling of infectious diseases and their application with validity are considered see [15–25], where different fractional operators are used to obtained the results, such as the Caputo, *CF*, and *AB*. The *CF* derivative was introduced in order to overcome the limitations of the singularities, and the *AB* derivative was introduced to overcome the deficiency of singular and nonsingular kernel. The authors in [15] recently studied the comparison of the *CF* and *AB* derivative for the dengue model and suggested that the *AB* derivative results are flexible and provide better behavior. The oxygen diffusion equation with different fractional operators is considered by the authors . The competition model for the comparison of bank data in Caputo and *AB* operators is formulated and analyzed in [16], and the appropriateness of *AB* operator for the real data is suggested. The dynamics of *TB* through *AB* derivative with real data is discussed in [17]. More recently, a Zika model with asymptomatic carriers was modeled through *AB* derivative in [26].

By the above interesting studies about the fractional modeling, we aim to consider a mathematical model that is used for the Zika virus with mutation in [27]. Here, in this study, we consider the published model [27] in two different fractional operators that is the Caputo–Fabrizio (*CF*) and the Atangana–Baleanu (*AB*) and investigate their dynamical analysis. Initially, we provide some background material of the fractional operators and then give the model details. The *CF* and *AB* model is formulated with detailed analysis, and finally the results are obtained with numerical approaches with comparisons.

11.2 Background of Fractional Operators

We provide here some background material of the fractional calculus that will be utilized in fractional modeling of *CF* and *AB* derivative for Zika model [28–30].

Definition 11.1 Suppose $w \in H^1(p,q)$, with $q > p$, and $0 \leq \theta \leq 1$, then the definition of Caputo–Fabrizio derivative is as follows:

$$D_t^{\theta}(w(t)) = \frac{Q(\theta)}{1-\theta}\int_a^t w'(y)\exp\left[-\theta\frac{t-y}{1-\theta}\right]dy, \tag{11.1}$$

where $Q(\theta)$ denotes the normalized function and holds $Q(0) = Q(1) = 1$. If $w \notin H^1(p,q)$s then the following is suggested as:

$$D_t^{\theta}(w(t)) = \frac{\theta\, Q(\theta)}{1-\theta}\int_a^t (w(t)-w(y))\exp\left[-\theta\frac{t-y}{1-\theta}\right]dy. \tag{11.2}$$

Remark 11.1 Let $\nu = \frac{1-\theta}{\theta} \in [0, \infty)$, $\theta = \frac{1}{1+\nu} \in [0, 1]$, then equation given by (11.2) can be expressed as follows:

$$D_t^\nu(w(t)) = \frac{Q(\nu)}{\nu} \int_a^t w'(y) \exp\left[-\frac{t-\theta}{\nu}\right] dy, \quad Q(0) = Q(\infty) = 1. \tag{11.3}$$

Furthermore,

$$\lim_{\nu \to 0} \frac{1}{\nu} \exp\left[-\frac{t-y}{\nu}\right] = \varphi(y-t). \tag{11.4}$$

Definition 11.2 Consider $\theta \in (0, 1)$ for a function $w(y)$, then we can write the integral of fractional-order θ is as follows:

$$I_t^\theta(w(t)) = \frac{2(1-\theta)}{(2-\theta)Q(\theta)} g(t) + \frac{2\theta}{(2-\theta)Q(\theta)} \int_0^t w(y)dy, t \geq 0. \tag{11.5}$$

Remark 11.2 In Eq. (11.2), the remainder of the Caputo-type noninteger – order integral of the function with order $\theta \in (0, 1)$ is a mean into w with integral of order 1. Thus, it requires

$$\frac{2}{2Q(\theta) - \theta Q(\theta)} = 1, \tag{11.6}$$

which implies that $Q(\theta) = \frac{2}{2-\theta}$, $\theta \in (0, 1)$. Based on Eq. (11.6), a new Caputo derivative is suggested with $\theta \in (0, 1)$ and is given by

$$D_t^\theta(w(t)) = \frac{1}{1-\theta} \int_0^t w'(y) \exp\left[-\theta \frac{t-y}{1-\theta}\right] dy. \tag{11.7}$$

Definition 11.3 Consider $w \in H^1(p, q)$, where q greater than p, and $0 \leq \theta \leq 1$, then we define the Atangana–Baleanu derivative in the following:

$${}_a^{ABC}D_t^\theta w(t) = \frac{Q(\theta)}{1-\theta} \int_a^t w'(y) E_\theta \left[-\theta \frac{(t-y)^\theta}{1-\theta}\right] dy. \tag{11.8}$$

Definition 11.4 The fractional integral for the Atangana–Baleanu derivative is expressed as follows:

$${}_a^{ABC}I_t^\theta w(t) = \frac{1-\theta}{Q(\theta)} w(t) + \frac{\theta}{Q(\theta)\Gamma(\theta)} \int_a^t f(y)(t-y)^{\theta-1} dy. \tag{11.9}$$

One can restore the original function for the case when $\theta = 0$.

11.3 Model Framework

The present section aims to formulate the dynamics of Zika virus with mutation. Initially, we divide the total population of human denoted by $N_h(t)$ into three epidemiologically meaningful compartments that is the susceptible human that attract the Zika virus, $S_h(t)$, individuals infected with Zika virus, $I_h(t)$, and the individuals recovered from Zika virus, $R_h(t)$; so, $N_h(t) = S_h(t) + I_h(t) + R_h(t)$. Similarly, the total population of vector Zika is denoted by $N_v(t) = S_m(t) + I_m(t)$, where $S_m(t)$ denotes discerptible mosquitoes and $I_m(t)$ is the infected mosquitoes due to Zika virus at any time t. Using these assumptions, the evolutionary dynamics of Zika virus is presented through the following system of differential equations:

$$\begin{aligned}
\frac{dS_h}{dt} &= \Lambda_h(S_h(t) + R_h(t)) + \phi R_h(t) - d_1 S_h(t) - \frac{\beta_h S_h(t) I_m(t)}{N_h(t)}, \\
\frac{dI_h}{dt} &= \frac{\beta_h S_h(t) I_m(t)}{N_h(t)} - (d_1 + \delta_h) I_h(t) + \Lambda_h I_h(t), \\
\frac{dR_h}{dt} &= \delta_h I_h(t) - (\phi + d_1) R_h(t), \\
\frac{dS_m}{dt} &= \Lambda_m(S_m(t) + I_m(t)) - d_2 S_m(t) - \frac{\beta_m S_m(t) I_h(t)}{N_h(t)}, \\
\frac{dI_m}{dt} &= \frac{\beta_m S_m(t) I_h(t)}{N_h(t)} - d_2 I_m(t),
\end{aligned} \tag{11.10}$$

with non-negative initial conditions. The parameter Λ_h represents the birth rate of human individuals and is consider here equal to the natural death rate of human individuals d_1. Susceptible humans are infected by Zika virus and generate infection through the rate β_h while the infected human interaction with susceptible mosquitoes is shown by β_m. The birth rate of the mosquitos is given by Λ_m whereas the natural death rate of the mosquitos is shown by d_m. The parameter ϕ shows the susceptibility of the human infected individuals with Zika virus while the Zika individuals infected the Zika virus are become infectious at a rate of δ_h. A fully vertical transmission in human population is considered. The inflow due to the vertical transmission in infected class is shown by $\Lambda_h I_h$. Furthermore, considering the birth and death rates are equal, we have the following reduced model:

$$\frac{dS_h}{dt} = R_h(\Lambda_H + \phi) - \frac{\beta_h S_h(t) I_m(t)}{N_h(t)},$$

$$\frac{dI_h}{dt} = \frac{\beta_h S_h(t) I_m(t)}{N_h(t)} - \delta_h I_h(t),$$

$$\frac{dR_h}{dt} = \delta_h I_h(t) - (\phi + d_1) R_h(t),$$

$$\frac{dS_m}{dt} = \Lambda_m I_m(t) - \frac{\beta_m S_m(t) I_h(t)}{N_h(t)},$$

$$\frac{dI_m}{dt} = \frac{\beta_m S_m(t) I_h(t)}{N_h(t)} - d_2 I_m(t). \tag{11.11}$$

Further making the following transformation,

$$S_H = \frac{S_h(t)}{N_h(t)}, \quad I_H = \frac{I_h(t)}{N_h(t)}, \quad R_H = \frac{R_h(t)}{N_h(t)}, \quad S_M = \frac{S_m(t)}{N_m(t)}, \quad I_M = \frac{I_m(t)}{N_m(t)}, \tag{11.12}$$

and following the relations $R_H = 1 - S_H - I_H$ and $S_M = 1 - I_M$, we finally have the following model:

$$\frac{dS_H}{dt} = (\Lambda_h + \phi)(1 - S_H(t) - I_H(t)) - \beta_h S_H(t) I_M(t),$$

$$\frac{dI_H}{dt} = \beta_h S_H(t) I_M(t) - \delta_h I_h(t),$$

$$\frac{dI_m}{dt} = \beta_m I_H(t)(1 - I_M(t)) - d_2 I_M(t), \tag{11.13}$$

where $\beta_h = m\beta_h' = (N_h/N_h)\beta_h'$. The biological feasible region for the given model is

$$\Pi = \left\{(S_H, I_H, I_M) | 0 \leq S_H + I_H \leq 1, 0 \leq I_M \leq 1\right\}, \tag{11.14}$$

which is positive for the model (11.13). The disease-free equilibrium of the model (11.10) is $E_1 = (1, 0, 0)$. The basic reproduction number for the Zika model is given by

$$\mathcal{R}_0 = \sqrt{\frac{\beta_h \beta_m}{\delta_h d_2}}.$$

11.4 A Fractional Zika Model with Different Fractional Derivatives

Here in this section, we present two different fractional derivatives that is the Caputo–Fabrizio derivative and the Atangana–Baleanu derivative. We apply these two derivatives on the Zika model described by the system (11.13) as follows:

$$
\begin{aligned}
{}_0^{CF}D_t^{\theta} S_H &= (\Lambda_h + \phi)(1 - S_H(t) - I_H(t)) - \beta_h S_H(t) I_M(t),\\
{}_0^{CF}D_t^{\theta} I_H &= \beta_h S_H(t) I_M(t) - \delta_h I_h(t),\\
{}_0^{CF}D_t^{\theta} I_M &= \beta_m I_H(t)(1 - I_M(t)) - d_2 I_M(t),
\end{aligned}
\tag{11.15}
$$

where the fractional order is shown by θ. The model with Atangana–Baleanu derivative can be written by following the definition described above is follows as:

$$
\begin{aligned}
{}_0^{ABC}D_t^{\theta} S_H &= (\Lambda_h + \phi)(1 - S_H(t) - I_H(t)) - \beta_h S_H(t) I_M(t),\\
{}_0^{ABC}D_t^{\theta} I_H &= \beta_h S_H(t) I_M(t) - \delta_h I_h(t),\\
{}_0^{ABC}D_t^{\theta} I_M &= \beta_m I_H(t)(1 - I_M(t)) - d_2 I_M(t).
\end{aligned}
\tag{11.16}
$$

11.5 Numerical Scheme for Caputo–Fabrizio Model

Here we present the numerical solution for the Caputo-Fabrizio model equation (11.15) by using the scheme presented [31]. The following steps are taken as specified in [31] for the solution of our Eq. (11.15).

$$
S_H(t) - S_H(0) = \frac{(1-\theta)}{Q(\theta)} \mathcal{J}_1(t, S_H) + \frac{\theta}{Q(\theta)} \int_0^t \mathcal{J}_1(\zeta, S_H) d\zeta. \tag{11.17}
$$

For $t = t_{n+1}$, $n = 0, 1, 2, \dots$, we obtain

$$
S_H(t_{n+1}) - S_{H0} = \frac{1-\theta}{Q(\theta)} \mathcal{J}_1(t_n, S_{Hn}) + \frac{\theta}{Q(\theta)} \int_0^{t_{n+1}} \mathcal{J}_1(t, S_H) dt. \tag{11.18}
$$

The successive terms difference is given as follows:

$$
S_{Hn+1} - S_{Hn} = \frac{1-\theta}{Q(\theta)} \{\mathcal{J}_1(t_n, S_{Hn}) - \mathcal{J}_1(t_{n-1}, S_{Hn-1})\} + \frac{\theta}{Q(\theta)} \int_{t_n}^{t_{n+1}} \mathcal{J}_1(t, S_H) dt. \tag{11.19}
$$

Over the close interval $[t_k, t_{(k+1)}]$, the function $\mathcal{J}_1(t, S_H)$ can be approximated by the interpolation polynomial

$$
P_k(t) \cong \frac{f(t_k, y_k)}{h}(t - t_{k-1}) - \frac{f(t_{k-1}, y_{k-1}).}{h}(t - t_k), \tag{11.20}
$$

where $h = t_n - t_{n-1}$. Calculating the integral in (11.19) using above polynomial approximation, we get

$$
\begin{aligned}
\int_{t_n}^{t_{n+1}} \mathcal{J}_1(t, S_H) dt &= \int_{t_n}^{t_{n+1}} \frac{\mathcal{J}_1(t_n, S_{Hn})}{h}(t - t_{n-1}) - \frac{\mathcal{J}_1(t_{n-1}, S_{Hn-1})}{h}(t - t_n) dt\\
&= \frac{3h}{2} \mathcal{J}_1(t_n, S_{Hn}) - \frac{h}{2} \mathcal{J}_1(t_{n-1}, S_{Hn-1}).
\end{aligned}
\tag{11.21}
$$

Putting (11.21) in (11.19) and after simplification, we obtained

$$S_{Hn+1} = S_{Hn} + \left(\frac{1-\theta}{Q(\theta)} + \frac{3h}{2Q(\theta)} \right) \mathcal{J}_1(t_n, S_{Hn}) - \left(\frac{1-\theta}{Q(\theta)} + \frac{\theta h}{2Q(\theta)} \right) \times \mathcal{J}_1(t_{n-1}, S_{Hn-1}). \tag{11.22}$$

In similar way for the rest of equations of system (11.15), we obtained the recursive formula as below

$$\begin{aligned} I_{Hn+1} &= I_{H0} + \left(\frac{1-\theta}{Q(\theta)} + \frac{3h}{2Q(\theta)} \right) \mathcal{J}_2(t_n, I_{Hn}) \\ &\quad - \left(\frac{1-\theta}{Q(\theta)} + \frac{\theta h}{2Q(\theta)} \right) \mathcal{J}_2(t_{n-1}, I_{Hn-1}), \\ I_{Mn+1} &= I_{M0} + \left(\frac{1-\theta}{Q(\theta)} + \frac{3h}{2Q(\theta)} \right) \mathcal{J}_3(t_n, I_{Mn}) \\ &\quad - \left(\frac{1-\theta}{Q(\theta)} + \frac{\theta h}{2Q(\theta)} \right) \mathcal{J}_3(t_{n-1}, I_{Mn-1})). \end{aligned} \tag{11.23}$$

11.5.1 Solutions Existence for the Atangana–Baleanu Model

The model presented given by system (11.16) shows the dynamics of Zika model which is composed in nonlinear differential equation for which their exact solution is not possible, therefore, the existence of their solution is important if we show with some conditions the solution existence of Zika model (11.16). We give the following model for the implication purpose and then apply the procedure to the given model (11.16):

$$\begin{cases} {}_0^{ABC}D_t^{\theta} y(t) = \mathcal{M}(t, y(t)), \\ y(0) = y_0, \ 0 < t < T < \infty, \end{cases} \tag{11.24}$$

where $y(t) = (S_H, I_H, I_M)$ shows the vector with state variables S_H, I_H, I_M and is a continuous vector function and can be defined as follows:

$$\mathcal{M} = \begin{pmatrix} \mathcal{M}_1 \\ \mathcal{M}_2 \\ \mathcal{M}_3 \end{pmatrix} = \begin{pmatrix} (\Lambda_h + \phi)(1 - S_H(t) - I_H(t)) - \beta_h S_H(t) I_M(t) \\ \beta_h S_H(t) I_M(t) - \delta_h I_h(t) \\ \beta_m I_H(t)(1 - I_M(t)) - d_2 I_M(t) \end{pmatrix}. \tag{11.25}$$

It can shown easily that the function $\mathcal{M}$ will satisfy the Lipschitz condition as shown below:

$$\| \mathcal{M}(t, y_1(t)) - \mathcal{M}(t, y_2(t)) \| \leq \mathcal{P} \| y_1(t) - y_2(t) \|. \tag{11.26}$$

Therefore, we have the existence and uniqueness results for the Zika model (11.16). The following result is presented as:

Theorem 11.1 *The Zika model in Atangana–Baleanu form (11.16) may have a unique solution with some conditions, if the following satisfy,*

$$\frac{(1-\theta)}{ABC(\theta)}\mathcal{P}+\theta\frac{T^{\theta}_{max}}{ABC(\theta)\Gamma(\theta)}\mathcal{P}<1. \tag{11.27}$$

Proof: The application of Atangana–Baleanu fractional integration on both sides of Eq. (11.24), we have

$$y(t)=y_0+\frac{1-\theta}{ABC(\theta)}\mathcal{M}(t,y(t))+\frac{\theta}{ABC(\theta)\Gamma(\theta)}\int_0^t(t-\eta)^{\theta-1}\mathcal{M}(\eta,x(\eta))d\eta. \tag{11.28}$$

Suppose $J=(0,T)$ and the operator $\Pi : C(J,R^3)\to C(J,R^3)$ defined by

$$\Pi[y(t)]=y_0+\frac{1-\theta}{ABC(\theta)}\mathcal{M}(t,y(t))+\frac{\theta}{ABC(\theta)\Gamma(\theta)}\int_0^t(t-\eta)^{\theta-1}\mathcal{M}(\eta,x(\eta))d\eta. \tag{11.29}$$

Then, we write Eq. (11.28) as follows:

$$y(t)=\Pi[y(t)]. \tag{11.30}$$

We have after applying the supremum norm on J,

$$\|y(t)\|_J=\sup_{t\in J}\|y(t)\|,\quad y(t)\in C. \tag{11.31}$$

Obviously, $C(J,R^3)$ and the norm $\|.\|_J$ is a Banach space. Using the operator (11.30), the following is presented as:

$$\|\Pi[y_1(t)]-\Pi[y_2(t)]\|_J\leq\left\|\frac{(1-\theta)}{ABC(\theta)}(\mathcal{M}(t,y_1(t))-\mathcal{M}(t,y_2(t))+\frac{\theta}{ABC(\theta)\Gamma(\theta)}\right.$$
$$\left.\times\int_0^t(t-\eta)^{\theta-1}(\mathcal{M}(\eta,y_1(\eta))-\mathcal{F}(\eta,y_2(\eta)))d\eta\right\|_J. \tag{11.32}$$

Using triangular inequality and Lipschitz condition presented in (11.26), we have

$$\|\Pi[y_1(t)]-\Pi[y_2(t)]\|_J\leq\left(\frac{(1-\theta)\mathcal{P}}{ABC(\theta)}+\frac{\theta}{ABC(\theta)\Gamma(\theta)}\mathcal{P}T^{\theta}\right)\|y_1(t)-y_2(t)\|_J. \tag{11.33}$$

So, we get

$$\|\Pi[y_1(t)]-\Pi[y_2(t)]\|_J\leq L\|y_1(t)-y_2(t)\|_J, \tag{11.34}$$

where

$$L = \frac{(1-\theta)\mathcal{P}}{ABC(\theta)} + \frac{\theta}{ABC(\theta)\Gamma(\theta)}\mathcal{P}/T^{\alpha}.$$

The operator Π will be contraction if the condition (11.27) holds. Thus, the Banach fixed point theorem ensures that a unique solution for the Zika model in Atangana–Baleanu form (11.16) exists. □

11.5.2 Numerical Scheme for Atangana–Baleanu Model

Here we give a numerical procedure for the solution of fractional Zika model (11.16). The procedure shown [32] is used here for our proposed model (11.16) by writing the model (11.16) after using the fundamental theorem of fractional calculus:

$$y(t) - y(0) = \frac{(1-\theta)}{ABC(\theta)}\mathcal{J}(t, y(t)) + \frac{\theta}{ABC(\theta)\times\Gamma(\theta)}\int_0^t \mathcal{J}(\phi, x(\phi))(t-\phi)^{\theta-1}d\phi. \tag{11.35}$$

At $t = t_{n+1}$, $n = 0, 1, 2, \dots$, we have

$$\begin{aligned} y(t_{n+1}) - y(0) &= \frac{1-\theta}{ABC(\theta)}\mathcal{J}(t_n, y(t_n)) \\ &\quad + \frac{\theta}{ABC(\theta)\times\Gamma(\theta)}\int_0^{t_{n+1}} \mathcal{J}(\phi, y(\phi))(t_{n+1}-\phi)^{\theta-1}d\phi, \\ &= \frac{1-\theta}{ABC(\theta)}\mathcal{J}(t_n, y(t_n)) \\ &\quad + \frac{\theta}{ABC(\theta)\times\Gamma(\theta)}\sum_{j=0}^{n}\int_{t_j}^{t_{j+1}} \mathcal{J}(\phi, y(\phi))(t_{n+1}-\phi)^{\theta-1}d\phi. \end{aligned} \tag{11.36}$$

The function $\mathcal{J}(\phi, y(\phi))$ can be approximated over $[t_j, t_{j+1}]$, using the interpolation polynomial

$$\mathcal{J}(\phi, y(\phi)) \cong \frac{\mathcal{J}(t_j, y(t_j))}{h}(t - t_{j-1}) - \frac{\mathcal{J}(t_{j-1}, y(t_{j-1}))}{h}(t - t_j). \tag{11.37}$$

Putting in Eq. (11.36), we get

$$\begin{aligned} y(t_{n+1}) &= y(0) + \frac{1-\theta}{ABC(\theta)}\mathcal{J}(t_n, y(t_n)) \\ &\quad + \frac{\theta}{ABC(\theta)\times\Gamma(\theta)}\sum_{j=0}^{n}\left(\frac{\mathcal{J}(t_j, y(t_j))}{h}\int_{t_j}^{t_{j+1}}(t-t_{j-1})(t_{n+1}-t)^{\theta-1}dt\right. \\ &\quad \left. - \frac{\mathcal{J}(t_{j-1}, y(t_{j-1}))}{h}\int_{t_j}^{t_{j+1}}(t-t_j)(t_{n+1}-t)^{\theta-1}dt\right). \end{aligned} \tag{11.38}$$

Simplification leads to the following:

$$y(t_{n+1}) = y(t_0) + \frac{1-\theta}{ABC(\theta)}\mathcal{J}(t_n, y(t_n)) + \frac{\theta}{ABC(\theta)}\sum_{j=0}^{n}$$
$$\times\left(\frac{h^\theta \mathcal{J}(t_j, y(t_j))}{\Gamma(\theta+2)}((n+1-j)^\theta(n-j+2+\theta) - (n-j)^\theta(n-j+2+2\theta))\right.$$
$$\left. - \frac{h^\theta \mathcal{J}(t_{j-1}, y(t_{j-1}))}{\Gamma(\theta+2)}((n+1-j)^{\theta+1} - (n-j)^\theta(n-j+1+\theta))\right). \quad (11.39)$$

We have the results for the Zika model:

$$S_H(t_{n+1}) = S_H(t_0) + \frac{1-\theta}{ABC(\theta)}\mathcal{J}_1(t_n, S_H(t_n)) + \frac{\theta}{ABC(\theta)}\sum_{j=0}^{n}$$
$$\times\left(\frac{h^\theta \mathcal{J}_1(t_j, S_H(t_j))}{\Gamma(\theta+2)}((n+1-j)^\theta(n-j+2+\theta)\right.$$
$$-(n-j)^\theta(n-j+2+2\theta))$$
$$\left. - \frac{h^\theta \mathcal{J}_1(t_{j-1}, S_H(t_{j-1}))}{\Gamma(\theta+2)}((n+1-j)^{\theta+1} - (n-j)^\theta(n-j+1+\theta))\right),$$

$$I_H(t_{n+1}) = I_H(t_0) + \frac{1-\theta}{ABC(\theta)}\mathcal{J}_2(t_n, I_H(t_n)) + \frac{\theta}{ABC(\theta)}\sum_{j=0}^{n}$$
$$\times\left(\frac{h^\theta \mathcal{J}_2(t_j, I_H(t_j))}{\Gamma(\theta+2)}((n+1-j)^\theta(n-j+2+\theta)\right.$$
$$-(n-j)^\theta(n-j+2+2\theta))$$
$$\left. - \frac{h^\theta \mathcal{J}_2(t_{j-1}, I_H(t_{j-1}))}{\Gamma(\theta+2)}((n+1-j)^{\theta+1} - (n-j)^\theta(n-j+1+\theta))\right),$$

$$I_M(t_{n+1}) = I_M(t_0) + \frac{1-\theta}{ABC(\theta)}\mathcal{J}_3(t_n, I_M(t_n)) + \frac{\theta}{ABC(\theta)}\sum_{j=0}^{n}$$
$$\times\left(\frac{h^\theta \mathcal{J}_3(t_j, I_M(t_j))}{\Gamma(\theta+2)}((n+1-j)^\theta(n-j+2+\theta)\right.$$
$$-(n-j)^\theta(n-j+2+2\theta))$$
$$\left. - \frac{h^\theta \mathcal{J}_3(t_{j-1}, I_M(t_{j-1}))}{\Gamma(\theta+2)}((n+1-j)^{\theta+1} - (n-j)^\theta(n-j+1+\theta))\right). \quad (11.40)$$

11.6 Numerical Results

Here we give a brief details of the numerical results obtained as a figures for the scheme presented for the Caputo–Fabrizio model (11.15) and the Atangana–Baleanu model (11.16). We have taken the unit for time in days up to 100. The parameters values considered in this simulations are $\Lambda_h = 0.04$, $\phi = 0.002$, $\theta_h = 0.001$, $\delta_h = 0.002$, $\theta_m = 0.1250$, and $d_2 = 0.007\,139\,37$. We consider different values of the fractional-order parameter θ and obtained graphical results for the Caputo–Fabrizio model (see Figures 11.5–11.8), Atangana–Baleanu model (see Figures 11.1–11.4), and their comparison see Figures (11.9 and 11.10). From these graphical results, we observe the Atangana–Baleanu derivative give rapid increase/decrease for the decreasing of the fractional-order parameter (see Figures 11.11 and 11.12).

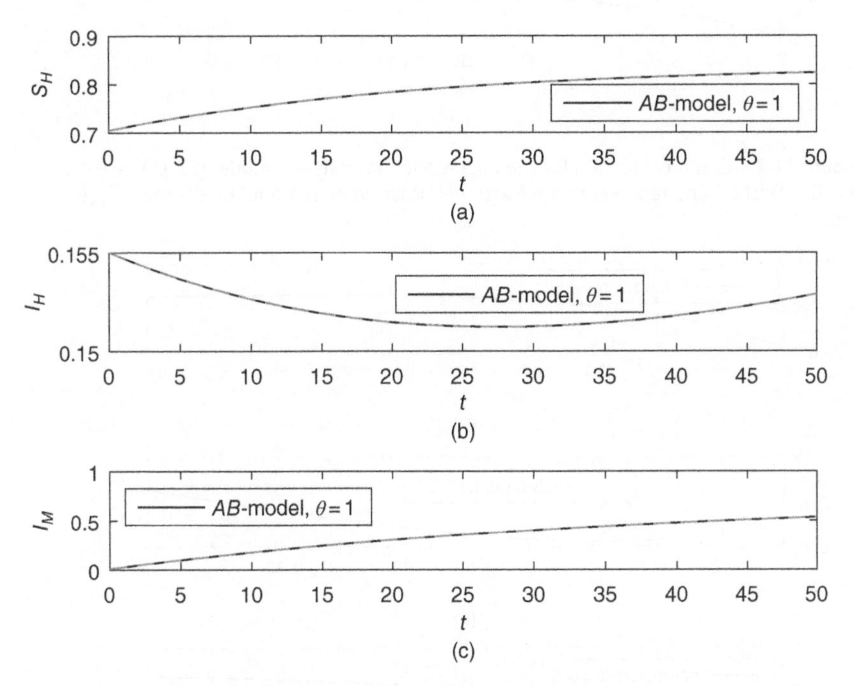

Figure 11.1 Graphical results for the Zika Atangana–Baleanu model (11.16), when $\theta = 1$. Dashed line represents the fractional order while the bold line is the integer order.

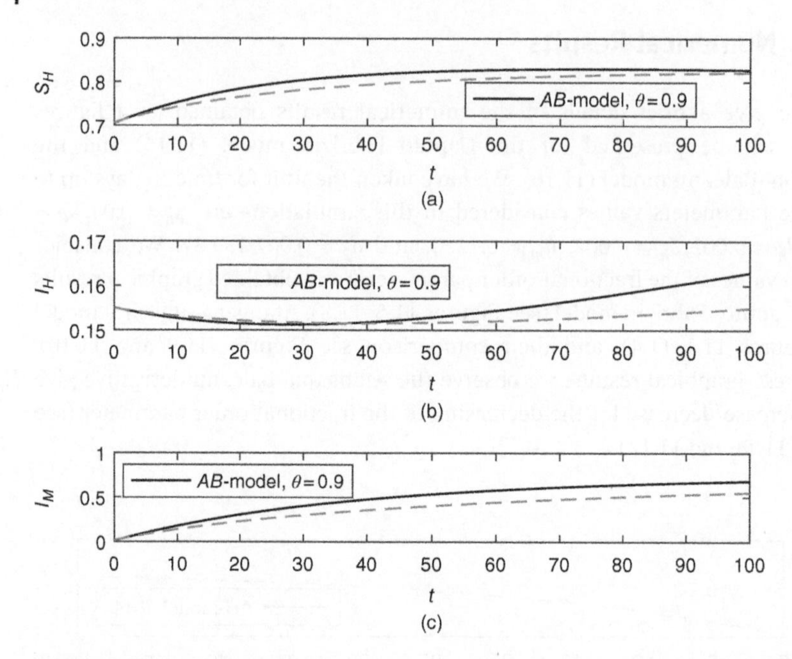

Figure 11.2 Graphical results for the Zika Atangana–Baleanu model (11.16), when $\theta = 0.9$. Dashed line represents the fractional order while the bold line is the integer order.

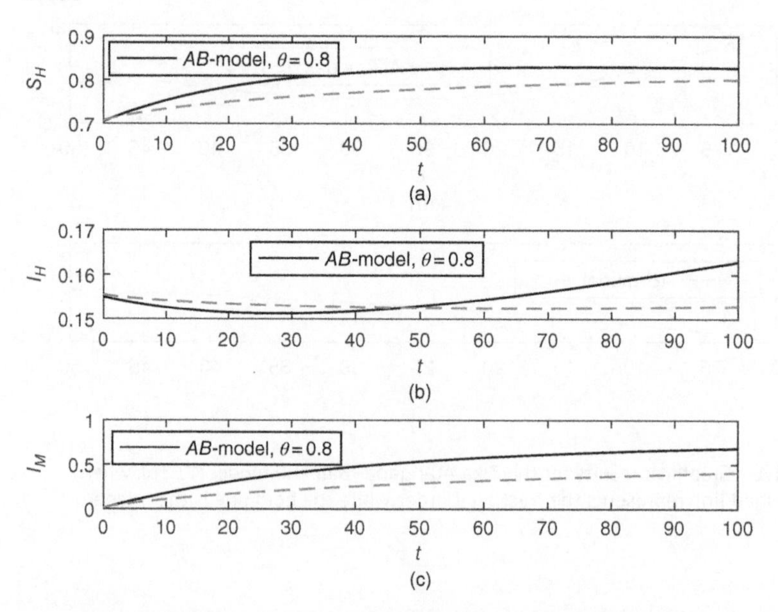

Figure 11.3 Graphical results for the Zika Atangana–Baleanu model (11.16), when $\theta = 0.8$. Dashed line represents the fractional order while the bold line is the integer order.

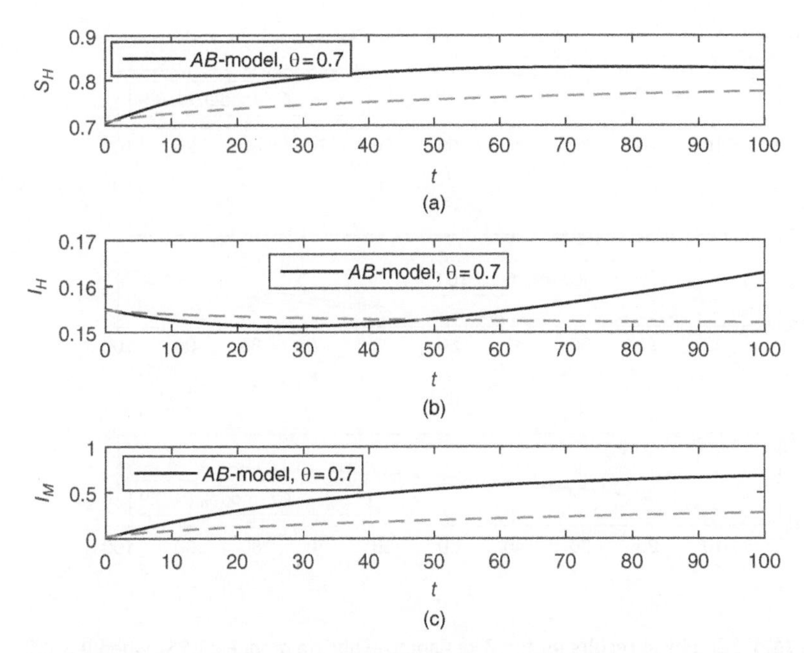

Figure 11.4 Graphical results for the Zika Atangana–Baleanu model (11.16), when $\theta = 0.7$. Dashed line represents the fractional order while the bold line is the integer order.

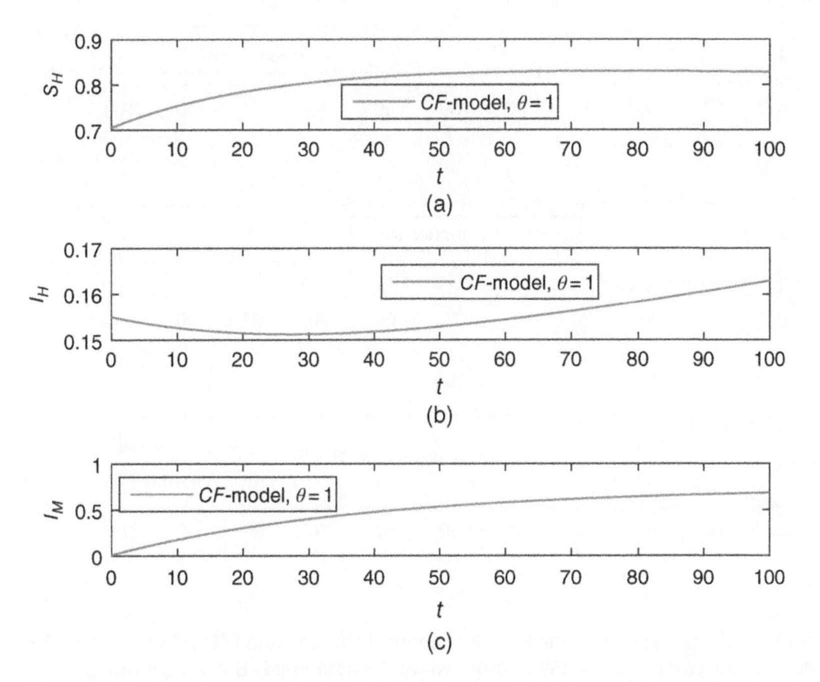

Figure 11.5 Graphical results for the Zika Caputo–Fabrizio model (11.15), when $\theta = 01$.

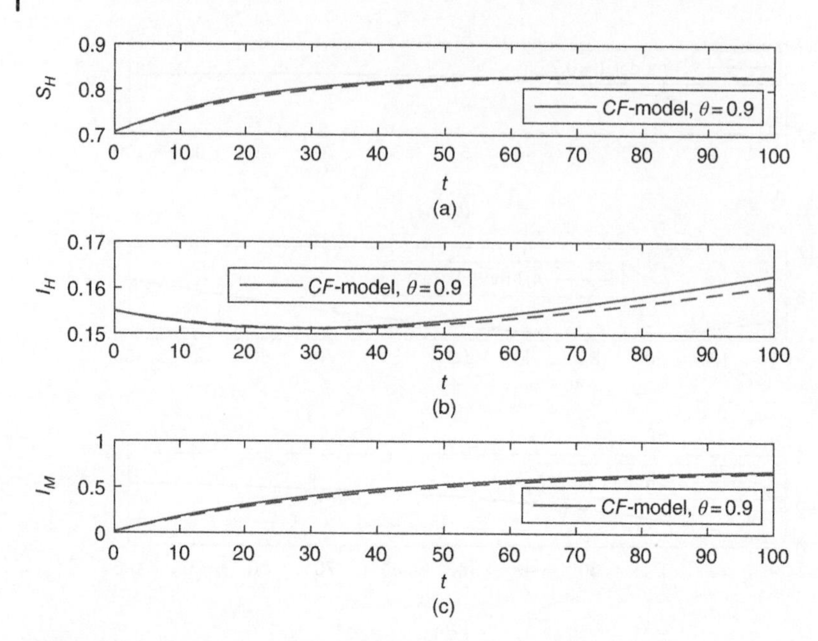

Figure 11.6 Graphical results for the Zika Caputo–Fabrizio model (11.15), when $\theta = 0.9$. Dashed line represents the fractional order while the bold line is the integer order.

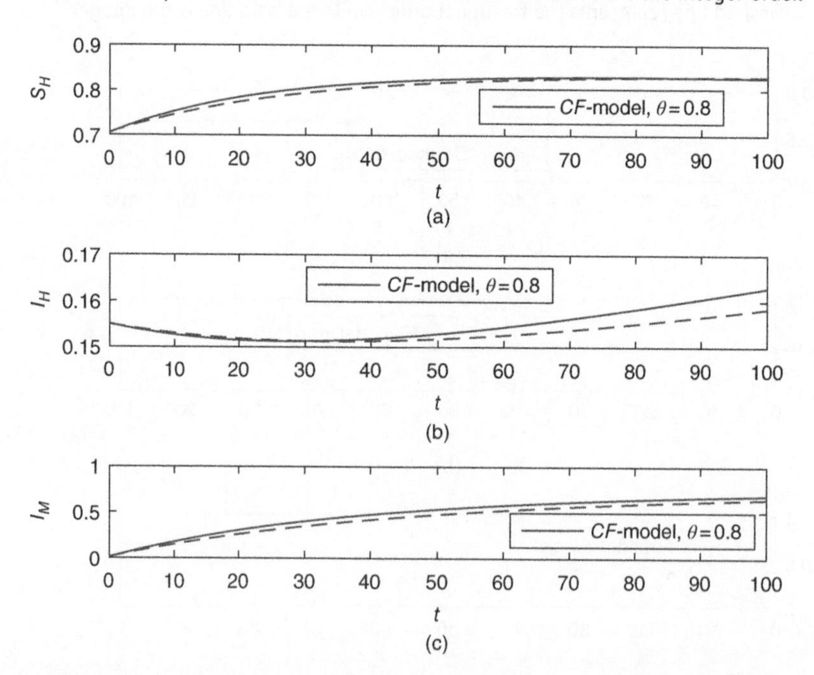

Figure 11.7 Graphical results for the Zika Caputo–Fabrizio model (11.15), when $\theta = 0.8$. Dashed line represents the fractional order while the bold line is the integer order.

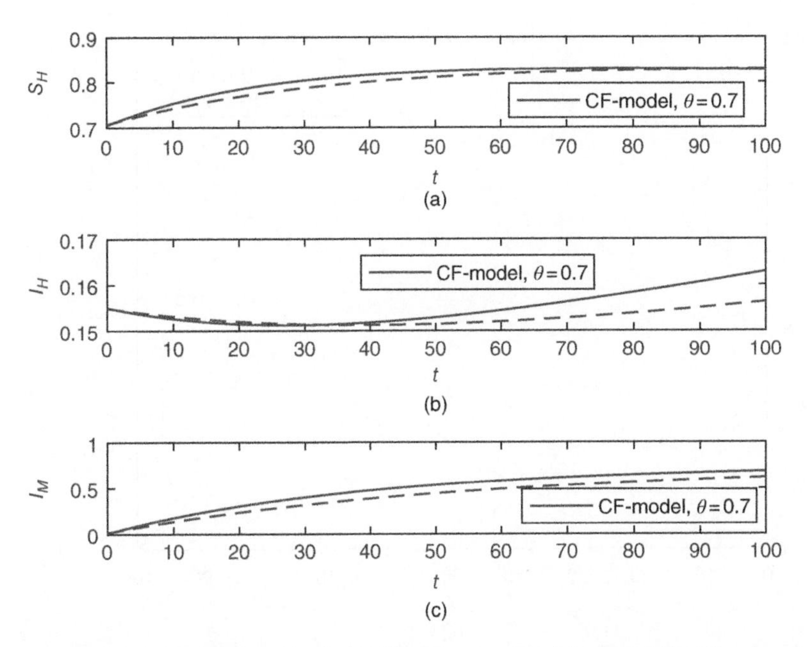

Figure 11.8 Graphical results for the Zika Caputo–Fabrizio model (11.15), when $\theta = 0.7$. Dashed line represents the fractional order while the bold line is the integer order.

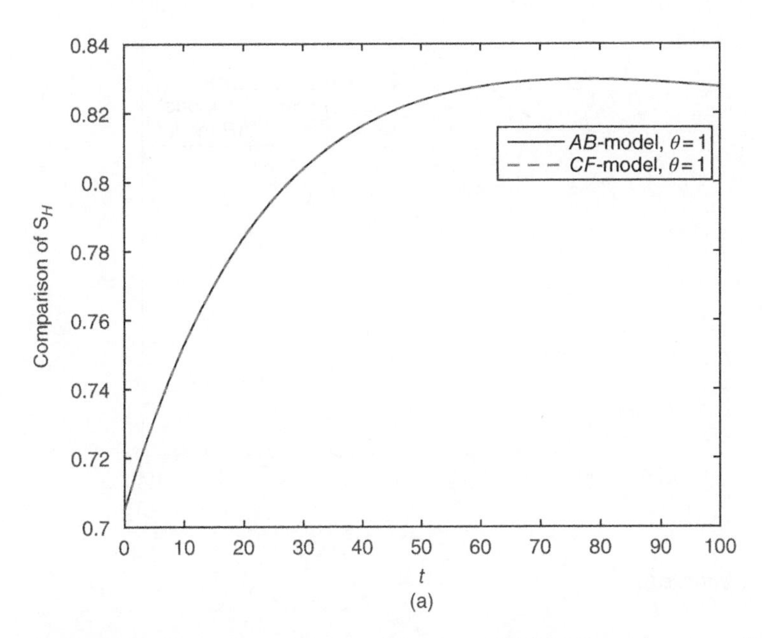

Figure 11.9 Comparison of Caputo–Fabrizio model (11.15) and Atangana–Baleanu model (11.16), when $\theta = 1$.

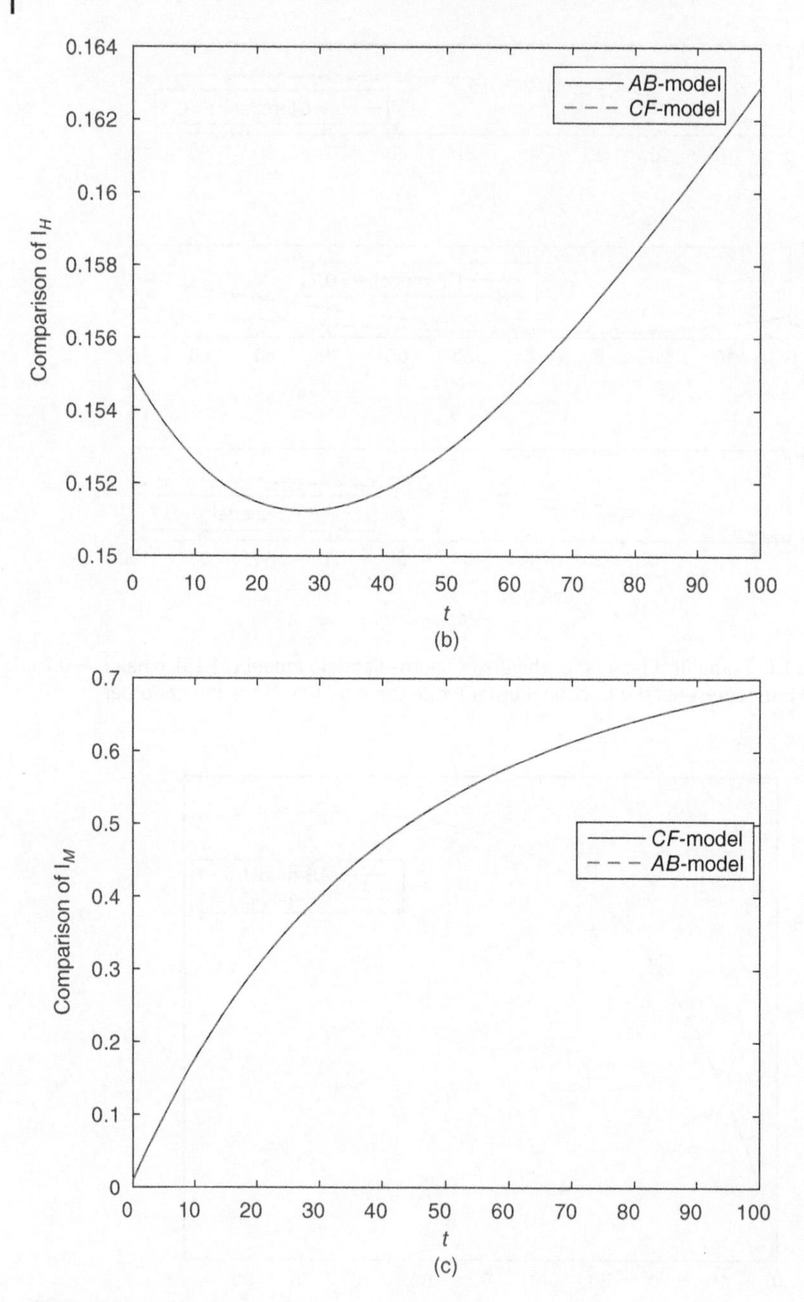

Figure 11.9 (*Continued*)

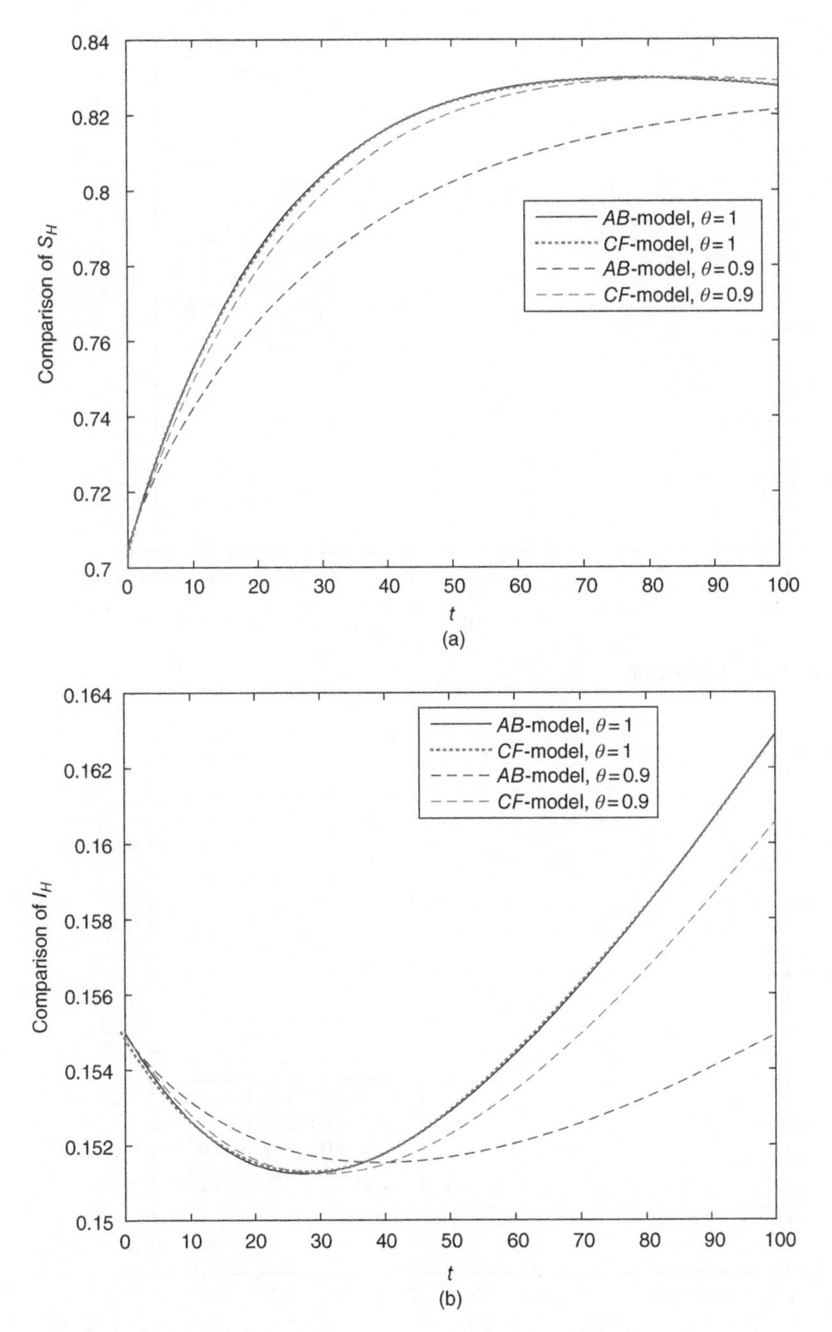

Figure 11.10 Comparison of Caputo–Fabrizio model (11.15) and Atangana–Baleanu model (11.16), when $\theta = 0.9$.

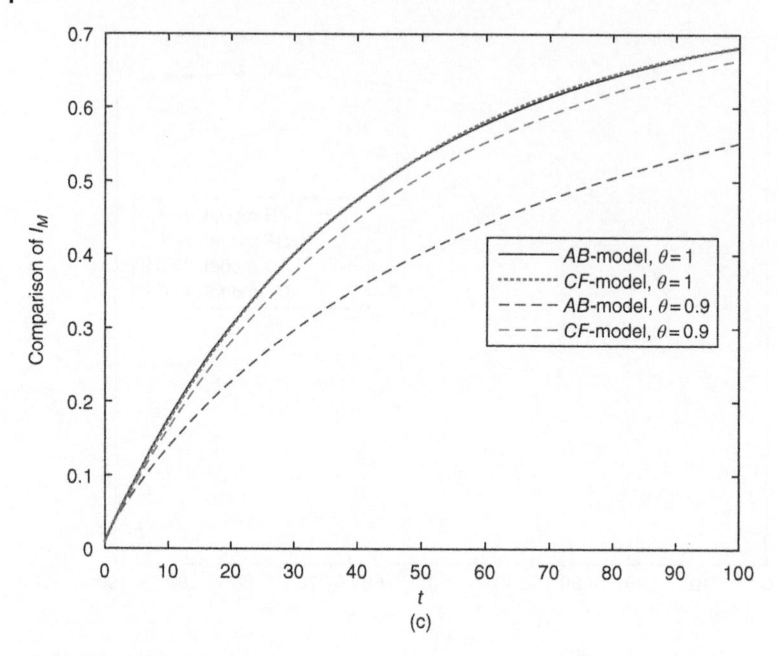

Figure 11.10 (*Continued*)

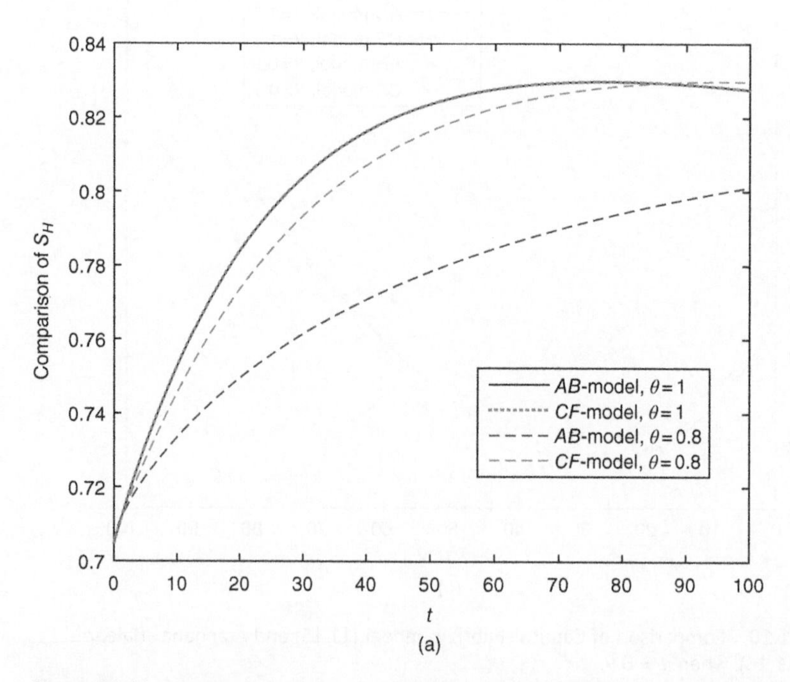

Figure 11.11 Comparison of Caputo–Fabrizio model (11.15) and Atangana–Baleanu model (11.16), when $\theta = 0.8$.

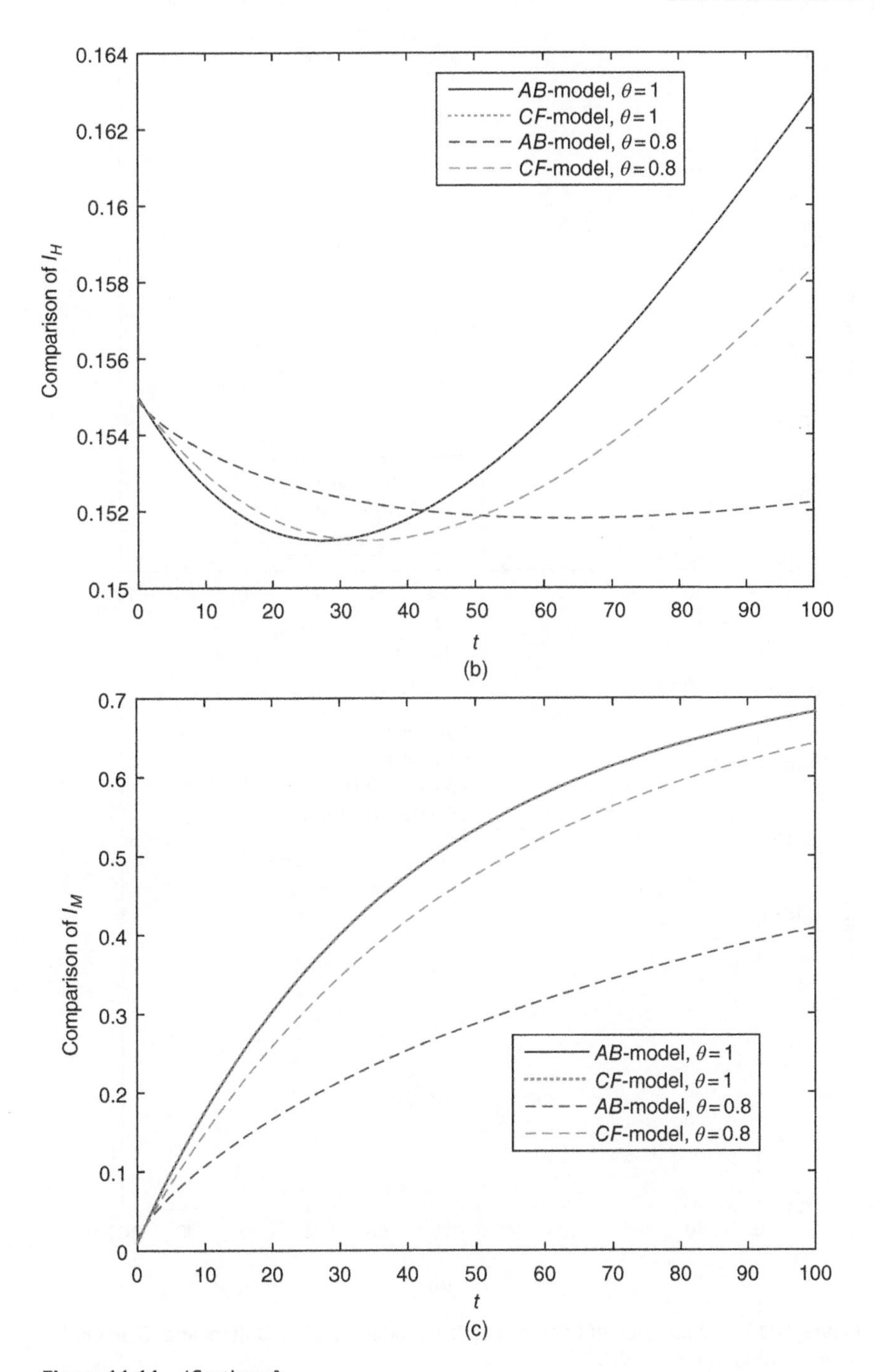

Figure 11.11 *(Continued)*

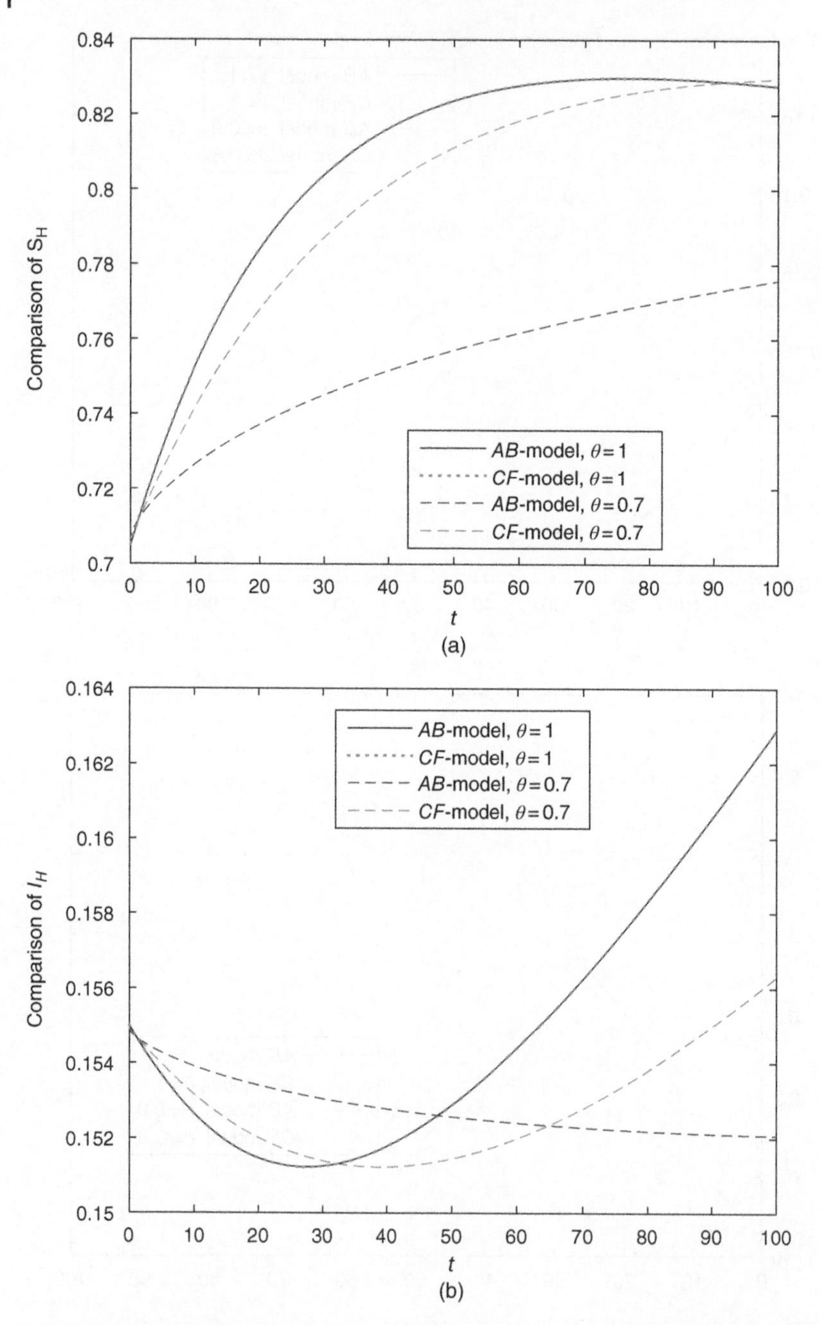

Figure 11.12 Comparison of Caputo–Fabrizio model (11.15) and Atangana–Baleanu model (11.16), when $\theta = 0.7$.

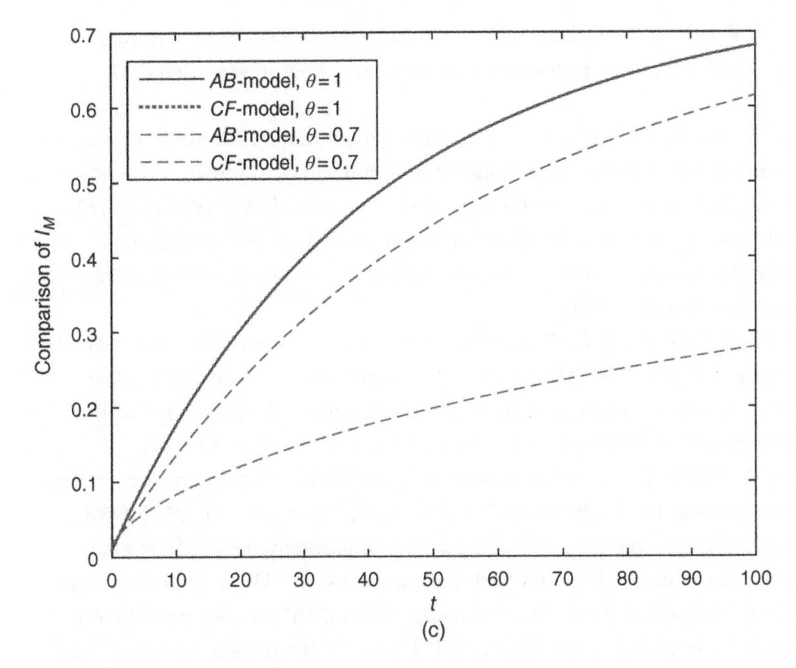

Figure 11.12 (*Continued*)

11.7 Conclusion

The present chapter investigated the dynamics of Zika model with Caputo–Fabrizio and Atangana–Baleanu derivative. Initially, we considered a Zika mathematical model from literature and obtained their mathematical results in details. Then, we apply the *CF* and *AB* operators and presented their analysis. Some useful numerical procedures for their simulations are provided and then detailed graphical results for each model with different fractional order and also their comparison were shown. We concluded that the results provided by Atangana–Baleanu derivative are more useful than that of Caputo – Fabrizio derivative.

Bibliography

1 Simpson, D.I.H. (1964). Zika virus infection in man. *Trans. R. Soc. Trop. Med. Hyg.* **58** (4): 335–337.

2 World Health Organization (WHO). Zika virus. https://www.who.int/emergencies/diseases/zika/en/(accessed 03 January 2019).

3 Duffy, M.R., Chen, T.-H., Hancock, W.T. et al. (2009). Zika virus outbreak on yap island, federated states of micronesia. *New Engl. J. Med.* **360** (24): 2536–2543.

4 Musso, D., Nhan, T., Robin, E. et al. (2014). Potential for Zika virus transmission through blood transfusion demonstrated during an outbreak in French Polynesia, November 2013 to February 2014. *Eurosurveillance* **19** (14): 20761.

5 Gao, D., Lou, Y., He, D. et al. (2016). Prevention and control of Zika as a mosquito-borne and sexually transmitted disease: a mathematical modeling analysis. *Sci. Rep.* **6**: 28070.

6 Foy, B.D., Kobylinski, K.C., Foy, J.L.C. et al. (2011). Probable non-vector-borne transmission of Zika virus, Colorado, USA. *Emerging Infect. Dis.* **17** (5): 880.

7 Bogoch, I.I., Brady, O.J., Kraemer, M.U.G. et al. (2016). Anticipating the international spread of zika virus from brazil. *Lancet* **387** (10016) 335–336.

8 Musso, D., Nilles, E.J., and Cao-Lormeau, V.-M. (2014). Rapid spread of emerging Zika virus in the Pacifc area. *Clin. Microbiol. Infect.* **20** (10): O595–O596.

9 Lee, E.K., Liu, Y., and Pietz, F.H. (2016). A compartmental model for Zika virus with dynamic human and vector populations. In: *AMIA Annual Symposium Proceedings*, vol. **2016**, 743. American Medical Informatics Association.

10 Agusto, F.B., Bewick, S., and Fagan, W.F. (2017). Mathematical model of Zika virus with vertical transmission. *Infect. Dis. Modell.* **2** (2): 244–267.

11 Suparit, P., Wiratsudakul, A., and Modchang, C. (2018). A mathematical model for Zika virus transmission dynamics with a time-dependent mosquito biting rate. *Theor. Biol. Med. Modell.* **15** (1): 11.

12 Ding, C., Tao, N., and Zhu, Y. (2016). A mathematical model of zika virus and its optimal control. In: *Control Conference (CCC), 2016 35th Chinese*, 2642–2645. IEEE.

13 Bonyah, E., Khan, M.A., Okosun, K.O., and Islam, S. (2017). A theoretical model for Zika virus transmission. *PLoS One* **12** (10): e0185540.

14 Khan, M.A., Shah, S.W., Ullah, S., and Gómez-Aguilar, J.F. (2019). A dynamical model of asymptomatic carrier Zika virus with optimal control strategies. *Nonlinear Anal. Real World Appl.* **50**: 144–170.

15 Jan, R., Khan, M.A., Kumam, P., and Thounthong, P. (2019). Modeling the transmission of dengue infection through fractional derivatives. *Chaos, Solitons Fractals* **127**: 189–216.

16 Morales-Delgado, V.F., Gómez-Aguilar, J.F., Saad, K.M. et al. (2019). Analytic solution for oxygen diffusion from capillary to tissues involving external force effects: a fractional calculus approach. *Physica A* **523**: 48–65.

17 Khan, M.A., Azizah, M., and Ullah, S. (2019). A fractional model for the dynamics of competition between commercial and rural banks in Indonesia. *Chaos, Solitons Fractals* **122**: 32–46.

18 Ullah, S., Khan, M.A., Farooq, M., and Alzahrani, E.O. (2020). A fractional model for the dynamics of tuberculosis (TB) using Atangana-Baleanu derivative. *Discrete Contin. Dyn. Syst. S* **13** (3): 937–956.

19 Khan, M.A. and Ullah, S., and Farhan, M. (2019). The dynamics of Zika virus with Caputo fractional derivative. *AIMS Math.* **4**: 134–146.

20 El-Dessoky, M.M. and Khan, M.A. (2019). Application of fractional calculus to combined modified function projective synchronization of different systems. *Chaos: Interdisciplin. J. Nonlinear Sci.* **29** (1): 013–107.

21 Khan, M.A., Hammouch, Z., and Baleanu, D. (2019). Modeling the dynamics of hepatitis E via the Caputo-Fabrizio derivative. *Math. Modell. Nat. Phenom.* **14** (3): 311.

22 Ullah, S., Khan, M.A., and Farooq, M. (2018). A fractional model for the dynamics of TB virus. *Chaos, Solitons Fractals* **116**: 63–71.

23 Ávalos-Ruiz, L.F., Gómez-Aguilar, J.F., Atangana, A., and Owolabi, K.M. (2019). On the dynamics of fractional maps with power-law, exponential decay and Mittag–Leffler memory. *Chaos, Solitons Fractals* **127**: 364–388.

24 Atangana, A. and Araz, S.I. (2019). Analysis of a new partial integro-differential equation with mixed fractional operators. *Chaos, Solitons Fractals* **127**: 257–271.

25 Solís-Pérez, J.E., Gómez-Aguilar, J.F., and Atangana, A. (2019). A fractional mathematical model of breast cancer competition model. *Chaos, Solitons Fractals* **127**: 38–54.

26 Gao, W., Ghanbari, B., and Baskonus, H.M. (2019). New numerical simulations for some real world problems with Atangana–Baleanu fractional derivative. *Chaos, Solitons Fractals* **128**: 34–43.

27 Aranda, D.F., González-Parrab, G., and Benincasaa, T. (2019). Mathematical modeling and numerical simulations of Zika in Colombia considering mutation. *Math. Comput. Simul.* **163**: 118.

28 Caputo, M. and Fabrizio, M. (2017). On the notion of fractional derivative and applications to the hysteresis phenomena. *Meccanica* **52**: 3043–3052.

29 Losada, J. and Nieto, J.J. (2015). Properties of the new fractional derivative without singular Kernel. *Progr. Fract. Differ. Appl.* **1**: 87–92.

30 Atangana, A. and Baleanu, D. (2016). New fractional derivatives with nonlocal and non-singular kernel: theory and application to heat transfer model. arXiv 2016, arXiv:1602.03408.

31 Atangana, A. and Owolabi, K.M. (2018). New numerical approach for fractional differential equations. *Math. Model. Nat. Phenom.* **13**: 121.

32 Toufik, M. and Atangana, A. (2017). New numerical approximation of fractional derivative with non-local and non-singular kernel: application to chaotic models. *Eur. Phys. J. Plus* **132**: 444.

Index

Fractional Order Analysis: Theory, Methods and Applications, First Edition.
Edited by Hemen Dutta, Ahmet Ocak Akdemir, and Abdon Atangana.

f

g

h

i

j

k

l

m

Printed and bound by CPI Group (UK) Ltd, Croydon, CR0 4YY